# 23
# Topics in Medicinal Chemistry

## Aims and Scope

Drug research requires interdisciplinary team-work at the interface between chemistry, biology and medicine. Therefore, the new topic-related series Topics in Medicinal Chemistry will cover all relevant aspects of drug research, e.g. pathobiochemistry of diseases, identification and validation of (emerging) drug targets, structural biology, drugability of targets, drug design approaches, chemogenomics, synthetic chemistry including combinatorial methods, bioorganic chemistry, natural compounds, high-throughput screening, pharmacological in vitro and in vivo investigations, drug-receptor interactions on the molecular level, structure-activity relationships, drug absorption, distribution, metabolism, elimination, toxicology and pharmacogenomics.

In general, special volumes are edited by well known guest editors.

In references Topics in Medicinal Chemistry is abbreviated Top Med Chem and is cited as a journal.

More information about this series at http://www.springer.com/series/7355

Dietmar Krautwurst
Editor

# Taste and Smell

With contributions by

M. Behrens · B. Boonen · S. Espinoza · J.N. Fletcher · R.R. Gainetdinov · C. Geithe · A.D. Kinghorn · D. Krautwurst · T. Kurahashi · P. Marcinek · W. Meyerhof · L. Pan · S. Prandi · J.B. Startek · H. Takeuchi · K. Talavera

Springer

*Editor*
Dietmar Krautwurst
Deutsche Forschungsanstalt fuer Lebensmittelchemie
Leibniz Institut
Freising, Germany

ISSN 1862-2461 ISSN 1862-247X (electronic)
Topics in Medicinal Chemistry
ISBN 978-3-319-48925-4 ISBN 978-3-319-48927-8 (eBook)
DOI 10.1007/978-3-319-48927-8

Library of Congress Control Number: 2016963069



Printed on acid-free paper

This Springer imprint is published by Springer Nature
The registered company is Springer International Publishing AG
The registered company address is: Gewerbestrasse 11, 6330 Cham, Switzerland

# Preface: The Chemical Senses Taste and Smell in Medicinal Chemistry

This thematic issue of "Topics in Medicinal Chemistry" highlights a selection of reviews on recent advances in modulating chemosensory receptors within the chemical senses taste and smell and beyond.

Discovery of new bioactives, i.e., therapeutic agents at the interface of medicinal chemistry, pharmacology, and various biological disciplines, typically starts with the identification of "hits," i.e., cognate compound/receptor pairs. The present special issue on "Taste and Smell" in "Topics on Medicinal Chemistry" needs thus to be target-centered. While there is a wealth of information about drugs and their side effects on our chemical senses [1–3], much less is known with respect to compounds designed to target and alter mechanisms of taste and smell perceptions.

In the case of compounds activating our chemical senses, olfaction and at least the umami, sweet, and bitter taste modalities, G protein-coupled receptors (GPCR) are the prime targets [4]. However, due to intrinsic properties of chemosensory GPCR and the heterologous test cell systems employed [5, 6], approximately 85% of our ~400 odorant receptors (OR) [9] and 16% of our 25 bitter taste receptors (TAS2R) [10] (chapter 'Taste Receptor Gene Expression Outside the Gustatory System' by Behrens et al.) are still, decades after their discovery [7, 8], orphan receptors with unknown specific agonists (Table 1). So far, little information on especially cognate odorant/OR pairs, validated by parameters of potency and efficacy and in vivo studies, is available.

Putative odorant and taste receptor agonists are small organic compounds of a specific biological activity, typically identified by bioassay-based approaches such as functional genomics, cellomics, and reverse pharmacology. Knowledge of the biological target molecules (receptor space) as well as of the biologically relevant chemical information on putative cognate agonists (stimulus space) will help to tackle complexity and to increase screening efficiency. Indeed, preselected, target-oriented compound libraries have been demonstrably advantageous in the identification process toward "hits" [21]. For example, only about 230 key food odorants out of 10,000 volatiles in food have been shown to be necessary and sufficient as a combinatorial toolbox to define most of today's food aromas [15]. Whether these

**Table 1** Human orphan receptors and GPCR with known agonists

| | *Human odorant GPCR* | *Human taste GPCR* | *Non-chemosensory GPCR* |
|---|---|---|---|
| Total | 391 [11] (413) [9] | 28 [8, 12, 13] | 356 [14] |
| With known agonists | 36 [15] (57) [16, 17] | 24 [10, 18, 19] | 262 |
| Without known agonists (orphans) | ~350 | 4 (TAS2R) | 94 [20] |

*GPCR* G protein-coupled receptors, *TAS2R* bitter taste receptors

key food odorants span the entire stimulus space of our sense of olfaction is however yet unclear. Similarly, a sensory-guided and diverse bitter compound library [22] is likely to include putative agonists, at least for the mostly broadly tuned bitter taste receptors in in vitro screening assays [18]. Indeed, target-oriented screening approaches have already delivered some valuable chemosensory information on cognate agonist/receptor pairs [10, 15]. Ideally, such screening endeavors embrace principles of reverse pharmacology, or employ collections of preselected, biological relevant, canonical activators of our chemical senses, for example, using validated foodborne key aroma and flavor compounds that are encountered by our chemical senses before a meal.

The receptors from our chemical senses taste and smell have evolved to detect chemicals of microbiological, plant, or animal origin, which carry chemosensory information. Thus, from a medicinal chemistry point of view, compounds that interact with our chemosensory receptors initially and ideally have to be naturals. In this volume, the medicinal chemistry of plant naturals as agonists/antagonists for taste GPCR is reviewed in the chapter "Medicinal Chemistry of Plant Naturals as Agonists/Antagonists for Taste Receptors" by Fletcher et al.

To put matters into a "medicinal chemistry" or pharmacognostic perspective:

1. More than 40% of marketed drugs target non-chemosensory GPCR [23, 24].
2. Ion channels represent the second largest target for existing drugs after GPCR [24].
3. Of the small molecules approved as drugs in 2010, more than 50% were natural products or directly derived therefrom [25].
4. An "ectopic" expression of the entire set of our ~430 chemosensory odorant and taste GPCR in nonolfactory, non-taste-related tissues and cells would about double the potential therapeutic GPCR target space (see Table 1).

Beyond GPCR, also ion channels of our chemical senses have advocated themselves as targets for a chemical modulation, for example, by odorants and tastants [26–28]. Ion channels are transmembrane proteins constituting ligand- or voltage-gated, water-filled pores to control active ion fluxes across membranes. The ion channel family is intimately involved in many aspects of cell physiology and signaling. For example, ion channels are prime effectors within the chemosensory receptors' signaling cascades, triggering frequency-encoded action potentials or transmitter release in olfactory sensory neurons or taste cells, respectively. In this volume, two chapters by Boonen et al. (chapter "Chemical Activation of Sensory TRP Channels") and by Takeuchi and Kurahashi (chapter "Olfactory Transduction Channels and Their Modulation by Varieties of Volatile Substances") review the effects of odorants or tastants on olfaction- or taste-related ion channels, opening a fresh view on an avenue of chemical intervention of these key effector molecules in the cellular signaling cascades of our chemical senses.

What could be the therapeutic potential of compounds targeting chemosensory receptors or ion channels? Is it to interfere hedonically with the regulation of appetite/craving to cut down or prevent calory intake in weight-challenged health risk groups [29–34]? Is it to develop bitter taste blockers to make bitter medicine palatable [35–37]? Or is it to identify allosteric modulators of our chemical senses, olfaction and taste [38], to boost appetite and hedonic experience of food for the chemosensory-challenged elderly, chronically ill, immune-deficient, and cancer patients [39–44]? Increasing evidence for an ectopic expression of chemosensory GPCR in tissues unrelated to taste and smell opens yet another, new perspective, in which any odorant or taste receptor agonists, beyond their function as adequate stimuli for our chemical senses, have to be considered bioactives in a variety of non-chemosensory cells, tissues, and organs [45–48]. In this volume, the reader will find these aspects reviewed by Behrens et al. for taste receptors (chapter "Taste Receptor Gene Expression Outside the Gustatory System") and by Marcinek et al. for odorant receptors, biogenic amine receptors, and taste receptors (chapter "Chemosensory G Protein-Coupled Receptors (GPCR) in Blood Leukocytes"), exemplified for their expression within the cellular immune system. Both chapters review evidence that, beyond our chemical senses smell and taste, some odorant and taste receptors are likely to emerge as genuine and relevant drug targets with a high chance of pharmacological intervention [49–51].

Trace amine-associated receptors are genuine olfactory receptors (at least in rodents [52–54]) with a non-yet-defined olfactory role in humans. In this volume, Espinoza and Gainetdinov review the neuronal function and emerging pharmacology of TAAR1 (chapter "Neuronal Functions and Emerging Pharmacology of TAAR1"), which is the best investigated human TAAR from a medicinal chemistry point of view [55]. Interestingly, for this receptor, there is increasing evidence for an ectopic expression in a variety of peripheral, nonolfactory tissues [56], such as the cellular immune system [57, 58], as reviewed in the chapter "Chemosensory G Protein-Coupled Receptors (GPCR) in Blood Leukocytes" by Marcinek et al. in this volume.

From a pharmacognostic point of view, there is an immense therapeutic potential of naturals [15, 59], peptides [60–62], metabolites [63], and drugs [1–3] targeting chemosensory receptors [64], ion channels [24, 27, 28, 65], or enzymes [66]. For example, natural compounds may act either as agonists [15, 59], antagonists [36, 37, 67–71], or modulators [38, 72] of the receptors of our chemical senses and, moreover, following a meal and after uptake via the gastrointestinal system, respiratory epithelia, or the skin [73–75], they may work as genuine bioactives in a variety of non-chemosensory tissues and organs via the same receptors "ectopically" expressed outside of our chemical senses taste and smell [46, 50, 51, 58, 76–79].

The challenges of medicinal chemistry and drug design on chemosensory receptors will become more complex as the knowledge on the variety of receptors, their cognate agonists, and their expressing cells and tissues increases.

Deutsche Forschungsanstalt fuer Lebensmittelchemie, Leibniz Institut
Freising, Germany

Dietmar Krautwurst

## References

1. Doty RL, Bromley SM (2004) Effects of drugs on olfaction and taste. Otolaryngol Clin North Am 37(6):1229–1254
2. Doty RL, Shah M, Bromley SM (2008) Drug-induced taste disorders. Drug Saf 31(3):199–215
3. Schiffman SS (2015) Influence of drugs on taste function. In: Handbook of olfaction and gustation. Wiley, pp 911–926
4. Silbering AF, Benton R (2010) Ionotropic and metabotropic mechanisms in chemoreception: 'chance or design'? EMBO Rep 11(3):173–179
5. Dalton RP, Lyons DB, Lomvardas S (2013) Co-opting the unfolded protein response to elicit olfactory receptor feedback. Cell 155(2):321–332
6. McClintock TS, Sammeta N (2003) Trafficking prerogatives of olfactory receptors. Neuroreport 14(12):1547–1552
7. Buck L, Axel R (1991) A novel multigene family may encode odorant receptors: a molecular basis for odor recognition. Cell 65(1):175–187
8. Chandrashekar J, Mueller KL, Hoon MA, Adler E, Feng L, Guo W, Zuker CS, Ryba NJ (2000) T2Rs function as bitter taste receptors. Cell 100(6):703–711
9. Olender T, Waszak SM, Viavant M, Khen M, Ben-Asher E, Reyes A, Nativ N, Wysocki CJ, Ge D, Lancet D (2012) Personal receptor repertoires: olfaction as a model. BMC Genom 13:414
10. Meyerhof W, Batram C, Kuhn C, Brockhoff A, Chudoba E, Bufe B, Appendino G, Behrens M (2010) The molecular receptive ranges of human TAS2R bitter taste receptors. Chem Senses 35(2):157–170
11. HORDE. The Human Olfactory Receptor Data Exploratorium (HORDE) (2011) The Weizmann Institute. http://bioportal.weizmann.ac.il/HORDE/. Accessed 16 Mar 2015
12. Nelson G, Hoon MA, Chandrashekar J, Zhang Y, Ryba NJ, Zuker CS (2001) Mammalian sweet taste receptors. Cell 106(3):381–390
13. Nelson G, Chandrashekar J, Hoon MA, Feng L, Zhao G, Ryba NJ, Zuker CS (2002) An amino-acid taste receptor. Nature 416(6877):199–202
14. IUPHAR/BPS. G protein-coupled receptors. IUPHAR/BPS Guide to PHARMACOLOGY (2015) http://www.guidetopharmacology.org/GRAC/FamilyDisplayForward?familyId=694. Accessed 16 Mar 2015
15. Dunkel A, Steinhaus M, Kotthoff M, Nowak B, Krautwurst D, Schieberle P, Hofmann T (2014) Nature's chemical signatures in human olfaction: a foodborne perspective for future biotechnology. Angew Chem Int Ed 53(28):7124–7143
16. Mainland JD, Keller A, Li YR, Zhou T, Trimmer C, Snyder LL, Moberly AH, Adipietro KA, Liu WL, Zhuang H, Zhan S, Lee SS, Lin A, Matsunami H (2014) The missense of smell: functional variability in the human odorant receptor repertoire. Nat Neurosci 17(1):114–120

17. Mainland JD, Li YR, Zhou T, Liu WL, Matsunami H (2015) Human olfactory receptor responses to odorants. Sci Data 2:150002
18. Behrens M, Meyerhof W (2013) Bitter taste receptor research comes of age: from characterization to modulation of TAS2Rs. Semin Cell Dev Biol 24(3):215–221
19. Behrens M, Meyerhof W, Hellfritsch C, Hofmann T (2011) Sweet and umami taste: natural products, their chemosensory targets, and beyond. Angew Chem Int Ed 50(10):2220–2242
20. Davenport AP, Alexander SP, Sharman JL, Pawson AJ, Benson HE, Monaghan AE, Liew WC, Mpamhanga CP, Bonner TI, Neubig RR, Pin JP, Spedding M, Harmar AJ (2013) International Union of Basic and Clinical Pharmacology. LXXXVIII. G protein-coupled receptor list: recommendations for new pairings with cognate ligands. Pharmacol Rev 65(3):967–986
21. Weidel E, Negri M, Empting M, Hinsberger S, Hartmann RW (2014) Composing compound libraries for hit discovery--rationality-driven preselection or random choice by structural diversity? Future Med Chem 6(18):2057–2072
22. Wiener A, Shudler M, Levit A, Niv MY (2012) BitterDB: a database of bitter compounds. Nucleic Acids Res 40 (Database issue):D413–D419
23. Garland SL (2013) Are GPCRs still a source of new targets? J Biomol Screen 18(9):947–966
24. Overington JP, Al-Lazikani B, Hopkins AL (2006) How many drug targets are there? Nat Rev Drug Discov 5(12):993–996
25. Newman DJ, Cragg GM (2012) Natural products as sources of new drugs over the 30 years from 1981 to 2010. J Nat Prod 75(3):311–335
26. Lin W, Ogura T, Margolskee RF, Finger TE, Restrepo D (2008) TRPM5-expressing solitary chemosensory cells respond to odorous irritants. J Neurophysiol 99(3):1451–1460
27. Chen TY, Takeuchi H, Kurahashi T (2006) Odorant inhibition of the olfactory cyclic nucleotide-gated channel with a native molecular assembly. J Gen Physiol 128(3):365–371
28. Gees M, Alpizar YA, Luyten T, Parys JB, Nilius B, Bultynck G, Voets T, Talavera K (2014) Differential effects of bitter compounds on the taste transduction channels TRPM5 and IP3 receptor type 3. Chem Senses 39(4):295–311
29. Scott TR (2011) Taste as a basis for body wisdom. Physiol Behav 104(1):57–63
30. Galindo MM, Schneider NY, Stahler F, Tole J, Meyerhof W (2012) Taste preferences. Prog Mol Biol Transl Sci 108:383–426
31. Sclafani A, Ackroff K (2012) Role of gut nutrient sensing in stimulating appetite and conditioning food preferences. Am J Physiol Regul Integr Comp Physiol 302(10):R1119–R1133
32. Kemps E, Tiggemann M, Bettany S (2012) Non-food odorants reduce chocolate cravings. Appetite 58(3):1087–1090
33. Frankort A, Roefs A, Siep N, Roebroeck A, Havermans R, Jansen A (2014) The craving stops before you feel it: neural correlates of chocolate craving during cue exposure with response prevention. Cereb Cortex 24(6):1589–1600
34. Rolls ET, McCabe C (2007) Enhanced affective brain representations of chocolate in cravers vs. non-cravers. Eur J Neurosci 26(4):1067–1076
35. Nagtegaal MJ, Swen JJ, Hanff LM, Schimmel KJ, Guchelaar HJ (2014) Pharmacogenetics of taste: turning bitter pills sweet? Pharmacogenomics 15(1):111–119
36. Ono N, Miyamoto Y, Ishiguro T, Motoyama K, Hirayama F, Iohara D, Seo H, Tsuruta S, Arima H, Uekama K (2011) Reduction of bitterness of antihistaminic drugs by complexation with beta-cyclodextrins. J Pharm Sci 100(5):1935–1943
37. Pydi SP, Sobotkiewicz T, Billakanti R, Bhullar RP, Loewen MC, Chelikani P (2014) Amino acid derivatives as bitter taste receptor (T2R) blockers. J Biol Chem 289(36):25054–25066
38. Servant G, Tachdjian C, Li X, Karanewsky DS (2011) The sweet taste of true synergy: positive allosteric modulation of the human sweet taste receptor. Trends Pharmacol Sci 32(11):631–636

39. Volkert D (2013) Malnutrition in older adults – urgent need for action: a plea for improving the nutritional situation of older adults. Gerontology 59(4):328–333
40. Murphy C (2008) The chemical senses and nutrition in older adults. J Nutr Elder 27(3–4):247–265
41. Mouritsen OG (2012) Umami flavour as a means of regulating food intake and improving nutrition and health. Nutr Health 21(1):56–75
42. Woolard S (2001) Wasting syndrome in HIV/AIDS patients. A guide to prevention and management. Adv Nurse Pract 9(9):42–46; quiz 47–48
43. de Pee S, Semba RD (2010) Role of nutrition in HIV infection: review of evidence for more effective programming in resource-limited settings. Food Nutr Bull 31(4):S313–S344
44. Boltong A, Keast R (2012) The influence of chemotherapy on taste perception and food hedonics: a systematic review. Cancer Treat Rev 38(2):152–163
45. Behrens M, Meyerhof W (2011) Gustatory and extragustatory functions of mammalian taste receptors. Physiol Behav 105(1):4–13
46. Clark AA, Liggett SB, Munger SD (2012) Extraoral bitter taste receptors as mediators of off-target drug effects. FASEB J 26(12):4827–4831
47. Feldmesser E, Olender T, Khen M, Yanai I, Ophir R, Lancet D (2006) Widespread ectopic expression of olfactory receptor genes. BMC Genom 7:121
48. Foster SR, Roura E, Thomas WG (2014) Extrasensory perception: odorant and taste receptors beyond the nose and mouth. Pharmacol Ther 142(1):41–61
49. Wellendorph P, Johansen LD, Brauner-Osborne H (2009) Molecular pharmacology of promiscuous seven transmembrane receptors sensing organic nutrients. Mol Pharmacol 76 (3):453–465. Epub 2009 Jun 2001
50. Clark AA, Nurmukhambetova S, Li X, Munger SD, Lees JR (2016) Odorants specifically modulate chemotaxis and tissue retention of CD4+ T cells via cyclic adenosine monophosphate induction. J Leukoc Biol 100:699–709
51. Malki A, Fiedler J, Fricke K, Ballweg I, Pfaffl MW, Krautwurst D (2015) Class I odorant receptors, TAS1R and TAS2R taste receptors, are markers for subpopulations of circulating leukocytes. J Leukoc Biol 97(3):533–545
52. Liberles SD (2009) Trace amine-associated receptors are olfactory receptors in vertebrates. Ann N Y Acad Sci 1170:168–172
53. Liberles SD, Buck LB (2006) A second class of chemosensory receptors in the olfactory epithelium. Nature 442(7103):645–650
54. Dewan A, Pacifico R, Zhan R, Rinberg D, Bozza T (2013) Non-redundant coding of aversive odours in the main olfactory pathway. Nature 497(7450):486–489
55. Jing L, Li JX (2015) Trace amine-associated receptor 1: a promising target for the treatment of psychostimulant addiction. Eur J Pharmacol 761:345–352
56. Babusyte A, Krautwurst D (2013) Tracing amines to their receptors: a synopsis in light of recent literature. Cell Biol Res Ther 2:1–3
57. Andersen G, Krautwurst D (2016) Trace amine-associated receptors in the cellular immune system. In: Farooqui T, Farooqui AA (eds) Trace amines and neurological disorders: potential mechanisms and risk factors. Academic Press Elsevier, pp 97–103
58. Babusyte A, Kotthoff M, Fiedler J, Krautwurst D (2013) Biogenic amines activate blood leukocytes via trace amine-associated receptors TAAR1 and TAAR2. J Leukoc Biol 93:387–394
59. Kinghorn AD, Pan L, Fletcher JN, Chai H (2011) The relevance of higher plants in lead compound discovery programs. J Nat Prod 74(6):1539–1555
60. Dotson CD, Geraedts MC, Munger SD (2013) Peptide regulators of peripheral taste function. Semin Cell Dev Biol 24(3):232–239

61. Kohl S, Behrens M, Dunkel A, Hofmann T, Meyerhof W (2013) Amino acids and peptides activate at least five members of the human bitter taste receptor family. J Agric Food Chem 61 (1):53–60
62. Schindler A, Dunkel A, Stahler F, Backes M, Ley J, Meyerhof W, Hofmann T (2011) Discovery of salt taste enhancing arginyl dipeptides in protein digests and fermented fish sauces by means of a sensomics approach. J Agric Food Chem 59(23):12578–12588
63. Tonack S, Tang C, Offermanns S (2013) Endogenous metabolites as ligands for G protein-coupled receptors modulating risk factors for metabolic and cardiovascular disease. Am J Physiol Heart Circ Physiol 304(4):H501–H513
64. Triballeau N, Van Name E, Laslier G, Cai D, Paillard G, Sorensen PW, Hoffmann R, Bertrand HO, Ngai J, Acher FC (2008) High-potency olfactory receptor agonists discovered by virtual high-throughput screening: molecular probes for receptor structure and olfactory function. Neuron 60(5):767–774
65. Bagal SK, Brown AD, Cox PJ, Omoto K, Owen RM, Pryde DC, Sidders B, Skerratt SE, Stevens EB, Storer RI, Swain NA (2013) Ion channels as therapeutic targets: a drug discovery perspective. J Med Chem 56(3):593–624
66. Ling G, Gu J, Genter MB, Zhuo X, Ding X (2004) Regulation of cytochrome P450 gene expression in the olfactory mucosa. Chem Biol Interact 147(3):247–258
67. Fletcher JN, Kinghorn AD, Slack JP, McCluskey TS, Odley A, Jia Z (2012) In vitro evaluation of flavonoids from Eriodictyon californicum for antagonist activity against the bitterness receptor hTAS2R31. J Agric Food Chem 59(24):13117–13121
68. Brockhoff A, Behrens M, Roudnitzky N, Appendino G, Avonto C, Meyerhof W (2011) Receptor agonism and antagonism of dietary bitter compounds. J Neurosci 31(41):14775–14782
69. Greene TA, Alarcon S, Thomas A, Berdougo E, Doranz BJ, Breslin PA, Rucker JB (2011) Probenecid inhibits the human bitter taste receptor TAS2R16 and suppresses bitter perception of salicin. PLoS One 6(5):e20123
70. Ley JP, Dessoy M, Paetz S, Blings M, Hoffmann-Lucke P, Reichelt KV, Krammer GE, Pienkny S, Brandt W, Wessjohann L (2012) Identification of enterodiol as a masker for caffeine bitterness by using a pharmacophore model based on structural analogues of homoeriodictyol. J Agric Food Chem 60(25):6303–6311
71. Slack JP, Brockhoff A, Batram C, Menzel S, Sonnabend C, Born S, Galindo MM, Kohl S, Thalmann S, Ostopovici-Halip L, Simons CT, Ungureanu I, Duineveld K, Bologa CG, Behrens M, Furrer S, Oprea TI, Meyerhof W (2010) Modulation of bitter taste perception by a small molecule hTAS2R antagonist. Curr Biol 20(12):1104–1109
72. Hillmann H, Mattes J, Brockhoff A, Dunkel A, Meyerhof W, Hofmann T (2012) Sensomics analysis of taste compounds in balsamic vinegar and discovery of 5-acetoxymethyl-2-furaldehyde as a novel sweet taste modulator. J Agric Food Chem 60:9974–9990
73. Colburn WA, Bekersky I, Blumenthal HP (1981) Dietary saccharin kinetics. Clin Pharmacol Ther 30(4):558–563
74. Genter MB, Kendig EL, Knutson MD (2009) Uptake of materials from the nasal cavity into the blood and brain: are we finally beginning to understand these processes at the molecular level? Ann N Y Acad Sci 1170:623–628
75. Genter MB, Krishan M, Prediger RD (2015) The olfactory system as a route of delivery for agents to the brain and circulation. In: Handbook of olfaction and gustation. Wiley, pp 453–484
76. Angelucci FL, Silva VV, Dal Pizzol C, Spir LG, Praes CE, Maibach H (2014) Physiological effect of olfactory stimuli inhalation in humans: an overview. Int J Cosmet Sci 36(2):117–123
77. Ekoff M, Choi JH, James A, Dahlen B, Nilsson G, Dahlen SE (2014) Bitter taste receptor (TAS2R) agonists inhibit IgE-dependent mast cell activation. J Allergy Clin Immunol 134 (2):475–478

78. Foster SR, Blank K, See Hoe LE, Behrens M, Meyerhof W, Peart JN, Thomas WG (2014) Bitter taste receptor agonists elicit G-protein-dependent negative inotropy in the murine heart. FASEB J 28(10):4497–4508
79. Lee RJ, Kofonow JM, Rosen PL, Siebert AP, Chen B, Doghramji L, Xiong G, Adappa ND, Palmer JN, Kennedy DW, Kreindler JL, Margolskee RF, Cohen NA (2014) Bitter and sweet taste receptors regulate human upper respiratory innate immunity. J Clin Invest 124(3):1393–1405

# Acknowledgments

To Professor Christian Harteneck, a friend and colleague who passed away before we could finish this volume, for his valuable advice and comments on some of the chapters.

To Dr. Herbert Wieser, who thoroughly and meticulously edited all contributions, and to Anneliese Stoiber, for the help in editing and administration of this volume.

To all the authors for their contributions, especially for their patience during the dynamic development of this volume.

To the staff of Springer, both in Germany and India, whose tireless efforts and patience have been critical to making this volume a success.

# Contents

Top Med Chem (2017) 23: 1–34
DOI: 10.1007/7355_2014_79

Published online: 26 October 2014

# Taste Receptor Gene Expression Outside the Gustatory System

**Maik Behrens, Simone Prandi, and Wolfgang Meyerhof**

**Abstract** The sense of taste facilitates the recognition of beneficial or potentially harmful food constituents prior to ingestion. For the detection of tastants, epithelial specializations in the oral cavity are equipped with taste receptor molecules that interact with sweet, umami (the taste of L-amino acids), salty, sour, and bitter-tasting substances. Over the past years, numerous tissues in addition to gustatory sensory tissue have been identified to express taste receptor molecules. These findings bear important implications for the roles taste receptors fulfill in vertebrates, which are currently envisioned much broader than thought previously. Taste receptive molecules are present in the brain, respiratory and gastrointestinal tracts, heart, male reproductive tissue, as well as other areas of the body just beginning to emerge. This review summarizes current knowledge on the occurrence and functional implications of taste receptive molecules outside the oral cavity.

**Keywords** Gastrointestinal tract, Gene expression, Respiratory epithelium, Taste receptor

## Contents



M. Behrens (✉), S. Prandi, and W. Meyerhof
Department of Molecular Genetics, German Institute of Human Nutrition
Potsdam-Rehbruecke, Arthur-Scheunert-Allee 114-116, 14558 Nuthetal, Germany
e-mail: behrens@dife.de

# 1 Introduction

The need of animals to feed their bodies constantly with energy-rich food items belongs to the most fundamental prerequisites to maintain life and well-being. As food sources are complex and consist of variable amounts of life-maintaining (macro-)nutrients as well as potentially harmful substances, a mechanism has evolved enabling animals evaluating food items prior to their consumption. The taste system present in the oral cavity furnishes animals with receptive proteins devoted to detect the building blocks of carbohydrates, proteins, and critical electrolytes as well as with sensors that identify potentially harmful substances. The activation of the oral taste receptors does not only provide information on the chemicals present in consumed food items, but it also evokes hedonic tones, e.g., perceived pleasantness or unpleasantness directly affecting consumption. Whereas sweet, umami, and low concentrations of sodium salts represent generally attractive taste stimuli promoting consumption, high salt concentrations and sour and bitter stimuli tend to elicit rejection.

In recent years, numerous reports on the extragustatory expression of taste receptors clearly suggested that their role is not limited to taste perception. Taste receptors have meanwhile been identified in the gastrointestinal (e.g., [1–26]) and respiratory tracts of mammals (e.g., [27–37]), in the male reproductive system [36, 38–43], as well as in the brain [44–49] and heart [50] to name just the areas having received the most attention during the recent years. Similar to the oral cavity, also the alimentary canal and the respiratory system represent epithelia exposed to potentially noxious substances originating from the environment. Hence, it appears conceivable that bitter taste receptors, devoted to protect organisms from the uptake of potentially toxic substances, play similar roles in these extragustatory organ systems as well. Another important activity of the intestinal mucosa is the monitoring of the nutritional value of the ingested food. This nutrient sensing allows for an efficient digestion and absorption of nutrients and allows fine tuning of metabolic parameters in response to demands and available resources. Thus, the occurrence of taste receptors, devoted to the detection of sweet and umami stimuli in the GI tract, is not surprising.

More surprising than finding taste receptors in the respiratory and gastrointestinal epithelia was the detection of these molecules in other organs such as the brain and heart. Since these organs are protected from direct contact to the outside world, the expression of TAS1R as well as TAS2Rs suggests that they are involved in the surveillance of molecules present in internal body fluids such as blood plasma or cerebrospinal fluid. It will be highly interesting to see in the future how rather low-affinity receptors, developed to detect concentrated taste stimuli in the oral cavity, fulfill their yet unknown roles in the brain and heart.

Another class of chemosensory receptors, the odorant receptors (ORs), has already been detected in sperm cells and associated with sperm motility and chemotaxis [51, 52]. The identification of taste receptors in the testis and sperm cells underscores the importance of chemoreceptor signaling for reproduction. However, in contrast to odorant receptors, which may actually serve a role in spermatozoa guidance, none of the recently published studies on taste receptors in the male reproductive tract put forward mechanisms indicating a similar function of these receptors [36, 38–43].

After a brief introduction into taste receptors and their associated signaling components, we discuss the expression of taste receptors and their functional implications in the respiratory and alimentary tract, the brain, the heart, and the reproductive system. For space constraints, we will not extensively discuss some so far rather isolated reports on taste receptor gene expression in additional tissues such as the bladder [53], adipose tissue [54, 55], and bone marrow stromal and vascular smooth muscle cells [56], as well as their emerging role in the regulation of autophagy [57] (Table 1).

## 2 Taste Receptors and Signaling Components

The gustatory system of mammals is equipped with taste receptors devoted to the detection of the five basic taste qualities: sweet, sour, salty, umami, and bitter. In addition to these well-accepted basic taste qualities, the perception of further stimuli such as fatty or metallic is discussed (for a recent review, see [58]). The recent generation of knockout mice, lacking the α-subunit of the epithelial sodium channel, ENaC, confirmed the role of this channel as receptor for attractive low concentrations of sodium in rodents [59]. Although numerous candidate molecules for sour taste receptors such as members of the polycystic-kidney-disease-like ion channels (PKD) had been proposed in the past (e.g., [60–62], but cf. [63]), the molecular identification of this receptor is still pending. The best characterized taste receptors are the G protein-coupled receptors responding to sweet [64–69], umami [70, 71], and bitter compounds [72–74], which belong to two different gene families (Fig. 1). The TAS1R family of taste receptors has only three members named TAS1R1, TAS1R2, and TAS1R3. Similar to other class C G protein-coupled receptors such as metabotropic glutamate receptors (mGluRs) [75] or γ-aminobutyric acid type B receptors ($GABA_B$) [76–79], TAS1Rs form oligomeric complexes. Whereas the two subunits TAS1R1 and TAS1R3 form the main receptor for the detection of umami-tasting L-amino acids (in human specifically L-glutamate and to a lesser degree L-aspartic acid) [70, 71], the subunits TAS1R2

**Table 1** List of taste receptors and taste-signaling components found in extraoral tissues and their putative functions

| System | Organs | Transduction molecules | Taste receptors | (Putative) functions |
|---|---|---|---|---|
| Central nervous system | Brain | α-Gustducin, Gβ3, Gγ13, PLCβ2, TRPM5 | *Mouse*: Tas1r1, Tas1r2, Tas1r3. *Rat*: Tas2r1, Tas2r4, Tas2r10, Tas2r38. *Human*: TAS2R4, TAS2R5, TAS2R10, TAS2R13, TAS2R14, TAS2R50 | Modulation of hunger/satiety, glucose homeostasis (Tas1rs); release of neuropeptides controlling food intake (Tas2rs) |
| Respiratory system | Nose | α-Gustducin, PLCβ2, TRPM5 | *Mouse*: Tas2r108, Tas2r119, Tas2r131. *Rat*: Tas1r2, Tas1r3, Tas2r13, Tas2r105, Tas2r107, Tas2r119, Tas2r121, Tas2r126, Tas2r123, Tas2r134. *Human*: TAS2R38 | Detection of irritants (Tas1rs); trigeminal-mediated control of respiration, mucociliary clearance and NO production (innate immunity) (Tas2rs) |
| | Vomeronasal organ | TRPM5 | *Mouse*: Tas2r131 | Recognition of pheromones (Tas2rs) |
| | Trachea, bronchi, lungs | α-Gustducin, PLCβ2, TRPM5 | *Mouse*: Tas1r3, Tas2r105, Tas2r107, Tas2r108. *Rat*: Tas1r1, Tas1r2, Tas1r3, Tas2r13, Tas2r105, Tas2r107, Tas2r119, Tas2r121, Tas2r123, Tas2r126, Tas2r134. *Human*: TAS2R1, TAS2R3, TAS2R4, TAS2R5, TAS2R7, TAS2R8, TAS2R9, TAS2R9, TAS2R10, TAS2R13, TAS2R14, TAS2R16, TAS2R19, TAS2R20, TAS2R30, TASR42, TAS2R45, TAS2R50, TAS2R46, TAS2R31 | Not investigated (Tas1rs); mucociliary clearance, bronchodilation (Tas2rs) |

(continued)

**Table 1** (continued)

| System | Organs | Transduction molecules | Taste receptors | (Putative) functions |
|---|---|---|---|---|
| Cardiovascular system | Heart | α-Gustducin, PLCβ2, TRPM5 | *Mouse*: Tas2r108, Tas2r120, Tas2r121, Tas2r126, Tas2r135, Tas2r137, Tas2r143. *Rat*: Tas1r1, Tas1r3, Tas2r108, Tas2r120, Tas2r121, Tas2r126, Tas2r135, Tas2r137, Tas2r147. *Human*: TAS1R3, TAS2R3, TAS2R4, TAS2R5, TAS2R9, TAS2R10, TAS2R13, TAS2R14, TAS2R19, TAS2R20, TAS2R30, TAS2R31, TAS2R43, TAS2R45, TAS2R46, TAS2R50 | Nutrient sensing and metabolic regulation (Tas1rs, Tas2rs) |
| | Blood vessel | – | *Mouse*: Tas2r116, Tas2r143. *Human*: TAS2R46 | Vasodilation (Tas2rs) |
| Gastrointestinal system | Stomach | α-Gustducin, transducin, PLCβ2, TRPM5 | *Mouse*: Tas1r1, Tas1r2, Tas1r3, Tas2r108, Tas2r119, Tas2r134, Tas2r138. *Rat*: Tas2r1, Tas2r2, Tas2r3, Tas2r5, Tas2r6, Tas2r7, Tas2r8, Tas2r9, Tas2r10, Tas2r12, Tas2r16, Tas2r34, Tas2r38. *Human*: TAS1R1, TAS1R3 | Nutrient sensing (Tas1rs); delay of gastric emptying, ghrelin secretion (Tas2rs) |
| | Small intestine | α-Gustducin, transducin, PLCβ2, TRPM5 | *Mouse*: Tas1r1, Tas1r2, Tas1r3, Tas2r108, Tas2r118, Tas2r119, Tas2r134, Tas2r138. *Rat*: Tas2r1, Tas2r2, Tas2r3, Tas2r4, Tas2r5, Tas2r6, Tas2r7, Tas2r8, Tas2r9, Tas2r12. *Human*: TAS1R1, TAS1R2, TAS1R3, TAS2R1 | Release of gastro-intestinal hormones from enteroendocrine cells and regulation of glucose absorption from enterocytes (Tas1rs, Tas2rs); detection and elimination of harmful compounds and bacterial molecules (Tas2rs) |

(continued)

**Table 1** (continued)

| System | Organs | Transduction molecules | Taste receptors | (Putative) functions |
|---|---|---|---|---|
| | Large intestine | α-Gustducin, PLCβ2, TRPM5 | *Mouse*: Tas1r1, Tas2r108, Tas2r118, Tas2r119, Tas2r131, Tas2r138. *Rat*: Tas2r16, Tas2r26. *Human*: TAS1R1, TAS1R2, TAS1R3, TAS2R1, TAS2R4, TAS2R5, TAS2R9, TAS2R10, TAS2R13, TAS2R38, TAS2R39, TAS2R40, TAS2R42, TAS2R43, TAS2R44, TAS2R45, TAS2R46, TAS2R47, TAS2R49, TAS2R50, TAS2R60 | Detection of nutrients and other intestinal components (Tas1rs); detection and elimination of harmful compounds and bacterial molecules (Tas2rs) |
| | Pancreas | α-Gustducin, PLCβ2, TRPM5 | *Mouse*: Tas1r2, Tas1r3 | Regulation of insulin release |
| Male reproductive system | Testis | α-Gustducin, Gγ13, PLCβ2, TRPM5 | *Mouse*: all Tas1rs and Tas2rs | Reproduction |
| Excretory system | Bladder | – | *Rat*: Tas1r2, Tas1r3. *Human*: TAS1R2, TAS1R3 | Control of bladder contraction |
| Mesenchymal tissues | Bone marrow | α-Gustducin, Gβ1, PLCβ2 | *Human*: TAS2R46 | Control of microenvironment |
| | Adipose tissue | – | *Mouse*: Tas1r2, Tas1r3 | Anti-adipogenic regulation |

and TAS1R3 combine to the universal sweet taste receptor [70, 80]. The bitter taste receptors belong to the TAS2R gene family. Vertebrates possess a highly variable number of putative functional TAS2R genes ranging from 2 to 3 in some bird species to more than 50 in some reptilian and amphibian species [81]. In mammals, an average number of 15 receptors in dogs to 36 in rats have been identified [81]. So far, the best characterized bitter taste receptor repertoire is represented by the 25 functional human TAS2Rs. For 21 of the 25 human TAS2Rs, activators have been identified by heterologous expression assays. It was shown that TAS2Rs differ remarkably in their breadth of tuning with few extremely broadly tuned "generalist" receptors, several narrowly tuned "specialist" receptors, as well as numerous receptors exhibiting intermediate tuning properties [82, 83].

Within the oral cavity, the G protein-coupled receptors are expressed in non-overlapping subsets of taste receptor cells dedicated to detect sweet, umami, or bitter stimuli [84]. Despite this segregation into specialized sweet, umami, and

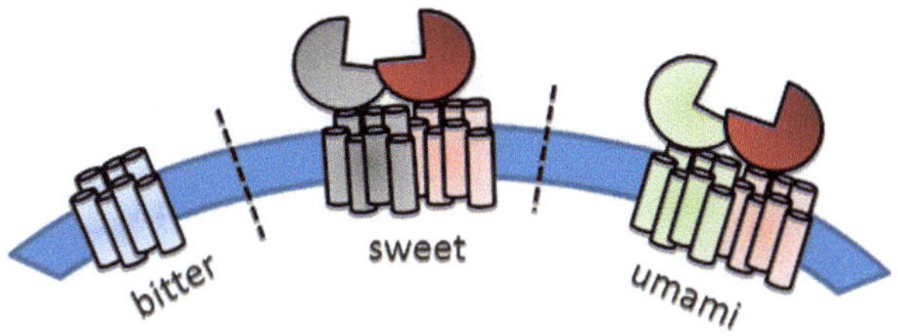


**Fig. 1** Schematic of the G protein-coupled taste receptors. Bitter taste receptors (*left*) do not possess long extracellular amino termini and function as monomers (or homooligomers), whereas sweet (*middle*) and umami (*right*) receptors exhibit large extracellular amino termini and form obligatory heterooligomers. Both receptors consist of a common subunit, TAS1R3, and a specific subunit, TAS1R2 and TAS1R1, for the sweet and umami receptor, respectively

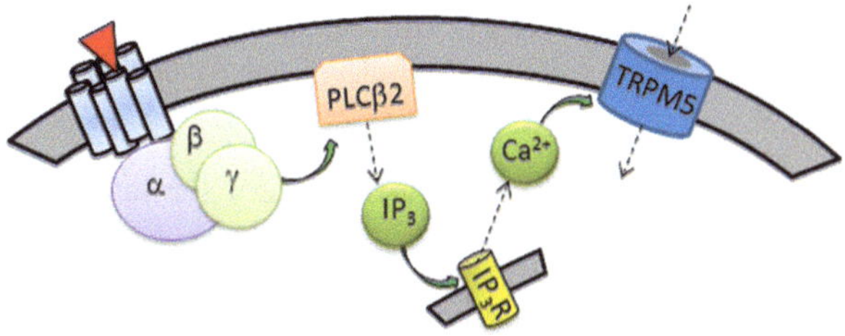


**Fig. 2** G protein-coupled taste receptor linked signal transduction. The activation of a taste GPCR by a taste stimulus (*red triangle*) leads to the activation of a heterotrimeric G protein complex which includes α-gustducin. The subsequent activation of phospholipase Cβ2 results in the generation of $IP_3$, which, in turn, causes release of calcium ions from intracellular stores. The elevated intracellular calcium ion level triggers the opening of the cation channel TRPM5 resulting in cell depolarization

bitter receptive cells, they share common downstream signaling components. These include heterotrimeric G proteins consisting of several alternative Gα-subunits, Gβ3 (Gβ1) and Gγ13 [85, 86]. Of these, the first identified and best characterized Gα-subunit is α-gustducin [87]. Upon receptor activation, the GTP-bound heterotrimeric G protein dissociates and transmits the signal to the membrane-bound phospholipase Cβ2 [88, 89] which, in turn, results in the generation of the second messenger molecule inositol-1,4,5-trisphosphate ($IP_3$). Next, $IP_3$ facilitates opening of the type III $IP_3$ receptor located in the membrane of the endoplasmatic reticulum of taste receptor cells [90–92]. The subsequent release of calcium ions into the cytoplasm activates the cation channel TRPM5 [89, 93–95] leading to membrane depolarization. Finally, the activated taste receptor cell devoid of synaptic contacts releases the neurotransmitter ATP [96] through channels [97–99] into the taste bud, where it eventually activates afferent nerve fibers. Signal termination after taste stimulation may involve RGS21 [100, 101], a protein accelerating the GTP hydrolysis of activated Gα-subunits found in taste receptor cells (Fig. 2).

Some of the signaling proteins, involved in taste signal transduction such as α-gustducin [102], have been identified prior to the taste receptors in non-gustatory

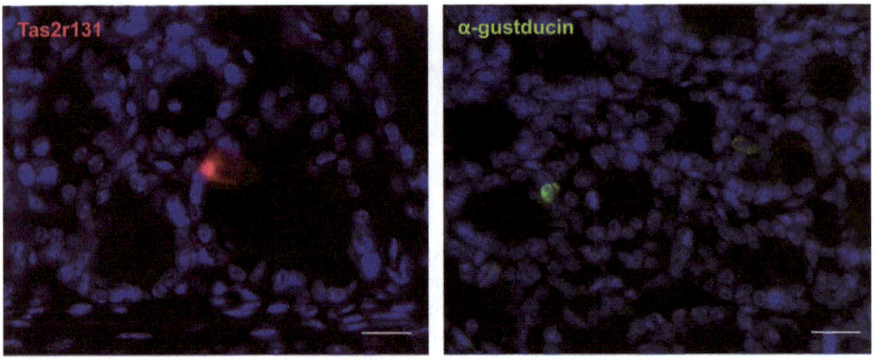


**Fig. 3** Expression of bitter taste receptors and taste-related signaling components in the GI tract. The *left panel* shows single cells in mouse colonic mucosa expressing a red fluorescent protein under the control of the bitter taste receptor Tas2r131 promoter. The *right panel* shows cells expressing α-gustducin, a component of the canonical taste transduction cascade. Scale bar; 20 μm

tissues and served, therefore, as surrogate markers for taste-specific signaling events outside the oral cavity (Fig. 3).

## 3 Brain

Perhaps, one of the most surprising tissues, reported to express taste receptor genes, is the brain. The brain is a highly vulnerable organ utterly safeguarded from direct contact to the environment and even only selectively accessible to substances present in the circulation by its protective blood-brain barrier [103]. One would think the brain might be the least place to find receptors tuned to recognize concentrated tastants present in the oral cavity. Nevertheless, recent reports indicate expression of genes coding for the sweet taste receptor subunits and several bitter taste receptors in human and rodent brains [44, 45, 47, 49]. The first report by Ren et al. [48] used RT-PCR expression profiling, in situ hybridization, and immunohistochemical detection methods to localize the Tas1r2 and Tas1r3 subunits necessary to form a functional sweet taste receptor in distinct areas of the mouse brain. Prominent expression was detected in the hippocampus, hypothalamus, cortex, and the paraventricular nucleus of the thalamus. Moreover, periventricular areas surrounding the third ventricle were found rich in presumably sweet-sensing cells as well. These brain areas do not only exhibit sweet taste receptors but also co-express components of the canonical taste-signaling cascade such as α-gustducin, Gβ3, and Gγ13. In vivo experiments, using food-deprived, hyperglycemic ob/ob mice and control mice fed *ad libitum,* demonstrated that the umami receptor-specific gene *Tas1r1*, but even more pronounced the sweet taste receptor-specific *Tas1r2* gene, is induced by starvation as well as in hyperglycemic mice as indicated by elevated corresponding mRNAs. This effect was restricted to hypothalamic tissue and not observed for cortical tissue suggesting a modulation of hunger/satiety regulatory circuits located in the hypothalamus. Since a selective decrease in Tas1r2 mRNA by elevated glucose levels in the culture medium was also observed in a mouse hypothalamic cell line, the authors concluded that the dynamic regulation of Tas1r1 and Tas1r2 mRNA might be a direct effect caused by local receptor activation rather than indirectly regulated by circulating hormones reporting the nutritional state of animals. Most importantly, the authors observed a similar effect on *Tas1r2* gene regulation also, if the non-metabolizable artificial sweetener sucralose is used instead of glucose excluding contributions by sweet taste receptor-independent glucose-sensing pathways.

A recent report on the effects of chronic treatment of mice with the artificial sweetener acesulfame K demonstrated a downregulation of the Tas1r3 subunit, common to both the sweet and the umami taste receptor, as well as a downregulation of sweet receptor-specific Tas1r2 transcripts in the hippocampus along with alterations in neurometabolic functions of the experimental animals [46]. Compared to wild-type mice, *Tas1r3* knockout animals were protected against negative cognitive effects of chronic acesulfame K treatment attesting to a putative role of *Tas1r* genes in learning tasks.

While taste receptors, devoted to the detection of sweet and umami compounds, may indicate an important endogenous role in energy surveillance of these receptors, the expression of bitter taste receptors in the brain is even more surprising. Nevertheless, Singh et al. [49] recently reported the expression of several *Tas2r* genes in the brainstem, cerebellum, cortex, and nucleus accumbens of rat brain. By RT-PCR, transcripts specific for the receptors Tas2r10, Tas2r4, and Tas2r38 were detected. Using a non-validated Tas2r4 antiserum, labeling was reported in neurons of the *nucleus tractus solitarius* (NTS) and the molecular layer of the cerebellum. Despite the apparent absence of bitter receptor gene expression in glial cells in vivo, the authors were able to detect several Tas2r-specific transcripts in cultured C6 glial cells, originating from experimentally induced rat gliomas [104], as well as in cultured primary neuronal cells. Tas2r4 may represent a functional ortholog of mouse Tas2r108, which is a low-sensitivity denatonium benzoate receptor [73], and/or of human TAS2R4, which has been reported to respond to a variety of bitter compounds, including denatonium benzoate [73, 83] and quinine [83, 105]. Based on this assumption functional calcium imaging experiments were performed. It turned out that, indeed, both cell lines responded in a dose-dependent fashion to rather high concentrations of the two tested stimuli. In another study performed in rat, the rat bitter taste receptor Tas2r1, along with signaling components involved in taste transduction such as α-gustducin, PLCβ2, and TRPM5, was detected in the brainstem tissue by RT-PCR experiments [47]. In contrast to the study by Singh et al [49], bitter receptor gene expression seemed to be rather restricted to few receptors in the brainstem and did not extend to the cerebellum. Further studies, using antisera specific for taste-signaling components, located these molecules in a subset of serotonergic neurons (=tryptophan hydroxylase immunoreactive) within the medullary raphe known to contain chemosensorically active cells [106]. Recent experimental evidence, obtained from human patients suffering from degenerative brain diseases, suggests that *TAS2R* gene expression as determined by quantitative RT-PCR analyses can be altered in cases of Parkinson's disease, Alzheimer's disease, progressive supranuclear palsy, and Creutzfeldt-Jakob disease [44, 45]. Whereas these data underscore the occurrence and possible functional role of brain-expressed bitter taste receptor genes, it is not known if the observed alterations in bitter taste receptor gene expression levels represent merely secondary events during disease progression or play a more active role in these processes.

# 4 Gastrointestinal Tract

The gastrointestinal (GI) system has received considerable attention over the past years, as it turned out to express components of the taste-signaling cascade. Early reports already suggested that cells with chemosensory ability might exist in the epithelial lining of the GI tract. Indeed, the presence of taste-signaling components was confirmed in many GI tissues, whereas the detection of G protein-coupled taste receptors on a cellular level was, depending on the receptor type, more difficult. These initial findings triggered the search for the physiological processes controlled by GI taste receptors.

## *4.1 Taste Receptors in the Gastrointestinal Tract*

The first G protein-coupled receptors described in the GI tract were 11 members of the Tas2r family, whose transcripts were detected by RT-PCR in gastric and duodenal mucosa of rodents [25]. Subsequent reports have described the identification of further bitter taste receptor transcripts in GI tract tissues of rat and mouse origin [3, 13, 14, 24] as well as in human cecum, colon [5, 14], and ileum [107]. From these reports, it appears that different GI organs and even functionally distinct parts of the same organ host different Tas2r subsets pointing to a complex role for Tas2rs in the GI system. A recent qRT-PCR study on the distribution and expression levels of a set of Tas2r along the mouse GI tract confirmed this by showing that (1) Tas2r distribution is rather heterogeneous, with some receptors being present in all GI organs analyzed and others being restricted to only few or single organs, and (2) Tas2rs in mouse GI tract are expressed at a very low level compared to gustatory tissue [19].

Also members of the *TAS1R* gene family of G protein-coupled receptors were identified in the GI system. Regarding the alimentary canal, it was found that mouse small intestine harbors all three subunits (TAS1R1, TAS1R2, and TAS1R3) of the TAS1R family as shown by both RT-PCR analyses and Western blotting [6]. Along the anteroposterior axis, the site of the highest expression is the jejunum in case of the sweet taste receptor subunits TAS1R2 and TAS1R3 (the latter being the most abundant). On the contrary, the umami-specific receptor subunit TAS1R1 is found mostly in the ileum [6]. Further studies took advantage of the available immunological tools for histological techniques to reveal the localization of TAS1R receptors. Sweet and umami taste receptor subunit expression was indeed shown mostly in mouse proximal small intestine in the mucosal layer and at all levels along the crypt-villus axis [2, 4, 16, 17, 108, 109].

In human, similar data to those from rodent models were obtained. In one study, an approximately tenfold higher expression of TAS1R3 with respect to TAS1R2 in small intestinal tissues was reported [26, 110]. Another study described an approximately 1,500 times higher expression of TAS1R3 compared with TAS1R2 in the same tissues [2]. In the same report, the authors measured the amount of TAS1R transcripts also in the large intestine, in which lower levels of expression compared with the small intestine were observed. The presence of TAS1R subunits in man has also been confirmed at the protein level using specific antibodies [11, 110].

In the upper GI tract, Tas1r3 was found in the corpus mucosa of the mouse stomach, both in adult [111, 112] and young postnatal mice [113]. The distribution of the Tas1r3-expressing cells is highly restricted, with the highest number in the region of the limiting ridge, at the border between the corpus and fundus regions [112].

Also other animal models were employed to investigate TAS1R expression in the GI tract: expression was revealed in the small intestine and stomach of pig [114, 115], dog, cat, and horse [116, 117] indicating a conserved and presumably essential role in the GI physiology of mammals for taste receptors. The identification of *Tas1r2* and *Tas1r3* gene expression in explanted pancreatic islets of mouse [18] and in human pancreas and liver tissues [118] suggests the conserved role of these taste receptors is not limited to those GI tissues that come into direct contact to the luminal food constituents.

## 4.2 Taste-Signaling Components in the GI Tract

All main components of the canonical taste transduction cascade (see Sect. 2) have been investigated in GI tissues. The first identified taste-related signaling molecule was α-gustducin in mucosal cells of the stomach and duodenum in rat [102] as well as in the pancreatic duct system [119]. That also other elements of the sensorial taste system are expressed in the GI tract became clear, when α-transducin, which is closely related to and can functionally replace α-gustducin [120], was found in intestinal epithelium and in gastric fundus and antrum [25, 121]. Recently, both G protein α-subunits were found to be expressed in ghrelin-producing cells of the oxyntic glands of the stomach [12]. Numerous reports confirmed and extended the initial findings of taste-signaling elements in the GI tract to most regions of the alimentary canal [2, 4, 16, 17, 23, 108, 117, 122], with the colon harboring the highest number of α-gustducin-positive cells [123]. Noteworthy is the finding of a numerically important α-gustducin-expressing cell population confined to a relative restricted region of the stomach, the limiting ridge [112]. In human, the presence of α-gustducin was demonstrated in the small intestine and in the colon [11, 20], both at mRNA and protein level.

The notion that taste-like mechanisms might work in specialized GI cells was further corroborated by the presence of PLCβ2. Single PLCβ2-positive cells in the small intestine and in colonic mucosa were demonstrated in rodent [2, 19, 122] and human tissue samples [11]. In the stomach, PLCβ2-expressing cells are found in separated populations in two distinct anatomical and functional regions of the organ, the corpus mucosa and the limiting ridge [112, 124].

Another taste-signaling element, the transient receptor potential subfamily M member 5 (TRPM5) channel, is expressed along the entire length of the GI tract from the stomach to colon [2, 11, 26, 110, 122, 125, 126]. In the gastric organ, as it is the case for other taste-related proteins, it is mostly expressed in the region of the limiting ridge [112, 113, 115, 124].

Taste transduction cascade elements are considered as surrogate markers when studying the occurrence of cells outside the sensorial systems with potential chemoresponsive abilities. By analogy with their role in taste cells, it is assumed that they would fulfill the same task also in non-taste cells. This would consequently imply the simultaneous presence of other components of the taste transduction cascade as well as upstream taste receptors. The good availability of experimental tools such as specific antisera and genetically modified mouse lines explains their extensive use as markers for potentially tastant-responsive cells. However, available data indicate that in GI tissues the link between the various taste-signaling components is not as tight as in the gustatory system. In fact, depending on the different regions of the GI tract and even within the same region, the degree of colocalization of the various transducing elements varies considerably. The entire population of TRPM5-GFP-labeled cells in duodenal villi and the colon co-expresses α-gustducin, while only one-third of TRPM5-positive duodenal cells show also PLCβ2 expression [2]. In the colon, PLCβ2-positive cells even represent a completely separated population with respect to TRMP5-labeled cells [2]. In the stomach, the situation is again different: within the gastric groove, all α-gustducin- and TRPM5-positive cells also express PLCβ2 [124]. Regarding the co-expression of taste transduction elements with taste receptors, it turned out that TRPM5-positive cells in mouse small intestine do neither express Tas1rs nor Tas2rs [2], while in the same regions α-gustducin exhibits a certain degree of colocalization with the Tas1r subunits [4, 11, 17]. Moreover, in the GI tract taste-related signaling elements may facilitate signal propagation of G protein-coupled receptors currently not considered as *bona fide* taste receptors such as fatty acid receptors and oleoylethanolamide and bile acid receptors which are co-expressed with α-gustducin in mouse colon [123].

## 4.3 Cell Types

The numerous studies performed so far have demonstrated that the GI tract contains a heterogeneous cell population expressing taste-related molecules. One of these identified cell types are brush cells (or tuft cells or caveolated cells) representing a minor differentiated cell type that manifest as single cells scattered throughout the mucosa. In mouse, they represent 0.4% of the intestinal epithelium and they possess a tapering cell body with a tuft of microvilli on top protruding into the intestinal lumen [127]. The structural features of brush cells have led to the speculation that they might sense the chemical content in the gut lumen through their apical pole [128]. The finding that brush cells express α-gustducin represented the first molecular clue that this cell type may indeed participate in intestinal chemoreception [102, 119, 121]. Subsequent studies have shown that mouse stomach brush cells, lining the border between fundus and corpus, the so-called limiting ridge, express also other taste transduction components like TRPM5, PLCβ2, and especially Tas1r3 [10, 124]. In the mucosa of the small and large intestine, brush cells express with a variable degree of overlap TRPM5, α-gustducin, PLCβ2, and Tas1rs [2, 122].

Another intestinal cell population, discovered to express taste-related molecules, are the enteroendocrine cells (1% of the lining epithelium) [6]. In their entirety, they express up to 20 gastrointestinal hormones that, together with their localization within the mucosa, define a complex class of enteroendocrine cells that function as important metabolic regulators [129]. Indeed, it was shown that human and mouse duodenal enteroendocrine L (expressing GLP-1, GLP-2, and PYY) and K (expressing GIP) cells harbor the sweet taste receptor-specific subunit Tas1r2, together with other taste-signaling components [11, 17]. In accordance with these findings, Tas1r2 was never found to be expressed in enteroendocrine I cells identified by the expression of CCK. On the contrary, the specific umami receptor subunit Tas1r1 is consistently expressed in I cells [4], indicating the segregation of the sweet and umami receptors in two different enteroendocrine cell populations that secrete different hormones and thus exert various functions such as induction of satiety, gastric emptying, intestinal motility, and stimulation of insulin secretion. Tas1r3 is expressed in both cell types, consistent with its role as binding partner in the formation of both functional sweet and umami taste receptors as well as in ghrelin-producing cells of the duodenum [4, 10]. In mouse jejunum and in human ileum, enterochromaffin (EC) cells producing serotonin were found to express α-gustducin [23], TAS1R3, and TAS2R1 [107]. In the colon, α-gustducin is expressed again in L and K cells but not in EC cells [20, 123]. Also in the stomach of mouse, the Tas1r3 subunit was found in gastrin-positive G cells [111]. Moreover, the Tas1r3 subunit was found to be expressed in the corpus of mouse stomach in ghrelin-positive cells [10] (Fig. 4).

Regarding the identification of cell type(s) expressing Tas2rs in situ, the data are much more limited. The main reasons for the scarcity of data might be a lack of reliable antisera as well as rather low expression levels of bitter taste receptor

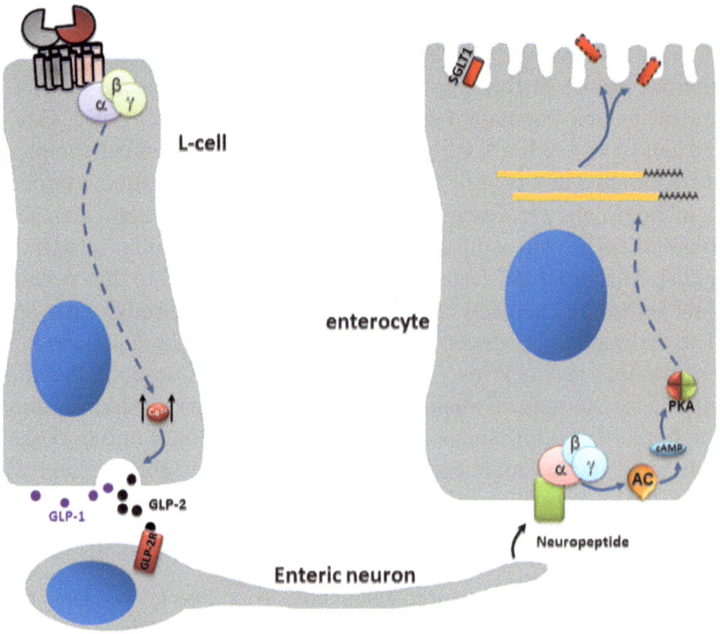


**Fig. 4** Model of the sweet receptor mediated pathway leading elevated intestinal glucose absorption. Activated sweet taste receptors localized on enteroendocrine L cells trigger the elevation of intracellular calcium levels via a pathway involving α-gustducin. The L cells secrete the peptide hormones GLP-1 and GLP-2. Secreted GLP-2 interacts with GLP-2 receptors located on neighboring enteric neurons, which by means of a so far unidentified neuropeptide stimulate enterocytes. Within the enterocytes, this signal results in an upregulation of SGLT1 mRNA and, finally, the incorporation of additional SGLT1 into the apical membrane. The elevation of SGLT-1 gene transcription is mediated by the cAMP-regulated protein kinase A (PKA). Redrawn and modified from Shirazi-Beechey et al. [21]

mRNAs in GI tissues, which limits the use of other histological methods such as in situ hybridization [19]. To date, only two reports directly addressed the in situ expression pattern of bitter taste receptors in the rodent GI tract. One study identified enteroendocrine cells in mouse duodenum, expressing bitter taste receptor genes by immunohistochemical experiments using a non-validated antiserum raised against the receptor Tas2r138 [13]. More recently, the difficulties to demonstrate the in situ expression of Tas2rs were overcome by using a knock-in mouse model, in which a fluorescent marker protein reports the activation of the Tas2r131 locus with high sensitivity and reliability. It was shown that the Tas2r131 gene is expressed in the GI tract with a proximal-to-distal gradient from the jejunum to colon and the Tas2r131-positive cells in the colonic mucosa were identified as a subset of goblet cells [19]. Curiously, these cells do not express α-gustducin or PLCβ2, so other signal transduction elements are likely to relay the signal downstream of the receptor [19].

## 4.4 Model Cell Lines

Over the years, several working groups used cell lines derived from gastrointestinal tissues to study taste receptor expression and function. GI taste receptors in enteroendocrine cell lines were firstly identified in the mouse STC-1 cell line [25], which was shown to express several bitter taste receptors as well as components of the canonical taste transduction cascade [3, 24, 25]. The same cell line also expresses the subunits Tas1r1 and Tas1r3 [4], and it is able to respond to tastants by releasing GI hormones such as CCK [3, 4]. Other enteroendocrine cell lines of gastrointestinal origin were used: the HuTu-80 cell line from the human intestine expresses TAS2Rs [20, 130], and the NCI-H716 of human origin is equipped with both sweet taste receptors [11] and bitter taste receptors [5, 20] that, upon activation, can trigger the secretion of GLP-1 [5, 11]. Despite these findings, the functional significance of the simultaneous presence of sweet and bitter taste receptors in the same cells still needs further clarification, as it appears counterintuitive that sweet and bitter taste receptors should trigger the same physiological responses.

## 4.5 Physiological Functions of GI Taste Receptors

There is an increasing body of evidence pointing at a role of sweet taste receptors in GI organs as regulators of glucose metabolism. It is well known that administration of glucose *per os* is more effective than a direct injection of glucose into the blood circulation to stimulate insulin release from the pancreas, which in turn promotes the cellular uptake of the circulating glucose [131]. This phenomenon is caused by the presence of sugars in the intestinal lumen triggering the secretion of potent insulin secretagogues, the incretin hormones GLP-1 and GIP, from enteroendocrine cells of the gut wall [132]. However, the sensor molecules, detecting luminal sugars and triggering the physiological responses, have been rather elusive. Jang et al. demonstrated that sweet taste receptors are expressed together with all components of taste transduction cascade in GLP-1- and GIP-secreting enteroendocrine L and K cells, respectively [11]. They observed that excised duodenal villi treated with glucose release GLP-1 and that in α-gustducin knockout animals upon glucose gavage the plasma levels of GLP-1, GIP, and insulin were lower than in wild-type littermates. They concluded that gustducin-coupled sweet taste receptors are the luminal sensors that detect sugars in the gut and trigger incretin hormone release from enteroendocrine cells. The physiological consequences of sweet taste receptor activation in gut enteroendocrine cells are not limited to the so-called incretin effect. Prolonged exposure to a carbohydrate-rich diet or to a diet enriched with artificial sweeteners causes an increase in the expression of the $Na^+$-glucose cotransporter 1 (SGLT-1) in the brush border membrane of enterocytes resulting in higher absorption of carbohydrates from the small intestine. This effect is abolished in α-gustducin and Tas1r3 knockout mice, suggesting that the sweet taste receptor in enteroendocrine cells senses luminal sugars to adjust the intestinal uptake capacity to the diet [17]. The authors proposed that a paracrine signal is released by the enteroendocrine cells resulting in an upregulation of the expression of SGLT-1 by enterocytes. The responsible signaling molecule might be GLP-2, which, after being secreted by enteroendocrine L cells [117, 133], activates enteric neurons expressing the receptor GLP-2R. This in turn causes the release of a yet unidentified neuropeptide that stimulates the basolateral membrane of enterocytes to upregulate SGLT-1 expression [21]. Also studies performed in human describe a situation in which sweet taste receptors are actively involved in the release of GLP-1 and PYY hormones from enteroendocrine cells of the gut upon intragastric and intraduodenal glucose infusions, as lactisole, a potent inhibitor of the human sweet taste receptor, decreased the quantity of measured hormones [22, 134]. Some in vivo studies involving animals and humans show conflicting results with regard to the specific involvement of sweet taste receptors in the measured response and pointed out that other carbohydrate sensors might operate in the gut. In some of these studies, artificial sweeteners did not increase incretin hormone levels from the gut of human volunteers and laboratory animals [135–137]. A recent study indeed showed that there are more mechanisms governing GLP-1 release upon glucose stimulation in the alimentary canal, and they differ depending on the localization along the GI tract [7]. Using Tas1r3 and Tas1r2 knockout mice, the authors showed that in an oral glucose tolerance test, the Tas1r3 subunit is fundamental for the control of plasma glucose and insulin levels, whereas Tas1r2 null mice do not show any difference with respect to WT animals. Experiments with small intestinal explants treated with glucose, fructose, and sucralose revealed that the Tas1r2 subunit is less effective in stimulating GLP-1 release than the Tas1r3 subunit, which can partially compensate for the absence of Tas1r2. Using colon explants, it was demonstrated that the main control mechanism of GLP-1 release is $K_{ATP}$ channel dependent and glucose specific, whereas taste-like mechanisms do not seem to be fundamental for eliciting incretin release from this part of the gut [7]. The observed phenotypical differences in Tas1r2 knockout versus Tas1r3 knockout mice, which both should render the sweet taste receptor nonfunctional, have been observed in other studies as well (e.g., see below [138, 139]). Whether this might be due to a residual function of homooligomeric Tas1r3 as low-affinity receptor for natural sweet compounds [140] and/or compensatory effects of other members of the class C GPCR family such as Tas1r1 [71], the calcium-sensing receptor [141], CaSR, and the amino acid-sensitive GPRC6A [142, 143], perhaps by the formation of alternative heterooligomers, remains to be determined.

There is evidence that other mechanisms in addition to the glucose metabolic pathway fine-regulate the release of insulin from pancreatic β-cells [144]. Indeed, it was observed that artificial sweeteners provoke both insulin release from isolated pancreatic islets and an augmented insulin secretion from MIN6 cells treated with glucose. A similar potentiation effect was demonstrated for fructose to increase insulin secretion from mouse and human islets [15, 18]. Some data suggest that pancreatic β-cells may sense sweet molecules mostly by means of a TAS1R3 homodimer, because (1) the TAS1R2 subunit is expressed at a very low level compared to TAS1R3 and (2) TAS1R2 mRNA knockdown does not affect the cellular response to artificial sweeteners, as it is the case with a knockdown of TAS1R3 [138, 139].

Concerning umami taste receptor function in the gut, fewer data are available. However, one study showed that the mouse STC-1 enteroendocrine cell line is able to release CCK in response to amino acid stimulation similar to small intestinal tissue. The response was shown to be umami taste receptor dependent as demonstrated by mRNA knockdown experiments with STC-1 cells and by treating intestinal explants with gurmarin, an inhibitor of rodent sweet [145, 146] and umami [147, 148] responses [117]. In particular in the gastric mucosa, L-amino acid sensing might be facilitated by mechanisms different from umami taste perception [149].

Also the potential role of bitter taste receptors in GI physiology is less well characterized compared to that of the sweet taste receptor. This seems to be mostly due to the limited knowledge on the cell type(s) expressing bitter taste receptors as well as the high pharmacological activity of bitter compounds, which, on top of that, possess an incredible variety of chemical structures [83]. A good example for the fact that GI responses must not necessarily involve (bitter) taste-related signaling elements has been obtained in rat pancreatic tissue. Here, Straub et al. [150] used the bitter substance denatonium benzoate to stimulate insulin release from clonal HIT-T15 β-cells as well as from isolated rat pancreatic islets. In the presence of denatonium, islets bathed in glucose solution start secreting insulin through a mechanism that does not involve gustducin or transducin as assessed in a trypsin digestion assay. It rather inhibits $K_{ATP}$ channel activity, leading to cell membrane depolarization that, in turn, opens the voltage-gated calcium channels allowing the influx of calcium from the extracellular milieu [150].

One of the hypothesized functions for Tas2r in murine GI tract is the delaying of gastric emptying. It is indeed intuitive that bitter compounds, which are frequently rather harmful for the organism [151], would trigger defensive reactions in case of accidental ingestion to prevent them from spreading throughout the body. The first reports on this issue showed contrasting evidences in experiments conducted in human subjects [152, 153]. Nevertheless, results obtained with rodent models pointed indeed to a modulatory role of ingested bitter compounds in the regulation of gastric motility. Intragastric infusions of 10 mmol/L denatonium significantly delayed the speed of gastric emptying in rats [8]. In another study, a mouse model was used to show that intragastric gavage of a mixture of bitter compounds provokes the secretion of ghrelin into the blood circulation leading to a short-term increase in food intake as assessed in GHS-R null mice [12]. Both ghrelin secretion and increase in food intake were proven to be dependent on α-gustducin. However, 4 h after the administration of the bitter mixture, a decrease in both gastric emptying and consequently in food intake was observed. These effects are independent of CCK and GLP-1 release from the small intestine but rather controlled by an unknown effect of bitter compounds on the smooth muscles of the stomach wall [12]. In view of these findings and due to the fact that rodents lack the vomiting reflex, the delay in gastric emptying appears to be an important adaptive mechanism that would slow down the rate of ingestion of further noxious compounds. In agreement with this, a report from Kaji et al. [14] showed that application of 6-*n*-propyl-2-thiouracil onto explanted colonic mucosa of human and rat elicits ion and fluid secretion. This would in turn be important to flush out harmful compounds that have reached the colon lumen. Jeon and coworkers suggest that murine Tas2r activity in the small intestine is regulated by diet composition presumably to prepare the intestinal lumen to a defensive response involving the xenobiotic transporter ABCB1 against ingested noxious compound [13, 154].

Part of the GI bitter-sensing mechanism may involve vagal nerve fibers, and it was proven that intragastric administration of bitter compounds in mice increases the number of activated neurons of the mid-NTS, the terminal station of the vagal afferents from the GI tract, through activation of $CCK_1$ and $Y_2$ receptors. This effect was shown by c-fos expression and it is abolished by subdiaphragmatic vagotomy. It was proposed that Tas2rs expressed in enteroendocrine cells trigger the release of CCK and PYY, which in turn activate the adjacent vagal fibers to relay the signal to the brain [9]. Further experiments with rats showed that other brain areas exhibit augmented c-fos expression after gavage with bitter compounds and that this correlates with an avoidance behavior upon stimulation with flavors previously paired with intragastrically administered bitter substances [155].

## 5 Respiratory System

Soon after the discovery of bitter taste receptor expression in the gastrointestinal tract [25], the respiratory system, another extraoral epithelium that is constantly challenged by potentially harmful substances present in the environment of organisms, was shown to express *Tas2r* genes [28]. The cells, which express not only Tas2r genes but also α-gustducin, PLCβ2 [28], and TRPM5 [125], show distinct morphological features identifying them as solitary chemosensory cells [28]. Within the respiratory epithelium of the rodents' nasal cavity, these bitter taste receptor-expressing cells are ideally suited to detect irritating noxious chemicals entering the organism through the inhaled air. Moreover, synaptic contact sites between solitary chemosensory cells in the nasal cavity with trigeminal nerve fibers have been identified [28]. Indeed, it was shown that stimulation of the cells with cognate bitter substances results in the activation of trigeminal nerve responses, which mediate depression of the respiratory rate [28]. More recently, bacterial quorum sensing molecules such as acyl-homoserine lactones were found to activate nasal solitary chemosensory cells as well suggesting that these cells not only serve a role in minimizing the amount of inhaled harmful xenobiotics but also are involved in defense mechanisms against pathogenic microorganisms [35, 156]. Recently, the presence of solitary chemosensory cells that express bitter receptors and corresponding signal transduction elements was demonstrated in human nasal respiratory epithelium as well suggesting the existence of a conserved protective mechanism [157].

Bitter taste receptor expression is not restricted to the upper airways but extends into the lower airways. Intriguingly, Shah et al. [32] demonstrated that ciliated cells of human airway epithelia express bitter taste receptors and respond to stimulation with bitter compounds with changes in their ciliary beat frequency. The authors believe that this mechanism allows the rapid elimination of noxious compounds in a cell-autonomous fashion. Recently, the direct involvement of the human bitter taste receptor TAS2R38 in the detection of bacterial quorum sensing molecules was proposed [30]. The authors found that human TAS2R38 is expressed in ciliated epithelial cells of the upper airways and that activation of this receptor by acyl-homoserine lactone results in an elevated ciliary beat frequency leading to improved mucociliary clearance as well as direct antibacterial effects [30]. Strikingly, a common genetic polymorphism of the TAS2R38 gene, which furnishes carriers of this genetic variant with a nonfunctional receptor protein, was shown to be associated with a higher incidence of sinonasal gram-negative bacterial infections [30]. A subsequent study, performed in knockout mice lacking taste-related signaling components such as PLCβ2, Trpm5, and α-gustducin, confirmed a role of bitter taste receptor signaling in the protection against airway infection caused by gram-negative bacteria. It was shown that nasal epithelial cells of mice signal in response to stimulation with acyl-homoserine lactone and that this response requires PLCβ2 and Trpm5 but not α-gustducin [158]. It should be noted that another study, performed in genetically modified mice expressing TRPM5-GFP, failed to identify taste-related cell types other than solitary chemosensory cells in the airways of mice; it remains to be seen how these contrasting results resolve [33].

Another site of bitter taste receptor expression in mammalian airways is smooth muscle cells of human airways [27]. Apparently counterintuitive, the stimulation of airway smooth muscle cells with bitter agonists did not result in bronchoconstriction consistent with a protective role, but rather resulted in the dilation of airways [27]. This report, although not undisputed (cf. [33, 159, 160]), generated considerable interest in the applicability of bitter tastants for the treatment of obstructive lung diseases. The bronchodilatory effect of bitter agonists observed by Deshpande et al. has been confirmed by a recent study performed by Zhang et al. [37] on cultured mouse airway smooth muscle cells and airway explants. Deshpande et al. [27] observed bitter tastant-evoked calcium increases in cultured airway smooth muscle cells, similar to the activity of established bronchoconstrictors, such as bradykinin or histamine, and suggested that local calcium increases via BKca channels could be responsible for the observed smooth muscle relaxation. In contrast, Zhang et al. suggested a different mechanism underlying this effect: the comparatively small increase of calcium ion levels in bitter compound-stimulated cultured airway smooth muscle cells and the failure to reproduce the local calcium events, observed by Deshpande et al. [27], led Zhang et al. to believe that bitter compound signaling exerts its bronchodilatory effect rather by antagonizing smooth muscle constriction. Indeed, the authors reported that bitter compounds inhibit the activation of voltage-dependent calcium channels by bronchoconstricting drugs and that this effect causes smooth muscle relaxation [37]. Although the exact mechanism, by which bitter taste receptor signaling causes bronchodilation, requires clarification, the finding that the large number of cognate bitter compounds may, in fact, represent a pool of potentially powerful drugs in the treatment of obstructive lung diseases is intriguing. The recent reports that *TAS2R* gene expression is increased in severe, therapy-resistant asthma in children and, hence, the potential target receptors for such alternative treatment strategies seem to be even upregulated [31] are promising prerequisites for a potential therapeutic value of cognate bitter substances.

Although the majority of reports about taste receptor expression and function in the airways focus on bitter taste receptors, which are believed to fulfill a protective role similar to their suspected function in the gustatory system, it should be noted that also *Tas1r* gene expression has been detected in the respiratory system of rodents [33]. Whether the detection of Tas1r3 in solitary chemosensory cells of the airways in rodents hints at the presence of sweet or umami receptors or both remains to be determined.

## 6 Heart

One of the most recent additions to the growing list of extraoral tissues expressing taste receptors is the heart [50]. By qRT-PCR analyses of rat neonatal whole heart cDNA, the two genes encoding the umami receptor subunits, Tas1r1 and Tas1r3, as well as seven bitter taste receptor genes were found to be expressed. Subsequently, samples of ventricular tissue of failing human hearts were tested and revealed the expression of more than half of all human *TAS2R* genes. Remarkably, the mRNA of the human bitter taste receptor TAS2R14, which is known to represent a very broadly tuned human TAS2R [83, 161], was found to be present at levels comparable to that of the $\beta_1$-adrenergic receptor [50]. Interestingly, a recent report suggested that this receptor responds to numerous clinically relevant drugs and that a considerable overlap between TAS2R14 agonists and small molecules, interacting with human ether-a-go-go related gene (hERG) potassium channels, exists [162]. As hERG potassium channels play an important role in cardiac physiology (e.g., [163]), more research on the impact of bitter compounds on cardiac tissue seems to be warranted. Using primary cell cultures derived from neonatal rat hearts, taste receptor gene expression as well as taste-related signaling components such as α-gustducin, phospholipase Cβ2, and TRPM5 was confirmed in cardiomyocytes. Concerning the localization of taste receptor mRNAs in situ, *Tas1r* and *Tas2r* gene expression was demonstrated in small subsets of cardiac cells by in situ hybridization. Further, Tas1r1 promoter-driven expression of fluorescent marker protein was observed in the myocardium of a knock-in mouse line [50]. Interestingly, the investigated taste receptor genes showed distinct developmental expression patterns and were regulated by experimentally induced nutrient deprivation [50].

# 7 Reproductive System

For successful fertilization, mammalian spermatozoa have to find their way along the female genital tract to eventually find the egg cell and fuse with it. The idea that spermatozoa are guided on their way by chemical cues is comprehensible, and in fact, chemotactic navigation of sperm cells has been demonstrated in invertebrates and vertebrates including mammals [164]. Among the first chemoreceptors suspected to serve a function in this process have been the odorant receptors (OR). Indeed, it has been shown that the human odorant receptor OR 17-4, which is located on human spermatozoa, is responsible for the attraction of sperm cells toward the compound bourgeonal [52]. Similarly, in mouse the odorant receptor MOR23 has been shown to mediate responsiveness of sperm cells to the agonist lyral to regulate sperm motility [51]. The exact role of ORs in mammalian sperm chemotaxis and the existence of relevant OR ligands in the female reproductive tract, however, remain to be determined [165]. Even though also the expression of taste receptor genes in mouse [67, 74] and human [166] testes was recognized quite some time ago, detailed analyses of expression patterns and putative functional roles have been performed only rather recently [39]. Similar to other extragustatory tissues, the first molecule that has been investigated to identify taste-related signaling in male reproductive tissue was the Gα-subunit, α-gustducin [87]. It was shown that α-gustducin expression occurs already in differentiating spermatids and is retained in mature spermatozoa [167]. Moreover, also other components of the gustatory signaling cascade, involved in signal transduction of taste GPCRs such as Gγ13, phospholipase Cβ2, and the transient receptor potential channel TRPM5, have been identified in mouse sperm cells [40]. Indeed, expression of all 3 *Tas1r* genes and all 35 putatively functional *Tas2r* genes has been detected in mouse testes [43]. The Tas1r3, the common subunit of sweet and umami receptors, is localized on the convex side of the head and the principle piece of the sperm flagellum [41]. By using genetically modified mice, which express a fluorescent marker protein under the control of the promoter of the umami receptor-specific *Tas1r1* gene, an overlapping expression pattern was observed indicating the presence of both subunits of the umami receptor. This was confirmed by double-labeling immunofluorescence experiments on human spermatozoa [41]. Intriguingly, sperm derived from Tas1r1-deficient mice showed differences in basal intracellular calcium ion as well as cyclic AMP levels suggesting a role of the umami receptor in sperm function [41]. Hence, umami receptor agonists and modulators of umami receptor responsiveness may play a, so far unanticipated, role for mammalian fertility including human. The latter may also apply for sweet-tasting substances and cognate modulators of the mammalian sweet taste receptor, as also the sweet taste receptor-specific *Tas1r2* gene is expressed in spermatozoa ([38, 42], however, cf. [41]).

Intriguingly, the simultaneous absence of *TAS1R3* and *α-gustducin* genes in genetically modified mice resulted in male-specific sterility confirming an important contribution of taste-signaling molecules for normal fertility [42]. Moreover, the authors of this study engineered mice with a humanized TAS1R3 subunit instead of the native murine Tas1r3 and combined it with an α-gustducin knockout mouse strain. This mouse line was susceptible to pharmacological induction of male sterility with the antilipid drug clofibrate, which blocks human TAS1R3-mediated signal transduction, but is not able to act via the mouse Tas1r3 ortholog [42]. As several drugs, but also environmental pollutants such as herbicides, have been identified to interfere with the activation of human TAS1R receptors, the authors indicate a potential link between some forms of male infertility and these compound classes. On the other hand the numerous TAS1R receptor agonists may show so far not anticipated treatment options for some forms of male infertility. Interestingly, in a double gene-targeted mouse line, it was shown that the umami receptor subunit in mouse sperm cells is colocalized with the bitter taste receptor Tas2r131, which is in sharp contrast to the situation in taste receptor cells [36]. As all 35 mouse bitter taste receptor genes are expressed in the testis and it was shown that mouse spermatids respond to several bitter compounds with increases of intracellular calcium levels [43], it would be highly interesting to see if also bitter taste receptor agonists and antagonists may affect male fertility. The generation of a mouse line, in which all cells that express the bitter taste receptor gene Tas2r105 are genetically ablated by the expression of diphtheria toxin A, revealed a considerable reduction in testicular size [40]. However, this mouse line still produces a reduced number of spermatids including Tas1r3-positive spermatids, suggesting a heterogeneous population of sperm cells. It would be very interesting to see whether bitter agonists and activators of Tas1rs influence sperm cell physiology in synergistic or opposing fashion or whether different subpopulations of sperm cells might be susceptible to different tastants or taste modulators.

# 8 Outlook

The finding of taste receptors and taste-related signaling components in non-gustatory tissues has received enormous, and ever-increasing, attention over the recent years. The investigation of taste-related signaling is no longer only relevant for researchers working in the field of chemoreception, but has attracted scientists coming from diverse areas such as respiratory, gastrointestinal, reproductive, and cardiovascular systems to name just a few. This has stimulated multidisciplinary research considerably and surely affected the way of how taste receptors are presently understood; they are clearly not only "taste" receptors anymore. However, this rapid expansion of knowledge, gained on potential roles for taste-related signaling systems outside the gustatory system, has raised numerous open questions, which need to be addressed, together with conflicting results that surfaced alongside, in the future.

One open question arises from the fact that numerous animal species have lost some, several, or even numerous taste receptor genes during evolution. For instance, cats [168] and chicken [169] lost their sweet taste receptor-specific

*Tas1r2* gene; the giant panda genome does not possess a functional umami taste receptor-specific *Tas1r1* gene [170]; and sea lions and bottlenose dolphins [171], as well as vampire bats [172, 173], do not possess any functional Tas1r subunit, and the dolphin genome may, in addition, not even contain *Tas2r* genes [171]. Assuming that taste receptors indeed fulfill an integral role in, e.g., brain function and fertility, how can the ever-growing number of animal species that have been demonstrated to lack taste receptor genes compensate for the absence of these molecules? The rather "benign" phenotypes, observed in the various taste receptor knockout mouse models, suggest that even animals, which have maintained their full taste receptor gene repertoire in the course of evolution, can obviously compensate the acute loss of extraoral taste receptors quite well.

Another issue arises from the fact that taste receptors in the oral cavity are rather low-affinity receptors devoted to detect food-derived compounds at high and nutritionally relevant concentrations. For taste receptors, expressed outside the oral cavity, the question arises whether the corresponding stimuli reach concentrations relevant to modulate the receptor's activities. For bitter receptors present in the brain, heart, and testis, this question appears most obvious: does one need to consider scenarios in which an orally consumed bitter substance reaches concentrations in the organism sufficiently high to activate Tas2rs in these tissues or would these receptors rather respond to yet undiscovered endogenous high-affinity ligands? If the latter was true, what would have been the major driving force for the development of Tas2rs during evolution, the endogenous ligand(s), or food-derived xenobiotics? Similarly, mammalian sweet taste receptors respond to natural sweet compounds in the mid- to high millimolar range [70], since blood glucose and even more so brain glucose levels are within the low millimolar concentration range [174]; under what circumstances would the sweet taste receptor will become activated?

In order to determine the physiological role of taste receptors in extragustatory tissues, it is important to identify the cell types that express the receptor gene as well as demonstrate unambiguously the involvement of the receptor in the physiological response observed upon stimulation with tastants. These two issues have rarely been addressed satisfactorily in the past. One of the reasons for the lack of data concerning the in vivo expression pattern of bitter taste receptors, e.g., in the gastrointestinal tract, is that specific antisera raised against these molecules are scarce. It would be beneficial to develop these tools in order to link the physiological activity of tastant molecules to taste receptor-dependent signaling more tightly. As it seems that some selective reagents such as an antiserum raised against the human bitter taste receptor TAS2R38 [175] start to become available, the chances that this situation will improve over time are good. Although requiring a considerable amount of effort, the generation of genetically modified mice, which strongly express marker molecules under the control of taste receptor gene promoters, represents a possible option for the identification of non-gustatory cell types producing Tas2rs [19].

Because tastants may activate cellular signals independent of taste receptors, physiological responses as a result of tastant stimulation are not necessarily sufficient evidence for the involvement of taste receptors. More convincing evidence for taste receptor-dependent activity could be obtained by knockdown/ knockout approaches in suitable cellular or animal systems. Alternatively, the use of taste receptor inhibitors such as, e.g., lactisole, a selective blocker of the human sweet taste receptor, as well as the use of receptor-matched agonist sets, consisting of known activators and non-activators, should help to clarify the putative involvement of taste receptors better. This of course depends heavily on the continuing success of the in vitro characterization of taste receptor responses to identify receptor activation patterns and inhibitors for a larger panel of taste receptors.

The intriguing finding that taste receptors are expressed in numerous non-gustatory tissues and fulfill within these tissues a variety of important physiological functions does not only have implications on the way we are looking at "taste" receptors but also their agonists, the tastants. Obviously, tastants may act via the activation of extraoral taste receptors in multiple ways on the physiology of vertebrates. In particular, bitter-tasting drugs may exert many off-target effects via the activation of extraoral bitter taste receptors as pointed out in a highly recommendable recent review article [176]. In the future it will be very important to investigate such potential effects in detail in order to develop strategies to avoid adverse side effects of bitter-tasting pharmaceuticals or even to identify novel drugs based on bitter "lead" structures.

**Acknowledgments** The authors thank the German Ministry of Education and Research (BMBF, #0315669, to WM and MB) for financial support.

## References

1. Behrens M, Meyerhof W (2011) Gustatory and extragustatory functions of mammalian taste receptors. Physiol Behav 105(1):4–13
2. Bezencon C, le Coutre J, Damak S (2007) Taste-signaling proteins are coexpressed in solitary intestinal epithelial cells. Chem Senses 32(1):41–49
3. Chen MC, Wu SV, Reeve JR Jr, Rozengurt E (2006) Bitter stimuli induce $Ca^{2+}$ signaling and CCK release in enteroendocrine STC-1 cells: role of L-type voltage-sensitive $Ca^{2+}$ channels. Am J Physiol Cell Physiol 291(4):C726–C739
4. Daly K, Al-Rammahi M, Moran A, Marcello M, Ninomiya Y, Shirazi-Beechey SP (2013) Sensing of amino acids by the gut-expressed taste receptor T1R1–T1R3 stimulates CCK secretion. Am J Physiol Gastrointest Liver Physiol 304(3):G271–G282. doi:10.1152/ajpgi.00074.2012
5. Dotson CD, Zhang L, Xu H, Shin YK, Vigues S, Ott SH, Elson AE, Choi HJ, Shaw H, Egan JM, Mitchell BD, Li X, Steinle NI, Munger SD (2008) Bitter taste receptors influence glucose homeostasis. PLoS One 3(12):e3974
6. Dyer J, Salmon KS, Zibrik L, Shirazi-Beechey SP (2005) Expression of sweet taste receptors of the T1R family in the intestinal tract and enteroendocrine cells. Biochem Soc Trans 33 (Pt 1):302–305
7. Geraedts MC, Takahashi T, Vigues S, Markwardt ML, Nkobena A, Cockerham RE, Hajnal A, Dotson CD, Rizzo MA, Munger SD (2012) Transformation of postingestive

glucose responses after deletion of sweet taste receptor subunits or gastric bypass surgery. Am J Physiol Endocrinol Metab 303(4):E464–E474
8. Glendinning JI, Yiin YM, Ackroff K, Sclafani A (2008) Intragastric infusion of denatonium conditions flavor aversions and delays gastric emptying in rodents. Physiol Behav 93 (4–5):757–765
9. Hao S, Sternini C, Raybould HE (2008) Role of CCK1 and Y2 receptors in activation of hindbrain neurons induced by intragastric administration of bitter taste receptor ligands. Am J Physiol Regul Integr Comp Physiol 294(1):R33–R38
10. Hass N, Schwarzenbacher K, Breer H (2010) T1R3 is expressed in brush cells and ghrelin-producing cells of murine stomach. Cell Tissue Res 339(3):493–504
11. Jang HJ, Kokrashvili Z, Theodorakis MJ, Carlson OD, Kim BJ, Zhou J, Kim HH, Xu X, Chan SL, Juhaszova M, Bernier M, Mosinger B, Margolskee RF, Egan JM (2007) Gut-expressed gustducin and taste receptors regulate secretion of glucagon-like peptide-1. Proc Natl Acad Sci U S A 104:15069–15074
12. Janssen S, Laermans J, Verhulst PJ, Thijs T, Tack J, Depoortere I (2011) Bitter taste receptors and alpha-gustducin regulate the secretion of ghrelin with functional effects on food intake and gastric emptying. Proc Natl Acad Sci U S A 108(5):2094–2099
13. Jeon TI, Zhu B, Larson JL, Osborne TF (2008) SREBP-2 regulates gut peptide secretion through intestinal bitter taste receptor signaling in mice. J Clin Invest 118(11):3693–3700
14. Kaji I, Karaki S, Fukami Y, Terasaki M, Kuwahara A (2009) Secretory effects of a luminal bitter tastant and expressions of bitter taste receptors, T2Rs, in the human and rat large intestine. Am J Physiol Gastrointest Liver Physiol 296(5):G971–G981
15. Kyriazis GA, Soundarapandian MM, Tyrberg B (2012) Sweet taste receptor signaling in beta cells mediates fructose-induced potentiation of glucose-stimulated insulin secretion. Proc Natl Acad Sci U S A 109(8):E524–E532. doi:10.1073/pnas.1115183109
16. Mace OJ, Affleck J, Patel N, Kellett GL (2007) Sweet taste receptors in rat small intestine stimulate glucose absorption through apical GLUT2. J Physiol 582(Pt 1):379–392
17. Margolskee RF, Dyer J, Kokrashvili Z, Salmon KS, Ilegems E, Daly K, Maillet EL, Ninomiya Y, Mosinger B, Shirazi-Beechey SP (2007) T1R3 and gustducin in gut sense sugars to regulate expression of $Na^+$-glucose cotransporter 1. Proc Natl Acad Sci U S A 104 (38):15075–15080
18. Nakagawa Y, Nagasawa M, Yamada S, Hara A, Mogami H, Nikolaev VO, Lohse MJ, Shigemura N, Ninomiya Y, Kojima I (2009) Sweet taste receptor expressed in pancreatic beta-cells activates the calcium and cyclic AMP signaling systems and stimulates insulin secretion. PLoS One 4(4):e5106. doi:10.1371/journal.pone.0005106
19. Prandi S, Bromke M, Hubner S, Voigt A, Boehm U, Meyerhof W, Behrens M (2013) A subset of mouse colonic goblet cells expresses the bitter taste receptor tas2r131. PLoS One 8(12): e82820. doi:10.1371/journal.pone.0082820
20. Rozengurt N, Wu SV, Chen MC, Huang C, Sternini C, Rozengurt E (2006) Colocalization of the alpha-subunit of gustducin with PYY and GLP-1 in L cells of human colon. Am J Physiol Gastrointest Liver Physiol 291(5):G792–G802
21. Shirazi-Beechey SP, Moran AW, Batchelor DJ, Daly K, Al-Rammahi M (2011) Glucose sensing and signalling; regulation of intestinal glucose transport. Proc Nutr Soc 70 (2):185–193. doi:10.1017/S0029665111000103
22. Steinert RE, Gerspach AC, Gutmann H, Asarian L, Drewe J, Beglinger C (2011) The functional involvement of gut-expressed sweet taste receptors in glucose-stimulated secretion of glucagon-like peptide-1 (GLP-1) and peptide YY (PYY). Clin Nutr 30(4):524–532. doi:10.1016/j.clnu.2011.01.007
23. Sutherland K, Young RL, Cooper NJ, Horowitz M, Blackshaw LA (2007) Phenotypic characterization of taste cells of the mouse small intestine. Am J Physiol Gastrointest Liver Physiol 292(5):G1420–G1428
24. Wu SV, Chen MC, Rozengurt E (2005) Genomic organization, expression, and function of bitter taste receptors (T2R) in mouse and rat. Physiol Genomics 22(2):139–149

25. Wu SV, Rozengurt N, Yang M, Young SH, Sinnett-Smith J, Rozengurt E (2002) Expression of bitter taste receptors of the T2R family in the gastrointestinal tract and enteroendocrine STC-1 cells. Proc Natl Acad Sci U S A 99(4):2392–2397
26. Young RL, Sutherland K, Pezos N, Brierley SM, Horowitz M, Rayner CK, Blackshaw LA (2009) Expression of taste molecules in the upper gastrointestinal tract in humans with and without type 2 diabetes. Gut 58(3):337–346
27. Deshpande DA, Wang WC, McIlmoyle EL, Robinett KS, Schillinger RM, An SS, Sham JS, Liggett SB (2010) Bitter taste receptors on airway smooth muscle bronchodilate by localized calcium signaling and reverse obstruction. Nat Med 16(11):1299–1304
28. Finger TE, Bottger B, Hansen A, Anderson KT, Alimohammadi H, Silver WL (2003) Solitary chemoreceptor cells in the nasal cavity serve as sentinels of respiration. Proc Natl Acad Sci U S A 100(15):8981–8986
29. Finger TE, Kinnamon SC (2011) Taste isn't just for taste buds anymore. F1000 Biol Rep 3:20
30. Lee RJ, Xiong G, Kofonow JM, Chen B, Lysenko A, Jiang P, Abraham V, Doghramji L, Adappa ND, Palmer JN, Kennedy DW, Beauchamp GK, Doulias PT, Ischiropoulos H, Kreindler JL, Reed DR, Cohen NA (2012) T2R38 taste receptor polymorphisms underlie susceptibility to upper respiratory infection. J Clin Invest 122(11):4145–4159. doi:10.1172/JCI64240
31. Orsmark-Pietras C, James A, Konradsen JR, Nordlund B, Soderhall C, Pulkkinen V, Pedroletti C, Daham K, Kupczyk M, Dahlen B, Kere J, Dahlen SE, Hedlin G, Melen E (2013) Transcriptome analysis reveals upregulation of bitter taste receptors in severe asthmatics. Eur Respir J 42(1):65–78. doi:10.1183/09031936.00077712
32. Shah AS, Ben-Shahar Y, Moninger TO, Kline JN, Welsh MJ (2009) Motile cilia of human airway epithelia are chemosensory. Science 325(5944):1131–1134
33. Tizzano M, Cristofoletti M, Sbarbati A, Finger TE (2011) Expression of taste receptors in solitary chemosensory cells of rodent airways. BMC Pulm Med 11:3. doi:10.1186/1471-2466-11-3
34. Tizzano M, Finger TE (2013) Chemosensors in the nose: guardians of the airways. Physiology (Bethesda) 28(1):51–60. doi:10.1152/physiol.00035.2012
35. Tizzano M, Gulbransen BD, Vandenbeuch A, Clapp TR, Herman JP, Sibhatu HM, Churchill ME, Silver WL, Kinnamon SC, Finger TE (2010) Nasal chemosensory cells use bitter taste signaling to detect irritants and bacterial signals. Proc Natl Acad Sci U S A 107 (7):3210–3215
36. Voigt A, Hubner S, Lossow K, Hermans-Borgmeyer I, Boehm U, Meyerhof W (2012) Genetic labeling of Tas1r1 and Tas2r131 taste receptor cells in mice. Chem Senses 37 (9):897–911
37. Zhang CH, Lifshitz LM, Uy KF, Ikebe M, Fogarty KE, ZhuGe R (2013) The cellular and molecular basis of bitter tastant-induced bronchodilation. PLoS Biol 11(3):e1001501. doi:10.1371/journal.pbio.1001501
38. Iwatsuki K, Nomura M, Shibata A, Ichikawa R, Enciso PL, Wang L, Takayanagi R, Torii K, Uneyama H (2010) Generation and characterization of T1R2-LacZ knock-in mouse. Biochem Biophys Res Commun 402(3):495–499
39. Li F (2013) Taste perception: from the tongue to the testis. Mol Hum Reprod 19(6):349–360. doi:10.1093/molehr/gat009
40. Li F, Zhou M (2012) Depletion of bitter taste transduction leads to massive spermatid loss in transgenic mice. Mol Hum Reprod 18(6):289–297
41. Meyer D, Voigt A, Widmayer P, Borth H, Huebner S, Breit A, Marschall S, de Angelis MH, Boehm U, Meyerhof W, Gudermann T, Boekhoff I (2012) Expression of Tas1 taste receptors in mammalian spermatozoa: functional role of Tas1r1 in regulating basal Ca(2)(+) and cAMP concentrations in spermatozoa. PLoS One 7(2):e32354. doi:10.1371/journal.pone.0032354
42. Mosinger B, Redding KM, Parker MR, Yevshayeva V, Yee KK, Dyomina K, Li Y, Margolskee RF (2013) Genetic loss or pharmacological blockade of testes-expressed taste

genes causes male sterility. Proc Natl Acad Sci U S A 110(30):12319–12324. doi:10.1073/pnas.1302827110
43. Xu J, Cao J, Iguchi N, Riethmacher D, Huang L (2013) Functional characterization of bitter-taste receptors expressed in mammalian testis. Mol Hum Reprod 19(1):17–28. doi:10.1093/molehr/gas040
44. Garcia-Esparcia P, Schluter A, Carmona M, Moreno J, Ansoleaga B, Torrejon-Escribano B, Gustincich S, Pujol A, Ferrer I (2013) Functional genomics reveals dysregulation of cortical olfactory receptors in Parkinson disease: novel putative chemoreceptors in the human brain. J Neuropathol Exp Neurol 72(6):524–539. doi:10.1097/NEN.0b013e318294fd76
45. Ansoleaga B, Garcia-Esparcia P, Llorens F, Moreno J, Aso E, Ferrer I (2013) Dysregulation of brain olfactory and taste receptors in AD, PSP and CJD, and AD-related model. Neuroscience 248C:369–382. doi:10.1016/j.neuroscience.2013.06.034
46. Cong WN, Wang R, Cai H, Daimon CM, Scheibye-Knudsen M, Bohr VA, Turkin R, Wood WH 3rd, Becker KG, Moaddel R, Maudsley S, Martin B (2013) Long-term artificial sweetener acesulfame potassium treatment alters neurometabolic functions in C57BL/6J mice. PLoS One 8(8):e70257. doi:10.1371/journal.pone.0070257
47. Dehkordi O, Rose JE, Fatemi M, Allard JS, Balan KV, Young JK, Fatima S, Millis RM, Jayam-Trouth A (2012) Neuronal expression of bitter taste receptors and downstream signaling molecules in the rat brainstem. Brain Res 1475:1–10. doi:10.1016/j.brainres.2012.07.038
48. Ren X, Zhou L, Terwilliger R, Newton SS, de Araujo IE (2009) Sweet taste signaling functions as a hypothalamic glucose sensor. Front Integr Neurosci 3:12
49. Singh N, Vrontakis M, Parkinson F, Chelikani P (2011) Functional bitter taste receptors are expressed in brain cells. Biochem Biophys Res Commun 406(1):146–151
50. Foster SR, Porrello ER, Purdue B, Chan HW, Voigt A, Frenzel S, Hannan RD, Moritz KM, Simmons DG, Molenaar P, Roura E, Boehm U, Meyerhof W, Thomas WG (2013) Expression, regulation and putative nutrient-sensing function of taste GPCRs in the heart. PLoS One 8(5):e64579. doi:10.1371/journal.pone.0064579
51. Fukuda N, Yomogida K, Okabe M, Touhara K (2004) Functional characterization of a mouse testicular olfactory receptor and its role in chemosensing and in regulation of sperm motility. J Cell Sci 117(Pt 24):5835–5845. doi:10.1242/jcs.01507
52. Spehr M, Gisselmann G, Poplawski A, Riffell JA, Wetzel CH, Zimmer RK, Hatt H (2003) Identification of a testicular odorant receptor mediating human sperm chemotaxis. Science 299(5615):2054–2058
53. Elliott RA, Kapoor S, Tincello DG (2011) Expression and distribution of the sweet taste receptor isoforms T1R2 and T1R3 in human and rat bladders. J Urol 186(6):2455–2462. doi:10.1016/j.juro.2011.07.083
54. Masubuchi Y, Nakagawa Y, Ma J, Sasaki T, Kitamura T, Yamamoto Y, Kurose H, Kojima I, Shibata H (2013) A novel regulatory function of sweet taste-sensing receptor in adipogenic differentiation of 3T3-L1 cells. PLoS One 8(1):e54500. doi:10.1371/journal.pone.0054500
55. Simon BR, Parlee SD, Learman BS, Mori H, Scheller EL, Cawthorn WP, Ning X, Gallagher K, Tyrberg B, Assadi-Porter FM, Evans CR, MacDougald OA (2013) Artificial sweeteners stimulate adipogenesis and suppress lipolysis independently of sweet taste receptors. J Biol Chem 288(45):32475–32489. doi:10.1074/jbc.M113.514034
56. Lund TC, Kobs AJ, Kramer A, Nyquist M, Kuroki MT, Osborn J, Lidke DS, Low-Nam ST, Blazar BR, Tolar J (2013) Bone marrow stromal and vascular smooth muscle cells have chemosensory capacity via bitter taste receptor expression. PLoS One 8(3):e58945. doi:10.1371/journal.pone.0058945
57. Wauson EM, Zaganjor E, Lee AY, Guerra ML, Ghosh AB, Bookout AL, Chambers CP, Jivan A, McGlynn K, Hutchison MR, Deberardinis RJ, Cobb MH (2012) The G protein-coupled taste receptor T1R1/T1R3 regulates mTORC1 and autophagy. Mol Cell 47(6):851–862. doi:10.1016/j.molcel.2012.08.001
58. Chaudhari N, Roper SD (2010) The cell biology of taste. J Cell Biol 190(3):285–296

59. Chandrashekar J, Kuhn C, Oka Y, Yarmolinsky DA, Hummler E, Ryba NJ, Zuker CS (2010) The cells and peripheral representation of sodium taste in mice. Nature 464(7286):297–301
60. Huang AL, Chen X, Hoon MA, Chandrashekar J, Guo W, Trankner D, Ryba NJ, Zuker CS (2006) The cells and logic for mammalian sour taste detection. Nature 442(7105):934–938
61. Ishimaru Y, Inada H, Kubota M, Zhuang H, Tominaga M, Matsunami H (2006) Transient receptor potential family members PKD1L3 and PKD2L1 form a candidate sour taste receptor. Proc Natl Acad Sci U S A 103(33):12569–12574
62. LopezJimenez ND, Cavenagh MM, Sainz E, Cruz-Ithier MA, Battey JF, Sullivan SL (2006) Two members of the TRPP family of ion channels, Pkd1l3 and Pkd2l1, are co-expressed in a subset of taste receptor cells. J Neurochem 98(1):68–77
63. Horio N, Yoshida R, Yasumatsu K, Yanagawa Y, Ishimaru Y, Matsunami H, Ninomiya Y (2011) Sour taste responses in mice lacking PKD channels. PLoS One 6(5):e20007. doi:10.1371/journal.pone.0020007
64. Bachmanov AA, Li X, Reed DR, Ohmen JD, Li S, Chen Z, Tordoff MG, de Jong PJ, Wu C, West DB, Chatterjee A, Ross DA, Beauchamp GK (2001) Positional cloning of the mouse saccharin preference (Sac) locus. Chem Senses 26(7):925–933
65. Hoon MA, Adler E, Lindemeier J, Battey JF, Ryba NJ, Zuker CS (1999) Putative mammalian taste receptors: a class of taste-specific GPCRs with distinct topographic selectivity. Cell 96 (4):541–551
66. Kitagawa M, Kusakabe Y, Miura H, Ninomiya Y, Hino A (2001) Molecular genetic identification of a candidate receptor gene for sweet taste. Biochem Biophys Res Commun 283 (1):236–242
67. Max M, Shanker YG, Huang L, Rong M, Liu Z, Campagne F, Weinstein H, Damak S, Margolskee RF (2001) Tas1r3, encoding a new candidate taste receptor, is allelic to the sweet responsiveness locus Sac. Nat Genet 28(1):58–63
68. Montmayeur JP, Liberles SD, Matsunami H, Buck LB (2001) A candidate taste receptor gene near a sweet taste locus. Nat Neurosci 4(5):492–498
69. Sainz E, Korley JN, Battey JF, Sullivan SL (2001) Identification of a novel member of the T1R family of putative taste receptors. J Neurochem 77(3):896–903
70. Li X, Staszewski L, Xu H, Durick K, Zoller M, Adler E (2002) Human receptors for sweet and umami taste. Proc Natl Acad Sci U S A 99(7):4692–4696
71. Nelson G, Chandrashekar J, Hoon MA, Feng L, Zhao G, Ryba NJ, Zuker CS (2002) An amino-acid taste receptor. Nature 416(6877):199–202
72. Adler E, Hoon MA, Mueller KL, Chandrashekar J, Ryba NJ, Zuker CS (2000) A novel family of mammalian taste receptors. Cell 100(6):693–702
73. Chandrashekar J, Mueller KL, Hoon MA, Adler E, Feng L, Guo W, Zuker CS, Ryba NJ (2000) T2Rs function as bitter taste receptors. Cell 100(6):703–711
74. Matsunami H, Montmayeur JP, Buck LB (2000) A family of candidate taste receptors in human and mouse. Nature 404(6778):601–604
75. Kunishima N, Shimada Y, Tsuji Y, Sato T, Yamamoto M, Kumasaka T, Nakanishi S, Jingami H, Morikawa K (2000) Structural basis of glutamate recognition by a dimeric metabotropic glutamate receptor. Nature 407(6807):971–977. doi:10.1038/35039564
76. Jones KA, Borowsky B, Tamm JA, Craig DA, Durkin MM, Dai M, Yao WJ, Johnson M, Gunwaldsen C, Huang LY, Tang C, Shen Q, Salon JA, Morse K, Laz T, Smith KE, Nagarathnam D, Noble SA, Branchek TA, Gerald C (1998) GABA(B) receptors function as a heteromeric assembly of the subunits GABA(B)R1 and GABA(B)R2. Nature 396 (6712):674–679. doi:10.1038/25348
77. Kaupmann K, Malitschek B, Schuler V, Heid J, Froestl W, Beck P, Mosbacher J, Bischoff S, Kulik A, Shigemoto R, Karschin A, Bettler B (1998) GABA(B)-receptor subtypes assemble into functional heteromeric complexes. Nature 396(6712):683–687. doi:10.1038/25360
78. Kuner R, Kohr G, Grunewald S, Eisenhardt G, Bach A, Kornau HC (1999) Role of heteromer formation in GABAB receptor function. Science 283(5398):74–77

79. White JH, Wise A, Main MJ, Green A, Fraser NJ, Disney GH, Barnes AA, Emson P, Foord SM, Marshall FH (1998) Heterodimerization is required for the formation of a functional GABA(B) receptor. Nature 396(6712):679–682. doi:10.1038/25354
80. Nelson G, Hoon MA, Chandrashekar J, Zhang Y, Ryba NJ, Zuker CS (2001) Mammalian sweet taste receptors. Cell 106(3):381–390
81. Dong D, Jones G, Zhang S (2009) Dynamic evolution of bitter taste receptor genes in vertebrates. BMC Evol Biol 9:12
82. Behrens M, Meyerhof W (2013) Bitter taste receptor research comes of age: from characterization to modulation of TAS2Rs. Semin Cell Dev Biol 24(3):215–221. doi:10.1016/j.semcdb.2012.08.006
83. Meyerhof W, Batram C, Kuhn C, Brockhoff A, Chudoba E, Bufe B, Appendino G, Behrens M (2010) The molecular receptive ranges of human TAS2R bitter taste receptors. Chem Senses 35(2):157–170
84. Chandrashekar J, Hoon MA, Ryba NJ, Zuker CS (2006) The receptors and cells for mammalian taste. Nature 444(7117):288–294
85. Huang L, Shanker YG, Dubauskaite J, Zheng JZ, Yan W, Rosenzweig S, Spielman AI, Max M, Margolskee RF (1999) Ggamma13 colocalizes with gustducin in taste receptor cells and mediates IP3 responses to bitter denatonium. Nat Neurosci 2(12):1055–1062
86. Rossler P, Boekhoff I, Tareilus E, Beck S, Breer H, Freitag J (2000) G protein betagamma complexes in circumvallate taste cells involved in bitter transduction. Chem Senses 25 (4):413–421
87. McLaughlin SK, McKinnon PJ, Margolskee RF (1992) Gustducin is a taste-cell-specific G protein closely related to the transducins. Nature 357(6379):563–569
88. Rossler P, Kroner C, Freitag J, Noe J, Breer H (1998) Identification of a phospholipase C beta subtype in rat taste cells. Eur J Cell Biol 77(3):253–261
89. Zhang Y, Hoon MA, Chandrashekar J, Mueller KL, Cook B, Wu D, Zuker CS, Ryba NJ (2003) Coding of sweet, bitter, and umami tastes: different receptor cells sharing similar signaling pathways. Cell 112(3):293–301
90. Clapp TR, Stone LM, Margolskee RF, Kinnamon SC (2001) Immunocytochemical evidence for co-expression of Type III IP3 receptor with signaling components of bitter taste transduction. BMC Neurosci 2:6
91. Hisatsune C, Yasumatsu K, Takahashi-Iwanaga H, Ogawa N, Kuroda Y, Yoshida R, Ninomiya Y, Mikoshiba K (2007) Abnormal taste perception in mice lacking the type 3 inositol 1,4,5-trisphosphate receptor. J Biol Chem 282:37225–37231
92. Miyoshi MA, Abe K, Emori Y (2001) IP(3) receptor type 3 and PLCbeta2 are co-expressed with taste receptors T1R and T2R in rat taste bud cells. Chem Senses 26(3):259–265
93. Liu D, Liman ER (2003) Intracellular $Ca^{2+}$ and the phospholipid PIP2 regulate the taste transduction ion channel TRPM5. Proc Natl Acad Sci U S A 100(25):15160–15165
94. Perez CA, Huang L, Rong M, Kozak JA, Preuss AK, Zhang H, Max M, Margolskee RF (2002) A transient receptor potential channel expressed in taste receptor cells. Nat Neurosci 5 (11):1169–1176
95. Zhang Z, Zhao Z, Margolskee R, Liman E (2007) The transduction channel TRPM5 is gated by intracellular calcium in taste cells. J Neurosci 27(21):5777–5786
96. Finger TE, Danilova V, Barrows J, Bartel DL, Vigers AJ, Stone L, Hellekant G, Kinnamon SC (2005) ATP signaling is crucial for communication from taste buds to gustatory nerves. Science 310(5753):1495–1499
97. Huang YJ, Maruyama Y, Dvoryanchikov G, Pereira E, Chaudhari N, Roper SD (2007) The role of pannexin 1 hemichannels in ATP release and cell–cell communication in mouse taste buds. Proc Natl Acad Sci U S A 104(15):6436–6441
98. Romanov RA, Rogachevskaja OA, Bystrova MF, Jiang P, Margolskee RF, Kolesnikov SS (2007) Afferent neurotransmission mediated by hemichannels in mammalian taste cells. EMBO J 26(3):657–667

99. Taruno A, Vingtdeux V, Ohmoto M, Ma Z, Dvoryanchikov G, Li A, Adrien L, Zhao H, Leung S, Abernethy M, Koppel J, Davies P, Civan MM, Chaudhari N, Matsumoto I, Hellekant G, Tordoff MG, Marambaud P, Foskett JK (2013) CALHM1 ion channel mediates purinergic neurotransmission of sweet, bitter and umami tastes. Nature 495(7440):223–226. doi:10.1038/nature11906
100. Cohen SP, Buckley BK, Kosloff M, Garland AL, Bosch DE, Cheng G Jr, Radhakrishna H, Brown MD, Willard FS, Arshavsky VY, Tarran R, Siderovski DP, Kimple AJ (2012) Regulator of G-protein signaling-21 (RGS21) is an inhibitor of bitter gustatory signaling found in lingual and airway epithelia. J Biol Chem 287(50):41706–41719. doi:10.1074/jbc.M112.423806
101. von Buchholtz L, Elischer A, Tareilus E, Gouka R, Kaiser C, Breer H, Conzelmann S (2004) RGS21 is a novel regulator of G protein signalling selectively expressed in subpopulations of taste bud cells. Eur J Neurosci 19(6):1535–1544
102. Hofer D, Puschel B, Drenckhahn D (1996) Taste receptor-like cells in the rat gut identified by expression of alpha-gustducin. Proc Natl Acad Sci U S A 93(13):6631–6634
103. Ribatti D, Nico B, Crivellato E, Artico M (2006) Development of the blood-brain barrier: a historical point of view. Anat Rec B New Anat 289(1):3–8. doi:10.1002/ar.b.20087
104. Benda P, Lightbody J, Sato G, Levine L, Sweet W (1968) Differentiated rat glial cell strain in tissue culture. Science 161(3839):370–371
105. Pydi SP, Bhullar RP, Chelikani P (2012) Constitutively active mutant gives novel insights into the mechanism of bitter taste receptor activation. J Neurochem 122(3):537–544
106. Hodges MR, Richerson GB (2010) Medullary serotonin neurons and their roles in central respiratory chemoreception. Respir Physiol Neurobiol 173(3):256–263. doi:10.1016/j.resp.2010.03.006
107. Kidd M, Modlin IM, Gustafsson BI, Drozdov I, Hauso O, Pfragner R (2008) Luminal regulation of normal and neoplastic human EC cell serotonin release is mediated by bile salts, amines, tastants, and olfactants. Am J Physiol Gastrointest Liver Physiol 295(2):G260–G272
108. Mace OJ, Lister N, Morgan E, Shepherd E, Affleck J, Helliwell P, Bronk JR, Kellett GL, Meredith D, Boyd R, Pieri M, Bailey PD, Pettcrew R, Foley D (2009) An energy supply network of nutrient absorption coordinated by calcium and T1R taste receptors in rat small intestine. J Physiol 587(Pt 1):195–210
109. Swartz TD, Duca FA, de Wouters T, Sakar Y, Covasa M (2012) Up-regulation of intestinal type 1 taste receptor 3 and sodium glucose luminal transporter-1 expression and increased sucrose intake in mice lacking gut microbiota. Br J Nutr 107(5):621–630. doi:10.1017/S0007114511003412
110. Young RL, Chia B, Isaacs NJ, Ma J, Khoo J, Wu T, Horowitz M, Rayner CK (2013) Disordered control of intestinal sweet taste receptor expression and glucose absorption in type 2 diabetes. Diabetes 62(10):3532–3541. doi:10.2337/db13-0581
111. Haid D, Widmayer P, Breer H (2011) Nutrient sensing receptors in gastric endocrine cells. J Mol Histol 42(4):355–364. doi:10.1007/s10735-011-9339-1
112. Hass N, Schwarzenbacher K, Breer H (2007) A cluster of gustducin-expressing cells in the mouse stomach associated with two distinct populations of enteroendocrine cells. Histochem Cell Biol 128(5):457–471
113. Sothilingam V, Hass N, Breer H (2011) Candidate chemosensory cells in the stomach mucosa of young postnatal mice during the phases of dietary changes. Cell Tissue Res 344(2):239–249. doi:10.1007/s00441-011-1137-2
114. Moran AW, Al-Rammahi MA, Arora DK, Batchelor DJ, Coulter EA, Daly K, Ionescu C, Bravo D, Shirazi-Beechey SP (2010) Expression of $Na^+$/glucose co-transporter 1 (SGLT1) is enhanced by supplementation of the diet of weaning piglets with artificial sweeteners. Br J Nutr 104(5):637–646
115. Widmayer P, Breer H, Hass N (2011) Candidate chemosensory cells in the porcine stomach. Histochem Cell Biol 136(1):37–45. doi:10.1007/s00418-011-0824-0

116. Batchelor DJ, Al-Rammahi M, Moran AW, Brand JG, Li X, Haskins M, German AJ, Shirazi-Beechey SP (2011) Sodium/glucose cotransporter-1, sweet receptor, and disaccharidase expression in the intestine of the domestic dog and cat: two species of different dietary habit. Am J Physiol Regul Integr Comp Physiol 300(1):R67–R75. doi:10.1152/ajpregu.00262.2010
117. Daly K, Al-Rammahi M, Arora DK, Moran AW, Proudman CJ, Ninomiya Y, Shirazi-Beechey SP (2012) Expression of sweet receptor components in equine small intestine: relevance to intestinal glucose transport. Am J Physiol Regul Integr Comp Physiol 303(2):R199–R208. doi:10.1152/ajpregu.00031.2012
118. Taniguchi K (2004) Expression of the sweet receptor protein, T1R3, in the human liver and pancreas. J Vet Med Sci 66(11):1311–1314
119. Hofer D, Drenckhahn D (1998) Identification of the taste cell G-protein, alpha-gustducin, in brush cells of the rat pancreatic duct system. Histochem Cell Biol 110(3):303–309
120. Ruiz-Avila L, McLaughlin SK, Wildman D, McKinnon PJ, Robichon A, Spickofsky N, Margolskee RF (1995) Coupling of bitter receptor to phosphodiesterase through transducin in taste receptor cells. Nature 376(6535):80–85
121. Hofer D, Asan E, Drenckhahn D (1999) Chemosensory perception in the gut. News Physiol Sci 14:18–23
122. Bezencon C, Furholz A, Raymond F, Mansourian R, Metairon S, Le Coutre J, Damak S (2008) Murine intestinal cells expressing Trpm5 are mostly brush cells and express markers of neuronal and inflammatory cells. J Comp Neurol 509(5):514–525. doi:10.1002/cne.21768
123. Li Y, Kokrashvili Z, Mosinger B, Margolskee RF (2013) Gustducin couples fatty acid receptors to GLP-1 release in colon. Am J Physiol Endocrinol Metab 304(6):E651–E660. doi:10.1152/ajpendo.00471.2012
124. Eberle JA, Richter P, Widmayer P, Chubanov V, Gudermann T, Breer H (2013) Band-like arrangement of taste-like sensory cells at the gastric groove: evidence for paracrine communication. Front Physiol 4:58. doi:10.3389/fphys.2013.00058
125. Kaske S, Krasteva G, Konig P, Kummer W, Hofmann T, Gudermann T, Chubanov V (2007) TRPM5, a taste-signaling transient receptor potential ion-channel, is a ubiquitous signaling component in chemosensory cells. BMC Neurosci 8:49. doi:10.1186/1471-2202-8-49
126. Kokrashvili Z, Rodriguez D, Yevshayeva V, Zhou H, Margolskee RF, Mosinger B (2009) Release of endogenous opioids from duodenal enteroendocrine cells requires Trpm5. Gastroenterology 137(2):598–606, 606 e591–592. doi:10.1053/j.gastro.2009.02.070
127. Gerbe F, Legraverend C, Jay P (2012) The intestinal epithelium tuft cells: specification and function. Cell Mole Life Sci 69(17):2907–2917. doi:10.1007/s00018-012-0984-7
128. Luciano L, Reale E, Ruska H (1968) On a glycogen containing brusc cell in the rectum of the rat. Z Zellforsch Mikrosk Anat 91(1):153–158
129. Sternini C, Anselmi L, Rozengurt E (2008) Enteroendocrine cells: a site of 'taste' in gastrointestinal chemosensing. Curr Opin Endocrinol Diabetes Obes 15(1):73–78
130. Le Neve B, Foltz M, Daniel H, Gouka R (2010) The steroid glycoside H.g.-12 from *Hoodia gordonii* activates the human bitter receptor TAS2R14 and induces CCK release from HuTu-80 cells. Am J Physiol Gastrointest Liver Physiol 299:G1368–G1375
131. McIntyre N, Holdsworth CD, Turner DS (1964) New interpretation of oral glucose tolerance. Lancet 2(7349):20–21
132. Baggio LL, Drucker DJ (2007) Biology of incretins: GLP-1 and GIP. Gastroenterology 132(6):2131–2157. doi:10.1053/j.gastro.2007.03.054
133. Sato S, Hokari R, Kurihara C, Sato H, Narimatsu K, Hozumi H, Ueda T, Higashiyama M, Okada Y, Watanabe C, Komoto S, Tomita K, Kawaguchi A, Nagao S, Miura S (2013) Dietary lipids and sweeteners regulate glucagon-like peptide-2 secretion. Am J Physiol Gastrointest Liver Physiol 304(8):G708–G714. doi:10.1152/ajpgi.00282.2012
134. Gerspach AC, Steinert RE, Schonenberger L, Graber-Maier A, Beglinger C (2011) The role of the gut sweet taste receptor in regulating GLP-1, PYY, and CCK release in humans. Am J Physiol Endocrinol Metab 301(2):E317–E325. doi:10.1152/ajpendo.00077.2011

135. Ford HE, Peters V, Martin NM, Sleeth ML, Ghatei MA, Frost GS, Bloom SR (2011) Effects of oral ingestion of sucralose on gut hormone response and appetite in healthy normal-weight subjects. Eur J Clin Nutr 65:508–513
136. Fujita Y, Wideman RD, Speck M, Asadi A, King DS, Webber TD, Haneda M, Kieffer TJ (2009) Incretin release from gut is acutely enhanced by sugar but not by sweeteners in vivo. Am J Physiol Endocrinol Metab 296(3):E473–E479
137. Ma J, Bellon M, Wishart JM, Young R, Blackshaw LA, Jones KL, Horowitz M, Rayner CK (2009) Effect of the artificial sweetener, sucralose, on gastric emptying and incretin hormone release in healthy subjects. Am J Physiol Gastrointest Liver Physiol 296(4):G735–G739
138. Nakagawa Y, Nagasawa M, Mogami H, Lohse M, Ninomiya Y, Kojima I (2013) Multimodal function of the sweet taste receptor expressed in pancreatic beta-cells: generation of diverse patterns of intracellular signals by sweet agonists. Endocr J 60(10):1191–1206
139. Nakagawa Y, Ohtsu Y, Nagasawa M, Shibata H, Kojima I (2014) Glucose promotes its own metabolism by acting on the cell-surface glucose-sensing receptor T1R3. Endocr J 61:119–131
140. Zhao GQ, Zhang Y, Hoon MA, Chandrashekar J, Erlenbach I, Ryba NJ, Zuker CS (2003) The receptors for mammalian sweet and umami taste. Cell 115(3):255–266
141. Brown EM, Gamba G, Riccardi D, Lombardi M, Butters R, Kifor O, Sun A, Hediger MA, Lytton J, Hebert SC (1993) Cloning and characterization of an extracellular Ca(2+)-sensing receptor from bovine parathyroid. Nature 366(6455):575–580. doi:10.1038/366575a0
142. Wellendorph P, Brauner-Osborne H (2004) Molecular cloning, expression, and sequence analysis of GPRC6A, a novel family C G-protein-coupled receptor. Gene 335:37–46. doi:10.1016/j.gene.2004.03.003
143. Wellendorph P, Hansen KB, Balsgaard A, Greenwood JR, Egebjerg J, Brauner-Osborne H (2005) Deorphanization of GPRC6A: a promiscuous L-alpha-amino acid receptor with preference for basic amino acids. Mol Pharmacol 67(3):589–597. doi:10.1124/mol.104.007559
144. Torres N, Noriega L, Tovar AR (2009) Nutrient modulation of insulin secretion. Vitam Horm 80:217–244. doi:10.1016/S0083-6729(08)00609-2
145. Imoto T, Miyasaka A, Ishima R, Akasaka K (1991) A novel peptide isolated from the leaves of Gymnema sylvestre–I. Characterization and its suppressive effect on the neural responses to sweet taste stimuli in the rat. Comp Biochem Physiol A Comp Physiol 100(2):309–314
146. Ninomiya Y, Imoto T (1995) Gurmarin inhibition of sweet taste responses in mice. Am J Physiol 268(4 Pt 2):R1019–R1025
147. Ninomiya Y, Nakashima K, Fukuda A, Nishino H, Sugimura T, Hino A, Danilova V, Hellekant G (2000) Responses to umami substances in taste bud cells innervated by the chorda tympani and glossopharyngeal nerves. J Nutr 130(4S Suppl):950S–953S
148. Yamamoto T, Matsuo R, Fujimoto Y, Fukunaga I, Miyasaka A, Imoto T (1991) Electrophysiological and behavioral studies on the taste of umami substances in the rat. Physiol Behav 49 (5):919–925
149. Nakamura E, Hasumura M, San Gabriel A, Uneyama H, Torii K (2010) New frontiers in gut nutrient sensor research: luminal glutamate-sensing cells in rat gastric mucosa. J Pharmacol Sci 112(1):13–18
150. Straub SG, Mulvaney-Musa J, Yajima H, Weiland GA, Sharp GW (2003) Stimulation of insulin secretion by denatonium, one of the most bitter-tasting substances known. Diabetes 52 (2):356–364
151. Behrens M, Meyerhof W (2010) Oral and extraoral bitter taste receptors. In: Meyerhof W, Beisiegel U, Joost H-G (eds) Sensory and metabolic control of energy balance, vol 52. Results and problems in cell differentiation. Springer, Heidelberg, pp 87-99. doi:10.1007/978-3-642-14426-4_8
152. Little TJ, Gupta N, Case RM, Thompson DG, McLaughlin JT (2009) Sweetness and bitterness taste of meals per se does not mediate gastric emptying in humans. Am J Physiol Regul Integr Comp Physiol 297(3):R632–R639

153. Wicks D, Wright J, Rayment P, Spiller R (2005) Impact of bitter taste on gastric motility. Eur J Gastroenterol Hepatol 17(9):961–965
154. Jeon TI, Seo YK, Osborne TF (2011) Gut bitter taste receptor signalling induces ABCB1 through a mechanism involving CCK. Biochem J 438(1):33–37
155. Hao S, Dulake M, Espero E, Sternini C, Raybould HE, Rinaman L (2009) Central Fos expression and conditioned flavor avoidance in rats following intragastric administration of bitter taste receptor ligands. Am J Physiol Regul Integr Comp Physiol 296(3):R528–R536
156. Sbarbati A, Tizzano M, Merigo F, Benati D, Nicolato E, Boschi F, Cecchini MP, Scambi I, Osculati F (2009) Acyl homoserine lactones induce early response in the airway. Anat Rec 292(3):439–448. doi:10.1002/ar.20866
157. Barham HP, Cooper SE, Anderson CB, Tizzano M, Kingdom TT, Finger TE, Kinnamon SC, Ramakrishnan VR (2013) Solitary chemosensory cells and bitter taste receptor signaling in human sinonasal mucosa. Int Forum Allergy Rhinol 3(6):450–457. doi:10.1002/alr.21149
158. Lee RJ, Chen B, Redding KM, Margolskee RF, Cohen NA (2013) Mouse nasal epithelial innate immune responses to *Pseudomonas aeruginosa* quorum-sensing molecules require taste signaling components. Innate Immun. doi:10.1177/1753425913503386
159. Belvisi MG, Dale N, Birrell MA, Canning BJ (2011) Bronchodilator activity of bitter tastants in human tissue. Nat Med 17(7):776. doi:10.1038/nm0711-776a, author reply 776–778
160. Morice AH, Bennett RT, Chaudhry MA, Cowen ME, Griffin SC, Loubani M (2011) Effect of bitter tastants on human bronchi. Nat Med 17(7):775. doi:10.1038/nm0711-775
161. Behrens M, Brockhoff A, Kuhn C, Bufe B, Winnig M, Meyerhof W (2004) The human taste receptor hTAS2R14 responds to a variety of different bitter compounds. Biochem Biophys Res Commun 319(2):479–485
162. Levit A, Nowak S, Peters M, Wiener A, Meyerhof W, Behrens M, Niv MY (2013) The bitter pill: clinical drugs that activate the human bitter taste receptor TAS2R14. FASEB J. doi:10.1096/fj.13-242594
163. Zhou PZ, Babcock J, Liu LQ, Li M, Gao ZB (2011) Activation of human ether-a-go-go related gene (hERG) potassium channels by small molecules. Acta Pharmacol Sin 32 (6):781–788. doi:10.1038/aps.2011.70
164. Yoshida M, Yoshida K (2011) Sperm chemotaxis and regulation of flagellar movement by $Ca^{2+}$. Mol Hum Reprod 17(8):457–465. doi:10.1093/molehr/gar041
165. Kaupp UB, Kashikar ND, Weyand I (2008) Mechanisms of sperm chemotaxis. Annu Rev Physiol 70:93–117. doi:10.1146/annurev.physiol.70.113006.100654
166. Behrens M, Bartelt J, Reichling C, Winnig M, Kuhn C, Meyerhof W (2006) Members of RTP and REEP gene families influence functional bitter taste receptor expression. J Biol Chem 281(29):20650–20659
167. Fehr J, Meyer D, Widmayer P, Borth HC, Ackermann F, Wilhelm B, Gudermann T, Boekhoff I (2007) Expression of the G-protein alpha-subunit gustducin in mammalian spermatozoa. J Comp Physiol A Neuroethol Sens Neural Behav Physiol 193(1):21–34
168. Li X, Li W, Wang H, Cao J, Maehashi K, Huang L, Bachmanov AA, Reed DR, Legrand-Defretin V, Beauchamp GK, Brand JG (2005) Pseudogenization of a sweet-receptor gene accounts for cats' indifference toward sugar. PLoS Genet 1(1):27–35
169. Shi P, Zhang J (2006) Contrasting modes of evolution between vertebrate sweet/umami receptor genes and bitter receptor genes. Mol Biol Evol 23(2):292–300
170. Li R, Fan W, Tian G, Zhu H, He L, Cai J, Huang Q, Cai Q, Li B, Bai Y, Zhang Z, Zhang Y, Wang W, Li J, Wei F, Li H, Jian M, Li J, Zhang Z, Nielsen R, Li D, Gu W, Yang Z, Xuan Z, Ryder OA, Leung FC, Zhou Y, Cao J, Sun X, Fu Y, Fang X, Guo X, Wang B, Hou R, Shen F, Mu B, Ni P, Lin R, Qian W, Wang G, Yu C, Nie W, Wang J, Wu Z, Liang H, Min J, Wu Q, Cheng S, Ruan J, Wang M, Shi Z, Wen M, Liu B, Ren X, Zheng H, Dong D, Cook K, Shan G, Zhang H, Kosiol C, Xie X, Lu Z, Zheng H, Li Y, Steiner CC, Lam TT, Lin S, Zhang Q, Li G, Tian J, Gong T, Liu H, Zhang D, Fang L, Ye C, Zhang J, Hu W, Xu A, Ren Y, Zhang G, Bruford MW, Li Q, Ma L, Guo Y, An N, Hu Y, Zheng Y, Shi Y, Li Z, Liu Q, Chen Y, Zhao J, Qu N, Zhao S, Tian F, Wang X, Wang H, Xu L, Liu X, Vinar T, Wang Y, Lam TW, Yiu SM,

Liu S, Zhang H, Li D, Huang Y, Wang X, Yang G, Jiang Z, Wang J, Qin N, Li L, Li J, Bolund L, Kristiansen K, Wong GK, Olson M, Zhang X, Li S, Yang H, Wang J, Wang J (2010) The sequence and de novo assembly of the giant panda genome. Nature 463 (7279):311–317. doi:10.1038/nature08696

171. Jiang P, Josue J, Li X, Glaser D, Li W, Brand JG, Margolskee RF, Reed DR, Beauchamp GK (2012) Major taste loss in carnivorous mammals. Proc Natl Acad Sci U S A 109 (13):4956–4961
172. Zhao H, Xu D, Zhang S, Zhang J (2012) Genomic and genetic evidence for the loss of umami taste in bats. Genome Biol Evol 4(1):73–79. doi:10.1093/gbe/evr126
173. Zhao H, Zhou Y, Pinto CM, Charles-Dominique P, Galindo-Gonzalez J, Zhang S, Zhang J (2010) Evolution of the sweet taste receptor gene Tas1r2 in bats. Mol Biol Evol 27 (11):2642–2650. doi:10.1093/molbev/msq152
174. Dunn-Meynell AA, Sanders NM, Compton D, Becker TC, Eiki J, Zhang BB, Levin BE (2009) Relationship among brain and blood glucose levels and spontaneous and glucoprivic feeding. J Neurosci 29(21):7015–7022. doi:10.1523/JNEUROSCI.0334-09.2009
175. Behrens M, Born S, Redel U, Voigt N, Schuh V, Raguse JD, Meyerhof W (2012) Immunohistochemical detection of TAS2R38 protein in human taste cells. PLoS One 7(7):e40304
176. Clark AA, Liggett SB, Munger SD (2012) Extraoral bitter taste receptors as mediators of off-target drug effects. FASEB J 26(12):4827–4831. doi:10.1096/fj.12-215087

Top Med Chem (2017) 23: 35–72
DOI: 10.1007/7355_2014_81

Published online: 28 January 2014

# Medicinal Chemistry of Plant Naturals as Agonists/Antagonists for Taste Receptors

**Joshua N. Fletcher, Li Pan, and A. Douglas Kinghorn**

**Abstract** The study of natural compounds that affect the perception of sweet or bitter tastes has a rich history. Coupled to this history is the recent discovery of the sweet and bitter taste receptors allowing for modern biological techniques to be used in taste research. Natural high potency sweeteners from stevia (*Stevia rebaudiana*) and monk fruit (*Siraitia grosvenorii*) have recently come to market, and several natural compounds find use as flavors to alter the perception of sweetness or bitterness. This chapter reviews these substances and several other natural products that alter the perception of bitterness or sweetness.

**Keywords** Bitter, Flavor, Natural substances, Sweet, Sweetener, Taste

## Contents



J.N. Fletcher (✉)
Tate and Lyle PLC, Hoffman Estates, IL 60192, USA
e-mail: Joshua.fletcher2@tateandlyle.com

L. Pan and A.D. Kinghorn
College of Pharmacy, The Ohio State University, Columbus, OH 43210, USA

## 1 Introduction

The perceptions of sweet and bitter tastes have long occupied a central role in guiding the human diet. At a basic level, sweetness indicates the presence of nutrients, while bitterness indicates the presence of a toxin to be avoided [1]. These basic taste sensations have also influenced agriculture. The domestication of the watermelon [*Citrullus lanatus* (Thunb.) Mansf.; Cucurbitaceae] has resulted in the transformation of a bitter-tasting predecessor crop with purgative properties into the sweet-tasting crop known today [2]. The removal of bitter (and in many cases toxic) secondary metabolites through plant breeding has occurred many times throughout human history (e.g., steroidal alkaloids in potatoes, quinolizidine alkaloids in lupins, and cyanogenic glycosides in almonds) [3–7].

The importance of sweetness to mankind can be seen in the yearly production of over 172 million metric tons of the most widely used sweetener, sucrose [8]. The desire for alternatives to sugar has led to the development of several artificial high-potency sweeteners including acesulfame K, alitame, aspartame, cyclamate, neotame, saccharin, and sucralose. The current trend towards natural sweeteners has allowed for significant advances in the commercialization of plant secondary metabolites as sweeteners. In addition to these more visible solutions, naturally occurring sweetness-enhancing and bitterness-masking agents are also being explored as can be seen in the recent FEMA GRAS lists (generally recognized as safe flavoring substances list provided by the Flavor and Extract Manufacturers Association). These natural products will be reviewed further later in this chapter.

While bitterness may be expected and even seen as desirable in some products such as beer, bitter melon, chocolate, coffee, grapefruit, lemon, and wine, in general, bitter tastants are avoided [4, 9]. There are also many methods to remove bitter compounds from foodstuffs including the use of transgenic organisms as in the removal of limonoids from oranges and chemical modification techniques such as the enzymatic and microbial transformation of flavones in orange juice [4]. The phenolic components of wine may be removed using a variety of techniques such as precipitation (as in the process of aging wines), adsorption onto proteins such as isinglass or casein, and the use of cyclodextrin beads [4]. Bitter components of foods may also be masked during food preparation by addition of fats, flavors, sweeteners, or salt [4].

## 2 Taste Receptor Assays and Natural Products

Perhaps the most significant recent breakthrough in taste research has been the discovery of the sweet and bitter taste receptors. There are at least 25 functioning bitterness receptor genes (TAS2Rs) encoded in the human genome, with an additional eight identified pseudogenes [10]. The sweet taste receptor on the other hand is generally accepted as being a heterodimer of two TAS1R subunits, with multiple binding sites for agonists [11, 12]. The use of this knowledge will be particularly beneficial in future attempts to elucidate structure-activity relationships between tastants that elicit the same taste response, as previous models attempted to dock all sweet (and sometimes all bitter) compounds in the same pocket [13, 14].

The discovery of the sweet and bitterness receptors has allowed for the development of cell-based assays with which to screen for taste-active compounds. A cell-based approach to new sweetener screening affords many benefits such as reproducibility, especially by removing taster bias and allowing for the collection of more data points, which leads to better quantification of the results obtained. Additionally, this approach renders possible high-throughput screening approaches to taste testing, using minimal amounts of samples, and bypasses most solubility issues and also reduces the need for safety testing to be carried out in conjunction with taste testing, until after compounds are established as potential leads [15, 16]. The minimal use of sample and the ability to test compounds, not readily soluble in water, are particularly beneficial to the potential discovery of new sweeteners and taste-modifying agents of natural origin. This becomes rapidly evident, when one considers the total amount of sample consumed over a series of chromatographic fractions in traditional taste tests, when purifying a sweet or other taste-modifying principle from a natural product extract. High-throughput screening methods in the field of taste have indeed been a long time in development, as one recent paper promotes a double-blinded screening method developed in 1966 as “state of the art” [17]. However, as with all in vitro techniques, lead compounds must be verified in in vivo sensory testing.

### *2.1 Natural High-Potency Sweeteners Used Commercially*

There are several naturally occurring high-potency sweeteners (nHPS) that find use in foods and beverages as extracts or highly purified compounds. However, there are only two widely used and approved structural classes of natural high-potency sweeteners used as such: those based on extracts of *Stevia rebaudiana* and those based on extracts of *Siraitia grosvenorii*. Recently, these sweeteners have attained what are perhaps the benchmarks of true sugar replacement, by both being produced in large scale and sold as tabletop sweeteners in sachets.

The majority of naturally occurring highly sweet compounds have not attained the same degree of success as mentioned above. However, there exist a considerable number of nHPS that find use as flavors or are used within limited geographical boundaries. Several of the nHPS flavors can be found in the Expert Panel of the Flavor and Extract Manufacturers Association of the United States' (FEMA) generally recognized as safe (GRAS) lists. While FEMA is not a government body, its GRAS lists are accepted in the United States, Australia, New Zealand, and the South American countries that make up Mercosur. In addition to these countries with codified acceptance of FEMA GRAS, several countries generally allow the use of FEMA GRAS flavors. Flavor levels set by FEMA for high-potency sweeteners are intended to be below their sweet taste thresholds, where they function as sweet taste enhancers.

## 2.2 Rebaudioside A and Related Compounds

Work on the sweet-tasting constituents of *Stevia rebaudiana* (Bertoni) Bertoni (Asteraceae) leaves began in the first half of the twentieth century with stevioside (**1**), the most abundant *ent*-kaurane diterpenoid (steviol) glycoside being crystallized over 80 years ago [18]. While intensely sweet, this compound suffers from a poor aftertaste. Rebaudioside A (**2**), another abundant steviol glycoside present in the leaves of *S. rebaudiana,* was first purified and determined structurally by the group of the late Prof. Osamu Tanaka at Hiroshima University in Japan in the 1970s [19]. This compound, while still possessing some bitterness, was found to have a superior taste quality to that of stevioside (**1**).

*Stevia rebaudiana* extracts have been used for decades as sweeteners, particularly in Japan. However, steviol glycosides did not obtain wide approval until 2008, after arduous safety screening was carried out, and their use was supported by a review by the FAO/WHO Joint Expert Committee on Food Additives (JECFA) [20]. Shortly after this approval, rebaudioside A (**2**)-based sweeteners were launched by several companies including collaborative efforts between Cargill and the Coca-Cola Company (Truvia®) and between PepsiCo and Merisant (PureVia®). Additionally, highly purified stevioside (**1**) can also be employed as a sweetener [21]. Very recently, a self-designation of GRAS was received by the US FDA for rebaudioside D (**3**) and rebaudioside X (**4**) (a compound with structure identical to that of the previously published rebaudioside M). These have been published as GRAS notices 456 and 473, respectively.

While not approved as sole sweetening agents per se, the steviol glycosides dulcoside A (**5**), rebaudioside B (**6**), rebaudioside C (**7**), rebaudioside D (**3**), rebaudioside F (**8**), rubusoside (**9**), and steviolbioside (**10**) may be present in significant quantities in *S. rebaudiana*-based sweeteners. According to the 73rd JEFCA recommendations, these minor glycosides are "generally present in preparations of steviol glycosides at levels lower than stevioside or rebaudioside A" (http://www.fao.org). For instance, FDA GRAS notice 275 was received with no questions asked in the United States and allows for a steviol glycoside sweetener with a content of 80% rebaudioside A (**2**) [22].

Enzyme-modified steviol glycosides, also known as "glucosyl steviol glycosides," "sugar-transformed steviol glycosides," or "enzyme-modified steviol glycosides," have been accepted in the United States for sweetening purposes [23]. Early work on glycosylated steviol glycosides was performed in the laboratory of Prof. Tanaka at Hiroshima University [24]. These compounds are 1,4-α-D-glucosylated forms of naturally occurring steviol glycosides and are produced from a *S. rebaudiana* extract that is reacted with a glucose-adding enzyme such as cyclodextrin glucosyltransferase, followed by an enzymatic glucose chain reduction [25]. GRAS notice 375 includes the mono- to penta-1,4-D-glucosylated derivatives of stevioside (**1**), rebaudioside A (**2**), rebaudioside B (**6**), rebaudioside C (**7**), dulcoside A (**5**), rubusoside (**9**), and steviolbioside (**10**).

## 2.3 *Mogrosides and Related Compounds*

The dried fruits of *Siraitia grosvenorii* (Swingle) C. Jeffrey ex A.M. Lu & Zhi Y. Zhang (Cucurbitaceae), commonly known as "lo han guo" or "monk fruit," produce a variety of sweet compounds, with the most abundant sweet constituent present being mogroside V (**11**). The initial discovery of a high-potency sweetener from this species was reported by Dr. Chi-Hang Lee in 1975, with mogroside V structurally characterized by Tsunematsu Takemoto and associates [26, 27]. Extractives of *S. grosvenorii* have been used as a sweetener in Japan for some time [28]. The use of monk fruit extracts with high levels of mogroside V for sweetening purposes was permitted recently in the United States [29], and Tate and Lyle commercialized a taste-improved monk fruit product with 50% mogroside V in 2011. This was followed in 2012 with the launch of the table top sweetener Nectresse™ by McNeil Nutritionals, LLC.

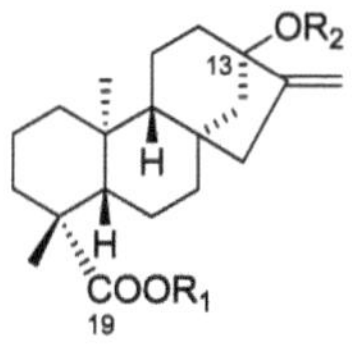


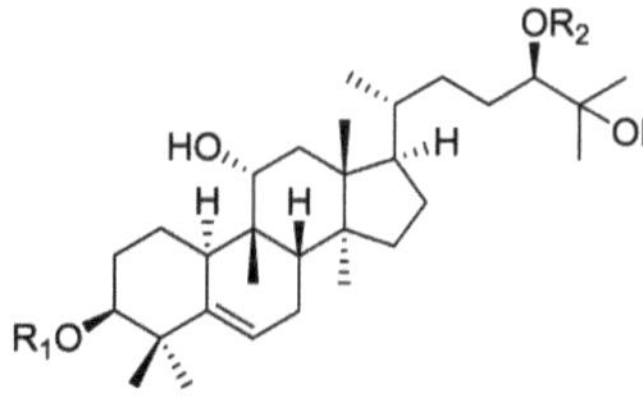


| | R1 | R2 |
|---|---|---|
| **1** | β-glc | β-glc$^{2}$-β-glc |
| **2** | β-glc | β-glc$^{2}$-β-glc<br>\|3<br>β-glc |
| **3** | β-glc$^{2}$-β-glc | β-glc$^{2}$-β-glc<br>\|3<br>β-glc |
| **4** | β-glc$^{2}$-β-glc<br>\|3<br>β-glc | β-glc$^{2}$-β-glc<br>\|3<br>β-glc |
| **5** | β-glc | β-glc$^{2}$-α-rha |

| | R1 | R2 |
|---|---|---|
| **6** | H | β-glc$^{2}$-β-glc<br>\|3<br>β-glc |
| **7** | β-glc | β-glc$^{2}$-α-rha<br>\|3<br>β-glc |
| **8** | β-glc | β-glc$^{2}$-β-xyl<br>\|3<br>β-glc |
| **9** | β-glc | β-glc |
| **10** | H | β-glc$^{2}$-β-glc |

| | R1 | R2 |
|---|---|---|
| **11** | β-glc$^{6}$-β-glc | β-glc$^{2}$-β-glc<br>\|6<br>β-glc |
| **48** | β-glc$^{4}$-β-glc | β-glc$^{2}$-β-glc<br>\|6<br>β-glc |

glc: D-glucopyranosyl
xyl: D-xylopyranosyl
rha: L-rhamnopyranosyl
glcA: D-glucuronopyranosyl

# 3 Natural Products Generally Used for Flavoring

## 3.1 Sweet Taste Enhancers

Many naturally occurring high-potency sweeteners find use as sweetness-enhancing agents when used at concentration levels below their sweet taste threshold. These compounds, when combined with a sweetener, result in an enhanced overall perception of sweetness. There is a second group of sweet taste enhancers referred to as "positive allosteric modulators" (PAMs). In taste research, this term generally refers to a compound that is not itself intensely sweet at any use level, but increases the potency of a sweetener. PAMs theoretically alter the configuration of the sweet taste receptor to allow for stronger binding of a sweetener to the sweet taste receptors. In pharmacological terms, subthreshold enhancers may act as ago-allosteric modulators [30], able to activate the sweet taste receptor at the allosteric site, and produce an increased response, when another sweetener is bound to the receptor.

Both monk fruit and stevia (*S. rebaudiana*) extracts may be used as flavors according to FEMA. Many steviol glycosides have been approved by FEMA as GRAS flavors and can generally find use as sweetness enhancers. These include stevioside (**1**), various purity levels of rebaudioside A (**2**) [31, 32], rebaudioside C (**7**) [33], and the previously mentioned glucosyl steviol glycosides, which are also known by the trade name NSF-02 [31]. Rubusoside (**9**), another steviol glycoside, a trace constituent in *S. rebaudiana* leaves but much more abundant in *Rubus suavissimus* S. Lee (Rosaceae), is also a FEMA-approved flavor as "sweet blackberry leaf extract." Sweet blackberry leaf extract is generally marketed as >70% rubusoside (**9**) [33].

The triterpenoid glycyrrhizin (**12**), or glycyrrhizic acid, is a sweet-tasting component of *Glycyrrhiza glabra* L. (Fabaceae) and other species with the common name "licorice." A crude plant extract containing glycyrrhizin was used as a sugar substitute in Japan in the early part of the twentieth century [34]. Although having a long history of human use, excessive glycyrrhizin consumption raises health concerns due to its propensity to cause pseudoaldosteronism [28, 35, 36]. Investigation into derivatives of glycyrrhizin with enhanced flavor profiles yielded both the monoglucuronide of glycyrrhetinic acid (accorded the abbreviation MGGR) (**13**) as well as the ammonium salt of glycyrrhizin (monoammonium glycyrrhizinate) [28, 37, 38]. Glycyrrhizin and its ammoniated derivative both possess a strong licorice flavor in addition to a lingering sweetness with a slow onset of taste. MGGR, which is produced using microbial transformation, was found to be sweeter than the parent compound and is used to flavor chocolate milk and soft drinks in Japan [28, 38]. Glycyrrhizin and monoammonium glycyrrhizinate have both attained GRAS status from the US FDA, when used as flavoring agents or surfactants (FDA 21 CFR Section: 184.1408). These were included in the first published list of FEMA-approved flavors [39].

The thaumatins are a group of closely related intensely sweet proteins (having at least 1,600 times the sweetness of sucrose on a weight basis) isolated from the fruits of the West African plant *Thaumatococcus danielli* Benth. (Marantaceae), with the major proteins thaumatin I and II isolated over 30 years ago [40]. The sweetness of the thaumatins was found to be dependent on the disulfide bonds present in these proteins, with six thaumatins, I, II, III, a, b, and c now having been isolated and characterized structurally thus far [40, 41]. Thaumatin is an approved sweetener in many countries and is a FEMA-approved flavor in the United States [28, 42].

The flavonoids and their synthetic derivatives have been particularly rich sources of sweet taste-enhancing compounds. Dihydrochalcones, flavans, flavanones, and synthetic derivatives of dihydrochalcones have all been approved by FEMA as flavors and possess sweetness-enhancing capabilities.

Neohesperidin dihydrochalcone (NHDC) (**14**) is a high-potency sweetener produced from the bitter-tasting flavonoid glycoside, neohesperidin. NHDC was originally produced in order to lower levels of the parent glycoside in the Seville orange, *Citrus aurantium* L. (Rutaceae) [43, 44]. NHDC displays a delayed onset in its sweetness and an aftertaste, which limit its use as a sweetener [43]. NHDC is approved as a sweetener and flavor in the European Union and is used as a flavoring agent in countries that follow FEMA recommendations [28]. Naringin dihydrochalcone (**15**), produced from naringin, was discovered concurrently with NHDC and has recently been given FEMA GRAS approval [31].

While the dihydrochalcones **14** and **15** have not been found in Nature to date, a structurally related natural dihydrochalcone, trilobatin (**16**), is a FEMA GRAS flavor (FEMA 25). This compound has been isolated from *Lithocarpus polystachyus* Rehder (Fagaceae) and *Malus trilobata* C.K. Schneid. (Rosaceae) and has long been known as a sweet-tasting natural product [45]. Phloretin (**17**), another FEMA-approved flavor [46], is the aglycone of trilobatin, which has been obtained previously from *Malus* sp. Both phloretin and trilobatin are patented for use as sweet taste enhancers [45, 47].

Phytochemical investigation of *Eriodictyon californicum* Decne. (Hydrophyllaceae) led to the determination that its constituent, hesperetin (**18**), is a sweetness enhancer [48]. Hesperetin was subsequently afforded FEMA GRAS status [46]. In an independent study of sweet taste-enhancing compounds from *E. californicum*, hesperetin and sakuranetin (**19**) were found to be active in the seven transmembrane domains of the sweet taste receptor [49]. However, the related compounds, homoeriodictyol (**20**) and naringenin (**21**), were not active in this assay [49]. There are conflicting reports on the sweet taste-enhancing activity of homoeriodictyol, and it appears that, if the compound is a sweetness enhancer, it only shows weak activity in this regard [49, 50]. An extract of *E. californicum* was also included in the first FEMA flavor list [39].

The sweet compound 7,3′-dihydroxy-4′-methoxyflavan (**22**) has recently been approved by FEMA as a sweet taste enhancer [33]. Although initially prepared by synthesis, the compound has also been identified as a constituent of *Terminalia argentea* Mart. (Combretaceae) [51, 52].

The pyridine-containing dihydrochalcone derivative 1-(2-hydroxyphenyl)-3-(pyridine-4-yl)propan-1-one (**23**) is a FEMA-approved flavor with sweet taste-enhancing capabilities [53]. Two compounds with similar structures, 1-(2-hydroxy-4-isobutoxyphenyl)-3-(pyridine-2-yl)propan-1-one (**24**) and 1-(2-hydroxy-4-methoxyphenyl)-3-(pyridine-2-yl)propan-1-one (**25**), are FEMA-approved flavors with patented use as umami enhancers [33, 54].

Phyllodulcin (**26**) is responsible for the sweet taste of a ceremonial Japanese tea (Buddha Tea) prepared from the leaves of *Hydrangea macrophylla* Seringe var. thunbergii (Siebold) Makino (Saxifragaceae) [28, 55]. While this compound is structurally an isocoumarin, it can be regarded as an isotere of the flavonoid class. The commercial use of phyllodulcin has been restricted due to its poor water solubility and less than optimal taste properties [28, 36]. In spite of these hurdles, Buddha Tea extract has recently been granted FEMA GRAS status [31].

Perillartine (**27**) has found limited use in Japan as a sweetener in maple syrup substitutes or as a licorice-type flavoring for tobacco [56, 57]. Perillartine may be prepared from perillaldehyde (**28**), a FEMA-approved flavor from *Perilla frutescens* (L.) Britton (Labiatae) [58]. Poor solubility and a bitter off-taste limit the use of perillartine [28, 57].

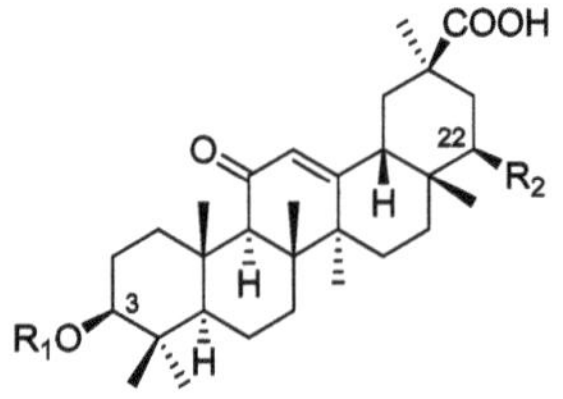


| | $R_1$ | $R_2$ |
|---|---|---|
| **12** | β-glcA²-β-glcA | H |
| **13** | β-glcA | H |
| **45** | β-glcA²-β-glcA | $OCOCH_3$ |

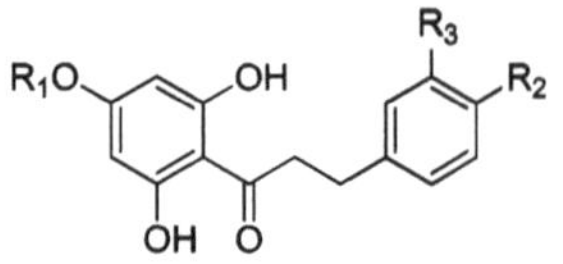


| | $R_1$ | $R_2$ | $R_3$ |
|---|---|---|---|
| **14** | β-glc²-α-rha | $OCH_3$ | OH |
| **15** | β-glc²-α-rha | OH | H |
| **16** | β-glc | OH | H |
| **17** | H | OH | H |

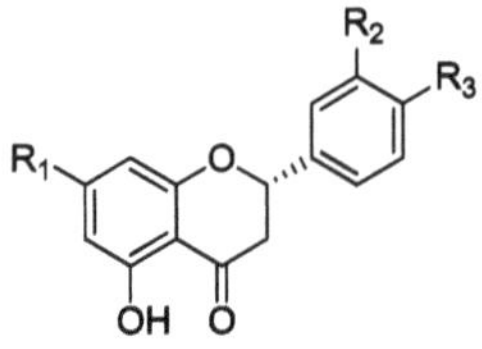


| | $R_1$ | $R_2$ | $R_3$ |
|---|---|---|---|
| **18** | OH | OH | $OCH_3$ |
| **19** | $OCH_3$ | H | OH |
| **20** | OH | $OCH_3$ | OH |
| **21** | OH | H | OH |

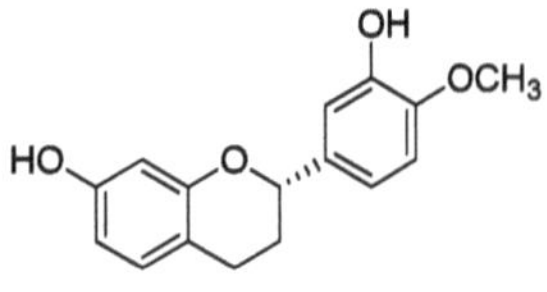


**22**

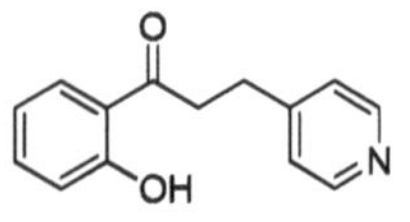


**23**

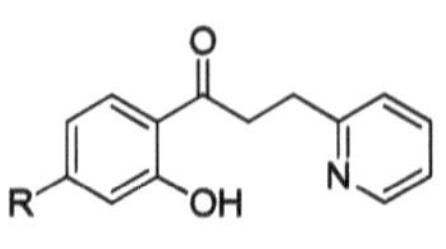


**24** R = 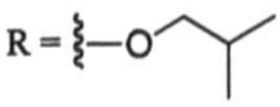


**25** $OCH_3$

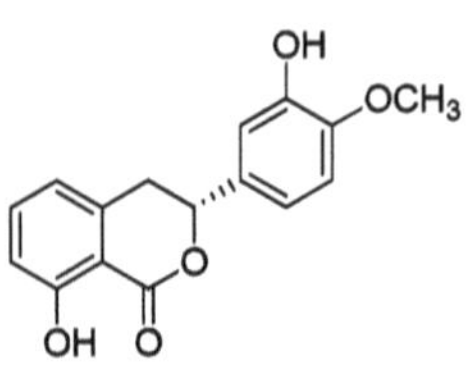


**26**

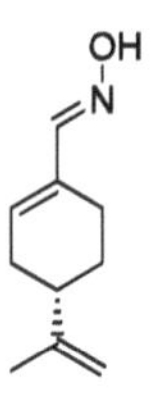


**27**

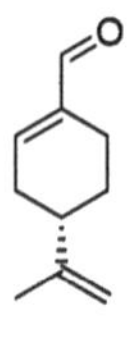


**28**

### 3.2 Naturally Occurring Sweeteners

In our earlier book chapter review, over 100 known highly sweet substances of natural origin reported by 2008 have been described in some detail [36]. These compounds were grouped into 20 major categories based on their chemical structural classes. The major categories of highly sweet-tasting principles derived from plant are terpenoids, flavonoids, and proteins. Other less common plant-derived sweeteners fall into the amino acid, benzo[*b*]indeno[1,2-*d*]pyran, isocoumarin, phenylpropanoid, proanthocyanidin, and steroidal saponin structural classes. The major classes of sweet-tasting terpenoids are diterpenoids and triterpenoids, with the most representative subtypes of diterpenoids being *ent*-kaurane- and labdane glucosides and of triterpenoids being cucurbitane, dammarane, and oleanane glycosides, respectively. The dihydrochalcone and the dihydroflavonol classes are reported to be responsible for the sweet taste of the flavonoids. Thus far, species representative of more than 25 separate plant families have been found to produce sweet-tasting compounds. Among these plant families, Asteraceae, Apiaceae, Lamiaceae, and Rosaceae are the major source of *ent*-kaurane and labdane glucosides, and Asclepiadaceae, Cucurbitaceae, Fabaceae, and Juglandaceae have been found to produce cucurbitane, dammarane, and oleanane glycosides [36, 59].

In this section of the chapter, plant-derived sweet-tasting compounds are updated based on our last review, with only those discovered since 2008 being mentioned. During the last 5 years, although many studies have led to the isolation and identification of a large number of structurally new derivatives of those known natural sweet principles, published reports focused on the organoleptic property evaluations, and related researches of these newly compounds are very limited. Among the new recently identified natural sweeteners, diterpenoids and triterpenoids are still the most typical compounds.

In 2010, a research group in Japan isolated and identified ten new minor steviol glycosides of the *ent*-kaurane type, namely, dulcoside B (SG4) (**29**) and rebaudiosides G-O (**30–35**,**4**,**36**,**37**), from a hot water (80°C) extract of the leaves of *Stevia rebaudiana* [60]. [It should be noted that the name dulcoside B, originally referred to a structure identical to rebaudioside C and was published as a new compound soon after the first report of rebaudioside C (near simultaneous discovery), has been somewhat abandoned and if mentioned is usually of the form "rebaudioside C (dulcoside B)" [61, 62].] All of these new compounds share the same aglycone (steviol) as rebaudioside A (**2**), with the structural differences between these substances occurring at the sugar chains at C13 and C19. No comprehensive organoleptic evaluations have been reported for these new steviol glycosides, with only rebaudiosides M (**4**) and N (**36**), each possessing a chain of three sugar units at C19, mentioned as having a sweetness equivalent to rebaudioside A (**2**). Volunteer human subjects reported a slightly latent sweet taste sensation, however. In a related patent application, based on this research [63], extracts and crystals, containing these new steviol glycoside mixtures, were evaluated by their taste quality. Besides the strong sweet taste ascribed to all the tested samples, the extracts, containing greater quantities and numbers of the new minor rebaudiosides than the crystalline samples, were claimed as having an improved "delicate" taste.

As a part of a continuing search for new natural sweeteners and to afford a better understanding of the physicochemical profiles of steviol glycosides, a research group at the Coca-Cola Company has produced several new semisynthetic *ent*-kaurane glycosides by catalytic hydrogenation of certain steviol glycosides of known composition [64, 65]. In this work, the C16/17 exocyclic double bond reduced derivatives of rubusoside (**38**), stevioside (**39**), and rebaudiosides A-D (**40–43**) were produced. Since the hydrogenation process is not stereoselective, the reduced product in each was a mixture of two epimers at C16. In the sensory evaluations conducted, all of the 16,17-dihydro derivatives were deemed to have lost sweet taste intensity to different degrees when compared with their respective parent compound. This study demonstrated that the C16/17 exocyclic double bond plays an important role in maintaining the sweetness of these steviol glycosides. This work confirms earlier observations that the C16/17 exocyclic double bond is important for the mediation of the sweetness of the steviol glucosides [66].

Using centrifugal partition chromatography, a new sweet-tasting oleanane triterpenoid glycoside, quercotriterpenoside I (**44**), was isolated from an ethanol-water extract of the heartwood of *Quercus petraea* (Matt.) Liebl. (Fagaceae), an oak tree grown in Caucasia [67]. In comparison of the structure of quercotriterpenoside I with other known oleanane-type sweeteners, the glucose moiety is positioned at C28, instead of having a more common sugar linkage at C3. In addition, the substitution of a galloyl group at C23 is also quite unusual. Interestingly, this discovery supports an observation made by some winemakers that aging in oak barrels has a sweetening effect on wine.

A glycyrrhizin derivative, 22β-acetoxyglycyrrhizin (**45**), was purified from an extract of the root of *Glycyrrhiza uralensis* Fisch. by Li and colleagues in 2007 [68]. This licorice triterpene was patented in Japan later as a natural sweetener, which was reported to be only ten times sweeter than that of sucrose, much less than the parent compound, glycyrrhizin [69]. The potency of this compound shows that an acetoxy group substitution at C22 dramatically decreases the sweet intensity of glycyrrhizin (**12**).

From an aqueous ethanolic extract of the aerial parts of *Mycetia balansae* Drake (Rubiaceae), two sweet-tasting dammarane-type glycosides, balansins A (**46**) and B (**47**), were isolated by Ley et al. of Symrise AG in Germany [70]. These two new compounds both possess a double bond between C20 and C22 and a hydroxy group at C21. In a sensory evaluation testing procedure, both balansins A and B were described as being intrinsically sweet with slightly bitterness, and at a concentration of 50 mg/l, the sweetness potencies of balansins A and B were reported to be equal to that of 0.1% and 0.2% sucrose, respectively.

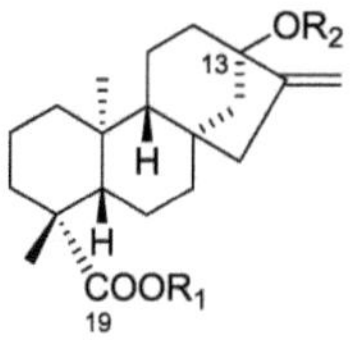


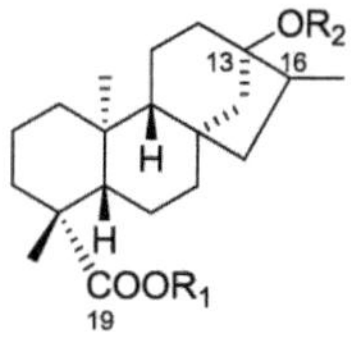


| | $R_1$ | $R_2$ |
|---|---|---|
| **29** | H | β-glc²-α-rha<br>│3<br>β-glc |
| **30** | β-glc | β-glc³-β-glc |
| **31** | β-glc | β-glc²-α-rha<br>│3 │3<br>β-glc β-glc |
| **32** | β-glc³-β-glc | β-glc²-β-glc<br>│3<br>β-glc |
| **33** | β-glc²-α-rha | β-glc²-β-glc<br>│3<br>β-glc |

| | $R_1$ | $R_2$ |
|---|---|---|
| **34** | β-glc²-β-glc | β-glc²-α-rha<br>│3<br>β-glc |
| **35** | β-glc | β-glc²-β-glc<br>│3 │6<br>β-glc β-glc |
| **36** | β-glc²-α-rha<br>│3<br>β-glc | β-glc²-β-glc<br>│3<br>β-glc |
| **37** | β-glc²-α-rha<br>│3 │3<br>β-glc β-glc | β-glc²-β-glc<br>│3<br>β-glc |

| | $R_1$ | $R_2$ |
|---|---|---|
| **38** | β-glc | β-glc |
| **39** | β-glc | β-glc²-β-glc |
| **40** | β-glc | β-glc²-β-glc<br>│3<br>β-glc |
| **41** | H | β-glc²-β-glc<br>│3<br>β-glc |
| **42** | β-glc | β-glc²-α-rha<br>│3<br>β-glc |
| **43** | β-glc²-β-glc | β-glc²-β-glc<br>│3<br>β-glc |

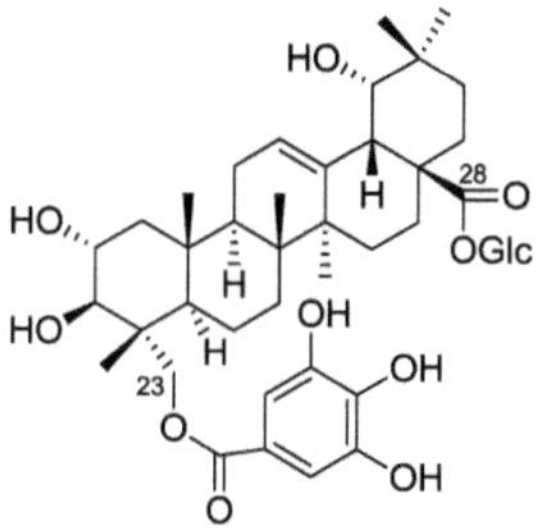


**44**

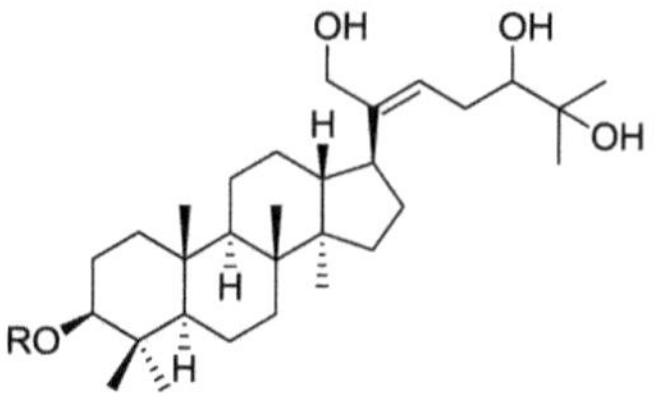


**46** R = β-glc²-β-glc
**47** R = β-glc²-β-glc<br>│6<br>β-glc

Iso-mogroside V (**48**) is a minor cucurbitane triterpenoid glycoside, isolated from the fruits of *Siraitia grosvenorii* (Cucurbitaceae), also known as "lo han guo" or monk fruit, by the late Zhonghua Jia and colleagues of Givaudan Flavors Corporation [71]. The structure of iso-mogroside V is quite similar to that of mogroside V (**11**), except that two glucose units at C3 in iso-mogroside V are joined by an α-1,4 glycosidic linkage instead of an α-1,6 glycosidic bond in mogroside V (**11**). In this report, the sweetness intensity of iso-mogroside V (**48**) was recorded as being even more potent than that of mogroside V and was rated as approximately 500 times than that of 5% sucrose.

As mentioned before, published research over the last few years on the discovery of structurally new naturally occurring sweet-tasting principles has been quite limited. On the other hand, considerable progress has been made to better understand the structure and function of the human taste receptors, which led to the successful application of receptor-based assay systems as a fast, sensitive, and selective method to facilitate the studies of interaction modes between the sweet-tasting ligands and the functional organoleptic receptors and to characterize the potential sweet properties of compounds with various structures.

The human sweet taste receptor (hT1R2/hT1R3), expressed on the surface of taste bud cells, is a heteromeric complex composed of two subunits, T1R2 and T1R3, which both belong to the C G-protein-coupled receptor family (C GPCRs) [72–75]. Each protein unit comprises a large extracellular amino-terminal domain (ATD), which also referred to the Venus flytrap module (VFTM), an extracellular cysteine-rich domain (CRD), a seven-helical transmembrane domain (TMD), and an intracellular carboxy-terminal domain [76, 77]. The human sweet taste receptor is able to respond to chemically diverse compounds ranging in molecular size from small naturally occurring and synthetic molecules to sweet-tasting proteins. In addition, this receptor exhibits stereoselectivity for certain molecules [78, 79], and thus, the illustration of the activation mechanism of the receptor is quite complex. Various research approaches, based on X-ray crystallographic analysis, molecular modeling, and site-directed mutagenesis, have been used to characterize the functional domains and the binding modes of C class GPCRs. The functional expression results indicate that, for the heteromeric complex of hT1R2/hT1R3, binding for most sweet ligands occurs on T1R2, which needs to be coexpressed with T1R3 to be functional [75, 80]. The study of the crystallized complexes of the homodimeric metabotropic glutamate 1 receptor (mGluR1) revealed the representative structural feature of C GPCRs. The ATD of C GPCRs was found to comprise two lobes, LB1 and LB2, of which the function is mediated through a Venus flytrap mechanism controlled by the equilibrium between an "open" and a "closed" conformation of these two lobes [81]. By using cross-species chimera approaches and homology modeling methods, multiple binding sites to different ligands are found on the sweet receptor T1R2/T1R3. Major artificial sweeteners such as aspartame or neotame and some naturally occurring sugar analogues were found to bind to the VFTM on T1R2 by functional expression of rat-human chimeric constructs [80, 82] as a result of spectroscopic analysis of the purified ligand-binding domains of T1R3 taste receptors [83]. Based on in silico docking experiments using homology models on mGluR1, sweet proteins such as brazzein, monellin, and thaumatin were suggested to bind to the open end of the VFTM with a postulated wedge model [11, 84]. However, this hypothesis has been challenged by several other studies. By using site-directed mutagenesis and generation of cross-species receptor chimera methods, the cysteine-rich domain of human T1R3 was found to be required for species-specific sensitivity to brazzein sweetness and was recognized as a new determinant of sweet receptor function [85]. Using calcium imaging analysis in HEK cells and a human-mouse chimeric receptor system, the amino-terminal domain (ATD) of T1R3 was identified as a new sweetener-binding region, which is essential for the receptor interaction of the sweet protein neoculin [86]. On evaluation of the sweetness of brazzein and its mutants, using a calcium-mobilization cell-based assay with heterologously expressed human sweet taste receptors, brazzein was found to interact with both T1R2 and T1R3. In addition, no expected decrease in brazzein response was observed to those mutations of receptor residues at putative functional sites predicted by wedge models. This study suggested that a series of multiple-step binding events might be involved during the interaction between brazzein and the sweet receptor [87]. These studies demonstrated the different functional roles of subunits on the heteromeric complex and indicated the complexity of cooperative interactions between the sweet-tasting ligands and the functional binding sites on the human sweet receptor, which supported the concept of allosteric modulation of the G protein-coupled receptor [88].

Studies of the interaction between naturally occurring small-molecule sweet-tasting ligands and the sweet receptors are lacking. However, cell-based functional taste receptor expression assays have been employed widely to evaluate the organoleptic properties of natural sweeteners in a systematic, comprehensive, and efficient manner. In a comparison of the human psychometric and taste receptor responses to selected steviol glycosides, derivatives including stevioside (**1**), rebaudiosides A-F (**2**, **3**, **6–8**), steviolbioside (**10**), dulcoside A (**5**), and rubusoside (**9**) were evaluated by an approach combining calcium imaging experiments conducted in HEK293 FlpIn T-Rex $G_{\alpha 15}G_{i3}$/hTAS1R2 cells and human sensory studies [89]. This investigation revealed that all steviol glycosides tested were able to activate the functionally expressed sweet taste receptor in a dose-dependent manner, and the concentrations preceded onset response in vitro were comparable to those obtained by human sensory testing. The rank order of potency of functionally expressed hTAS1R2/ hTAS1R3 for the individual steviol glycoside was also similar with the ranking based on the in vivo evaluation results. However, except for the consistent organoleptic profiles obtained based on the cell-based receptor assay and the in vivo test results, interesting differences were also observed between these two different evaluation systems. The maximal relative sweet intensities of the test compounds varied considerably in the in vivo test, while the potency of these compounds was very consistent in in vitro assay, which implies that the distinct sweetness elicited by these steviol glycosides seems not be mediated at the sweet taste receptor level. In addition, the sweet taste sensation, elicited by stevioside (**1**), rebaudioside A (**2**), and rubusoside (**9**), began to decrease after passing through the maximal concentrations in the human sensory experiments, while in the functional receptor assay, the sweet receptor response evoked by these three compounds did not decline beyond the maximum. The authors deduced that these attenuating effects in vivo of the three steviol glycosides could be caused by a cross-model suppressing effect of the associated intrinsic bitter taste.

A recombinant human taste receptor T1R2/T1R3-dependent cell-based system has been used in high-throughput screening of molecules for sweeteners and sweetness enhancers or modifiers. In this bioassay, the heterodimeric human taste receptors (T1R2/T1R3) are transfected with a multicistronic plasmid vector such as pTrix-Eb-R2R3 in HEK293 cells that stably express $G_{\alpha}15$, a promiscuous phospholipase C-linked G protein. Binding of the sweet-tasting stimuli with T1R2/T1R3 ligands induces transient increases in the intracellular calcium levels in $G_{\alpha}15$ cells, which can be evaluated by measuring the fluorescence values of cells labeled with the calcium-sensitive fluorescent dye Fluo-4 AM [90–95]. By using similar sweetness evaluation methods and protocols, several known natural secondary metabolites were claimed recently for their sweetness or/and sweetness enhancer properties [90–95]. Most of these compounds are derivatives of plant origin, which include (i) two flavonoids, nobiletin (**49**), a substance isolated from *Citrus* spp. [96], and afzelin 2″-*O*-gallate (**50**) found from *Calliandra haematocephala* Hassk. (Fabaceae) [97] and *Eugenia hyemalis* Cambess. (Myrtaceae) [98]; (ii) a triterpenoid, tormentic acid (**51**), which was initially isolated from *Potentilla tormentilla* Gilib. (Rosaceae) [99]; (iii) a benzyl glycoside, benzylprimeveroside (**52**), first isolated from *Alangium platanifolium* (Siebold & Zucc.) Harms. var. *trilobum* (Alangiaceae) [100]; and (iv) a phenylpropanoid derivative, *trans*-melilotoside (**53**), isolated from sweet clovers (*Melilotus* spp.) [101]. Besides these compounds from plants, microcarpalide (**54**), an alkyl-substituted nonenolide, originally isolated from the fermentation broth of an unidentified endophytic fungus growing on the bark of *Ficus microcarpa* Wight et Miq. [102], was also identified as a sweet substance of natural origin [95]. However, none of these compounds appears to have been reported to have a sweet taste as assessed by human volunteers. The sweet-tasting potential of these natural-origin constituents was identified only based on their activating effect of the expression of the hT1R2/hT1R3 sweet taste receptor in the cell-based functional taste receptor assay.

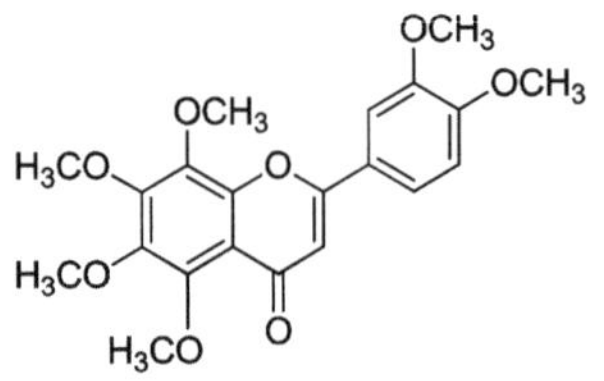


**49**

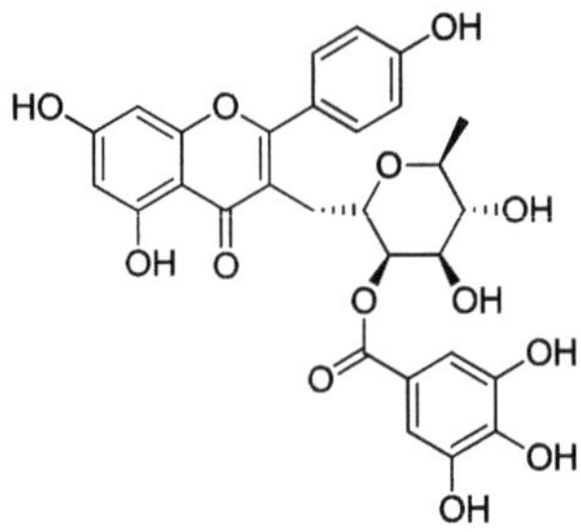


**50**

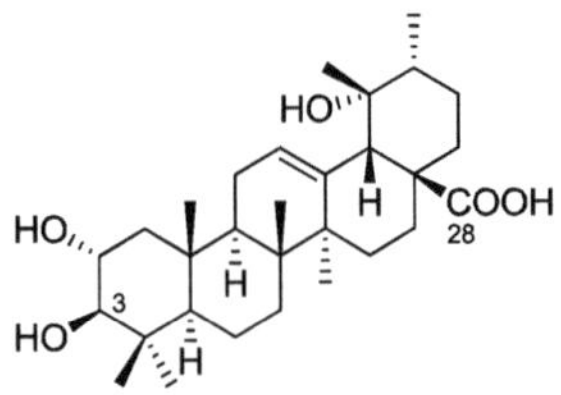


**51**

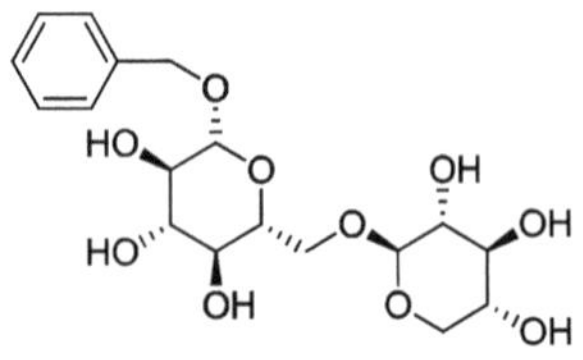


**52**

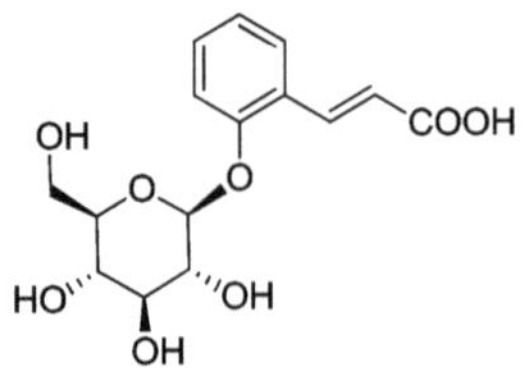


**53**

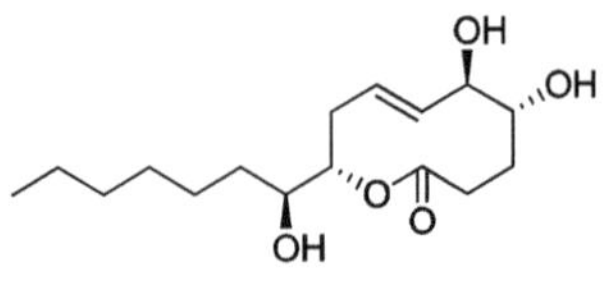


**54**

### *3.3 Naturally Occurring Sweet Taste Enhancers*

Several compounds of natural origin are known that are indicated as being of use for sweetness enhancement, but these have no indication of commercial utility. Aside from the sometimes obtuse language used, claims presented in the patent literature may not be supported by statistically significant experimental studies. Typical taste experiments feature a small number of panelists, and when a sweetness potency value is given, the enhancement seen is often within the published limits of error. As noted above, sweeteners used below their sweet taste threshold may be considered as sweet taste enhancers. However, currently, there is not a standard industry method to determine the sweet taste threshold. Therefore, one should be cautious when drawing conclusions based on the taste properties disclosed in the patent literature. For example, three out of five panelists found afzelin 2″-*O*-gallate (**50**) at a concentration of 20 μmol/l to increase the sweetness of a 4% fructose solution [91]. Benzylprimeveroside (**52**), a naturally occurring benzyl glycoside, was also found to enhance the sweetness of 4% fructose by three out of five panelists [93]. If treated as equal samples, the probability of at least three of five respondents, choosing the test sample as sweeter than the control fructose solution, is 50%. These patent applications, however, also employed cell-based assays, as mentioned in the previous section, which indicated that the compounds both activated the receptors and had a greater than additive effect with fructose.

In addition to the compounds previously mentioned, several phenylpropanoids have been disclosed as sweet taste enhancers. These include the structurally related flavans, 5,7,3′-trihydroxy-4′-methoxyflavan (**55**) and 4′,7-dihydroxyflavan (**56**), which are described as less active analogues of the FEMA-approved 7,3′-dihydroxy-4′-methoxyflavan (**57**) [103]. The dihydrochalcones, eriodictyol dihydrochalcone (**58**) and 4,2′-dihydroxy-3-methoxydihydrochalcone (**59**), have been disclosed as less active analogues of phloretin (**17**) [47]. The flavonoid nobiletin (**49**) has been claimed as a sweet taste enhancer for fructose [90]. 3,2′-Dihydroxy-4,4′,6′-trimethoxychalcone (**60**) was also determined to be a potential sweet taste enhancer in a cell-based assay [49]. This chalcone is a constituent of *Merrillia caloxylon* Swingle (Rutaceae) [104]. Rosmarinic acid (**61**), a constituent of several herbs, also has been disclosed to be a sweet taste enhancer [105]. In addition, the previously mentioned *trans*-melilotoside (**53**) has also been claimed to have this same biological property [94].

The phenolic compounds (+)-catechin (**62**), (+)-gallocatechin (**63**), (−)-epicatechin (**64**), and (−)-epigallocatechin (**65**) have been patented as sweetness enhancers [106]. These results were not replicated by independent research at the Coca-Cola Company [107]. However, the patent on these compounds deals mainly with time-intensity of sweetness in chewing gum, and thus, the compounds disclosed as sweet taste enhancers may be acting through mechanism other than direct interaction with the sweet taste receptor.

Microcarpalide (**54**) is a metabolite isolated from an unidentified endophytic fungus from *Ficus microcarpa* as an antimicrofilament agent. Subsequent investigation of microcarpalide in a cell-based sweet taste assay revealed the compound as an in vitro activator of the sweet taste receptor [95]. In a five-person panel, four panelists responded that 37 μmol/l of the compound enhanced the sweetness of a 4% fructose solution.

The surfactins, a group of cyclic peptides of microbial origin, have a patented use as sweet taste enhancers [108]. A mixture of surfactins from *Bacillus subtilis* with surfactin C (**66**) as the major component was screened for activity in a sweet receptor cell-based assay. This surfactin mixture was active in the assay and showed a super-additive effect with fructose at some concentrations of surfactin. This result indicates that surfactin may act as both a sweetener and a sweet taste enhancer for fructose.

Aladapcin (**67**), a bacterial metabolite from a *Nocardia* species, has been demonstrated as a sweetness enhancer using a cell-based assay [109]. The method used in this assay evaluated the compound at various concentrations with and without the addition of fructose. The results obtained showed an increase in response to fructose with increasing amounts of aladapcin. However, there was no response to aladapcin alone.

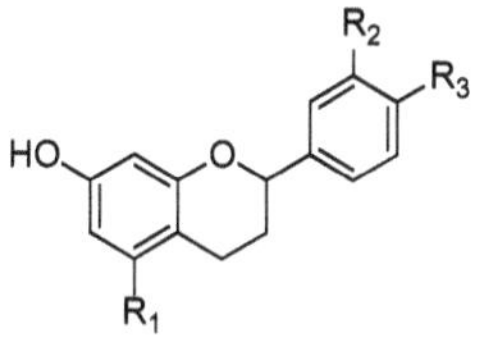

| | $R_1$ | $R_2$ | $R_3$ |
|---|---|---|---|
| **55** | OH | OH | $OCH_3$ |
| **56** | H | H | OH |
| **57** | H | OH | $OCH_3$ |

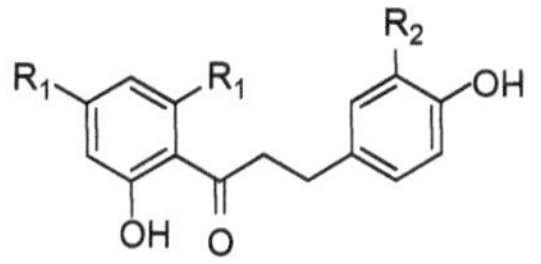

| | $R_1$ | $R_2$ |
|---|---|---|
| **58** | OH | OH |
| **59** | H | $OCH_3$ |

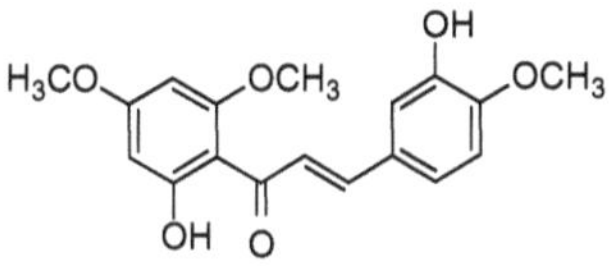

**60**

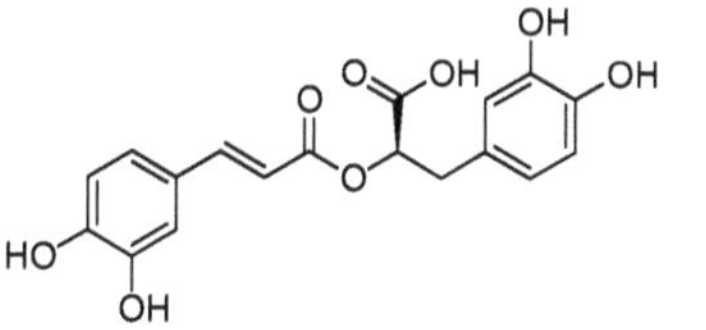

**61**

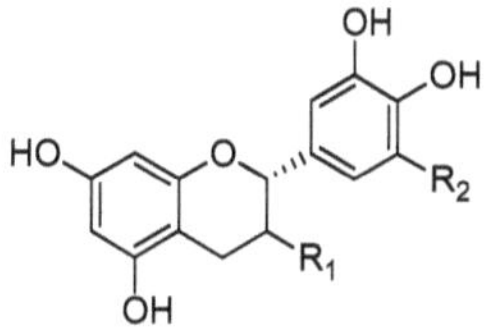

| | $R_1$ | $R_2$ |
|---|---|---|
| **62** | β-OH | H |
| **63** | β-OH | OH |
| **64** | α-OH | H |
| **65** | α-OH | OH |

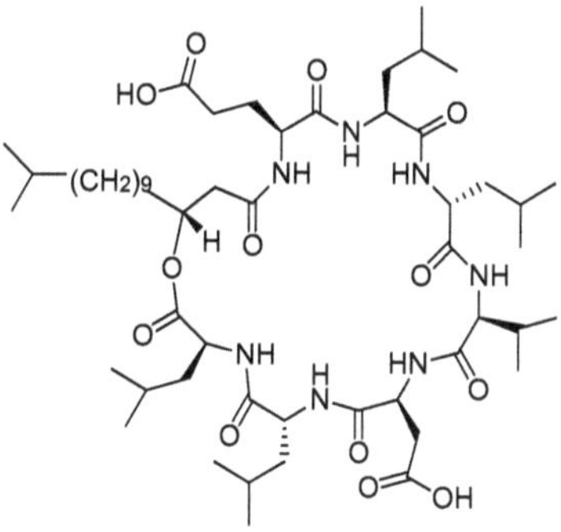

**66**

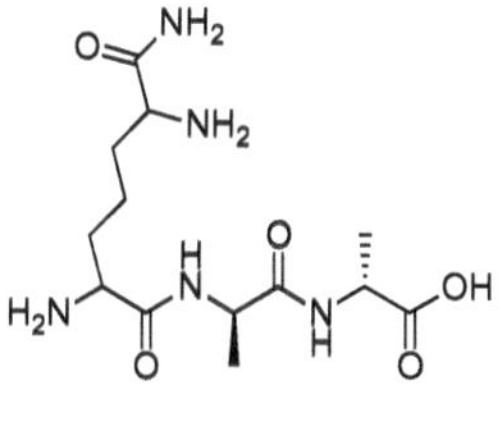

**67**

Tormentic acid (**51**) from *Potentilla tormentosa* has a patented use as a sweetener and a sweet taste enhancer [92]. The main evidence in support of this was from a cell-based assay. When evaluated by five panelists at 20 μmol/l, the compound was reported as having a raspberry-like taste with no strong sweetness. Perplexingly, the patent claims this taste experiment "show(s) that tormentic acid, beneficially, may be utilized as a sweetener, without contributing off tastes."

### 3.4 Naturally Occurring Sweet Taste Inhibitors

In addition to sweet taste-enhancing compounds, several sweet taste-inhibitory natural products are known. Although sweet taste inhibition may seem to be more of a novelty than a useful phenomenon, there are commercial applications for sweet taste inhibition. For instance, the sweetness inhibitor lactisole (**68**) is used in jellies and jams, where the physical properties of sugar are more important than the sweetness [50]. Additionally, sweetness inhibitors could be of use in lactose-free dairy products, which are sweeter than their lactose-containing counterparts [110]. Aside from food uses, there has also been some interest recently in non-gustatory taste receptors and the possibility of sweet taste inhibition to modify metabolic activity [110].

The above-mentioned lactisole (**68**) was discovered in a screening procedure for synthetic compounds active at the sweet taste receptor [111]. This compound was then discovered to be a naturally occurring component of coffee beans (*Coffea* spp.) [50]. Lactisole, while active against primate sweet taste receptors, does not impact rodent sweetness detection [112]. Using a series of human-mouse chimeric receptors, it was determined that lactisole interacts with the transmembrane portion of Tas1R3 [112]. Lactisole is a FEMA-approved flavor [113].

The most prevalent class of sweet taste inhibitors is represented by the triterpenoid glycosides and a great deal of work on these substances has been performed by investigators in Japan. These compounds have been recently reviewed elsewhere [36]; therefore, an abridged review will be given in the present chapter. The three major plant sources of natural antisweet compounds are *Gymnema sylvestre* (Retz.) Schult. (Asclepiadaceae), *Hovenia dulcis* Thunb. (Rhamnaceae), and *Ziziphus jujuba* Mill. (Rhamnaceae) [36].

*G. sylvestre* produces gymnemic acid I (**69**), which is the benchmark, against which other sweet taste-inhibitory compounds are generally compared [36]. The taste properties of this plant have a long history of study with the first phytochemical investigation occurring in the late nineteenth century [114]. The antisweet properties of *G. sylvestre* leaves were known long before this study, however, and the Hindi name for the plant translates as "sugar destroyer" [110]. Currently, gymnemic acids I–XVIII (**69–86**) have been isolated from *G. sylvestre*, of which all that were tested displayed antisweet activity [115–120]. Also, gymnemasaponins I–V (**87–91**) have been isolated from this same plant source, with compounds III–V displaying antisweet activity [121]. By comparing the gymnemasaponins, it appears that the sugar chain length plays an important role in the taste properties of these compounds. Additionally, the sodium salt of alternoside II (**92**), 21β-*O*-benzoylsitakisogenin 3-*O*-β-D-glucopyranosyl (1→3)-β-D-glucuronopyranoside (**93**), the potassium salt of longispinogenin 3-*O*-β-D-glucopyranosyl(1→3)-β-D-glucuronopyranoside (**94**), and the potassium salt of 29-hydroxylongispinogenin 3-*O*-β-D-glucopyranosyl(1→3)-β-D-glucuronopyranoside (**95**) were isolated from the leaves of *G. sylvestre*, with the former two compounds possessing antisweet properties [122]. In addition to these triterpenoids, the protein gurmarin from the plant has been found to suppress the detection of sweeteners in rodents but not in humans [110, 123]. *G. sylvestre* has been found to alter absorption of glucose and is used against diabetes mellitus as a traditional medicine [110].

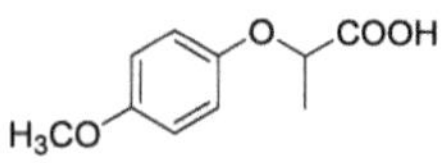


**68**

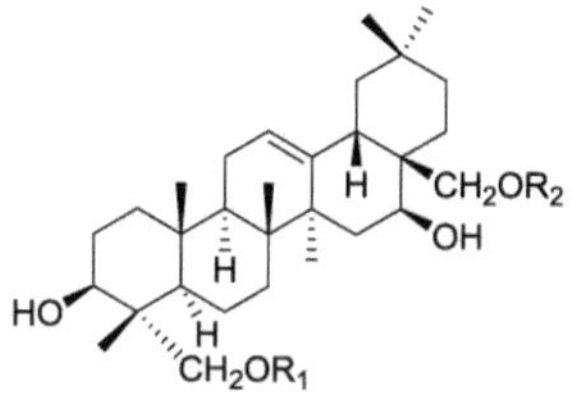


| | $R_1$ | $R_2$ |
|---|---|---|
| **87** | H | $\beta$-glc |
| **88** | $\beta$-glc | $\beta$-glc |
| **89** | $\beta$-glc | $\beta$-glc$^6$-$\beta$-glc |
| **90** | $\beta$-glc$^6$-$\beta$-glc | $\beta$-glc |
| **91** | $\beta$-glc$^6$-$\beta$-glc | $\beta$-glc$^6$-$\beta$-glc |

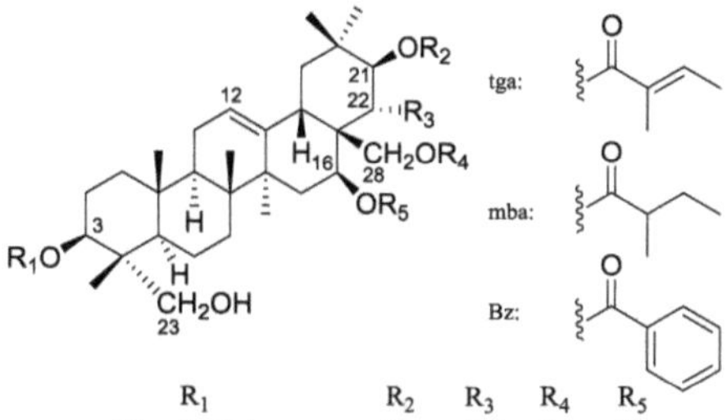


| | $R_1$ | $R_2$ | $R_3$ | $R_4$ | $R_5$ |
|---|---|---|---|---|---|
| **69** | $\beta$-glcA | tga | OH | Ac | H |
| **70** | $\beta$-glcA | mba | OH | Ac | H |
| **71** | $\beta$-glcA | mba | OH | H | H |
| **72** | $\beta$-glcA | tga | OH | H | H |
| **73** | $\beta$-glcA | tga | Otga | H | H |
| **74** | $\beta$-glcA$^3$-$\beta$-glc | tga | OH | H | H |
| **75** | $\beta$-glcA | H | H | H | H |
| **76** | $\beta$-glcA$^3$-$\beta$-OG | mba | OH | H | H |
| **77** | $\beta$-glcA$^3$-$\beta$-OG | tga | OH | H | H |
| **78** | $\beta$-glcA | H | OH | Ac | H |
| **79** | $\beta$-glcA | tga | OH | tga | H |
| **80** | $\beta$-glcA$^3$-$\beta$-glc | tga | OH | Ac | H |
| **81** | $\beta$-glcA | H | OH | mba | H |
| **82** | $\beta$-glcA | H | OH | tga | H |
| **83** | $\beta$-glcA | mba | Otga | H | H |
| **84** | $\beta$-glcA | H | Otga | H | tga |
| **85** | $\beta$-glcA | Bz | OH | H | H |
| **86** | $\beta$-glcA | H | OH | Bz | H |

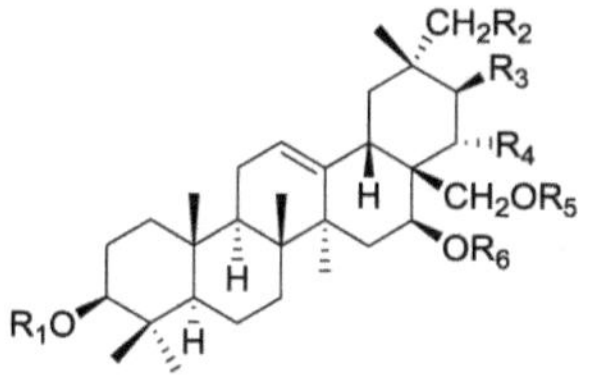


| | $R_1$ | $R_2$ | $R_3$ | $R_4$ | $R_5$ | $R_6$ |
|---|---|---|---|---|---|---|
| **92** | $\beta$-glcA$^3$-$\beta$-glc<br>6<br>$Na^+$ | H | H | OH | $\alpha$-rha | Ac |
| **93** | $\beta$-glcA$^3$-$\beta$-glc<br>6<br>$K^+$ | H | OBz | H | H | H |
| **94** | $\beta$-glcA$^3$-$\beta$-glc<br>6<br>$K^+$ | H | H | H | H | H |
| **95** | $\beta$-glcA$^3$-$\beta$-glc<br>6<br>$K^+$ | OH | H | H | H | H |

Several sweet taste-inhibitory dammarane derivatives have been identified from the leaves of *Hovenia dulcis*. Hodulosides I–V (**96–100**) and VII–IX (**101–103**) were all found to have antisweet properties along with hovenoside I (**104**), jujuboside B (**105**), and compounds named saponins $C_2$ (**106**), E (**107**), and H (**108**) [124, 125]. Hodulosides VI (**109**) and X (**110**) were not found to have antisweet activity indicating the importance of the substitution at C3 in sweet taste modification [124].

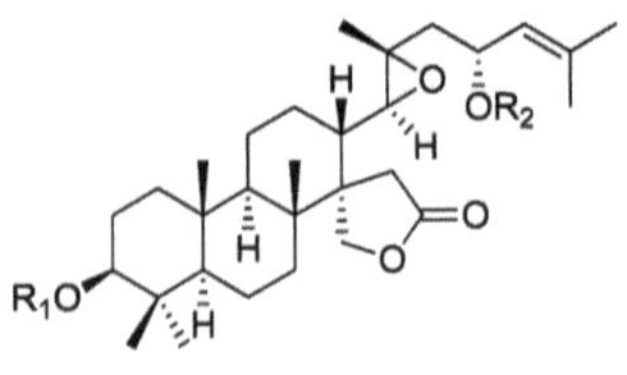

| | $R_1$ | $R_2$ |
|---|---|---|
| **96** | β-glc²-α-rha | β-glc |
| **97** | β-glc²-α-rha (³ β-glc) | H |
| **107** | β-glc²-α-rha | H |
| **108** | β-glc | H |

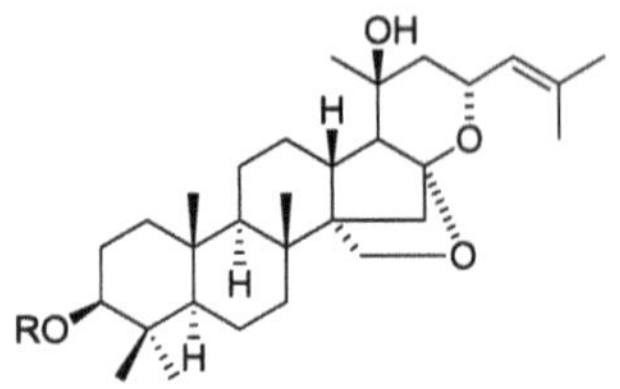

| | R |
|---|---|
| **98** | R = α-ara²-β-qui (³ β-glc) |
| **99** | R = α-ara²-β-glc (³ β-glc) |
| **100** | R = β-glc²-α-rha (³ β-glc) |
| **104** | R = α-ara²-β-xyl (³ β-glc) |
| **105** | R = α-ara²-α-rha (³ β-glc²-β-xyl) |
| **106** | R = α-ara²-α-rha (³ β-glc) |

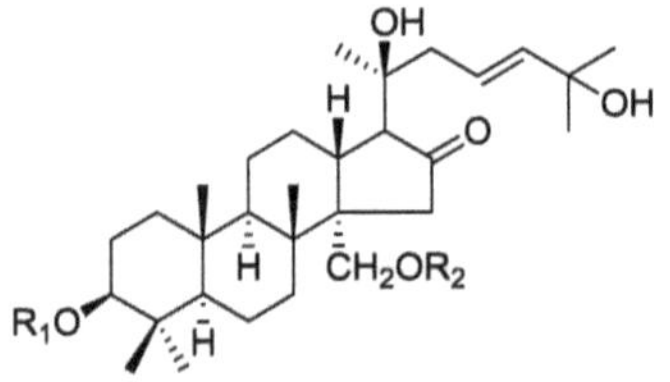

| | $R_1$ | $R_2$ |
|---|---|---|
| **101** | α-ara²-α-rha | β-glc |
| **102** | α-ara | β-glc⁶-β-xyl |
| **103** | α-ara²-α-rha | β-glc⁶-β-xyl |
| **109** | α-ara | β-glc |
| **110** | α-ara²-α-rha (³ β-glc) | β-glc |

Several dammaranes from the Chinese jujube, *Ziziphus jujuba*, possess antisweet activity, namely, jujubasaponins II–VI (**111–115**), ziziphin (**116**), and zizyphus saponins I–III (**117–119**). Initially, jujubasaponins I–III (**120**, **111**, **112**) were isolated [126, 127]. Since jujubasaponin I does not possess antisweet activity and jujubasaponins II and III do, the conclusion that acylation was important for the mediation of this activity was made [126]. However, isolation of the additional compounds has somewhat confounded a rigid structure-activity analysis.

Somewhat surprisingly, the domesticated chicken (*Gallus gallus domesticus*) produces a sweet taste-inhibiting protein, riboflavin-binding protein (RBP) [128]. It had previously been discovered that lysozyme from the eggs of various species including the chicken are high-potency sweeteners [129]. However, because egg white itself is not sweet, it was posited that there must also be a sweet taste inhibitor present [128]. RBP was found to selectively reduce the sweetness of the protein sweeteners thaumatin, monellin, and lysozyme [128]. RBP had no effect on the perceived sweetness of aspartame, glycine, D-phenylalanine, stevioside (**1**), saccharin, cyclamate, or sucrose [128]. Detailed studies have shown that RBP does not directly interact with sweet-tasting proteins indicating that its effect is mediated through interaction with the sweet taste receptor [128].

In addition to these straightforward sweet taste inhibitors, there exist several compounds in Nature that produce the related "sweet-water" effect. This type of effect is seen, when a compound is sampled in the mouth and no sweetness is perceived, but upon rinsing, a sweet taste is observed [130]. This effect is believed to be caused by the constitutive activity of the sweet taste receptor [130]. In other words, the sweet taste receptor is believed to exist in equilibrium between active and resting states, and the sweetness from the active state receptors is not usually perceived. However, when a sweet taste inhibitor is encountered, this background activity is suppressed. When the inhibitor is subsequently removed, the background constitutive activity can then be observed.

The ability of the artichoke, *Cynara scolymus* L. (Asteraceae), to produce the "sweet-water" effect resulted in phytochemical investigation of the plant [131]. The major active compounds were determined to be the related compounds chlorogenic acid (**121**) and cynarin (**122**). While this "sweet-water" effect was not attributed to blocking the sweet taste, there is evidence in vitro that lactisole also exhibits a "sweet-water" activity [130].

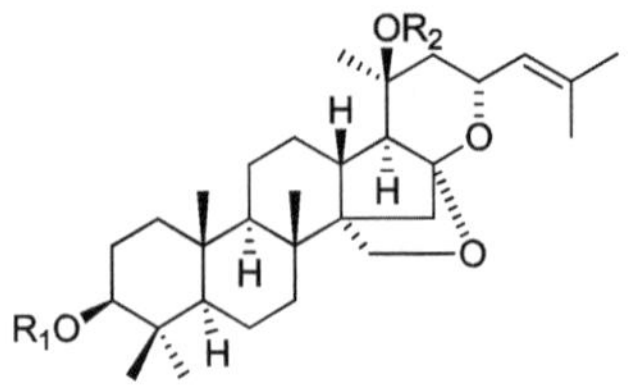


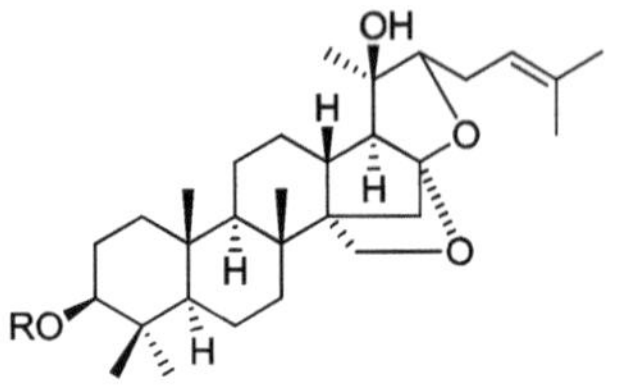


| | $R_1$ | $R_2$ |
|---|---|---|
| **111** | α-ara²-α-rha | α-rha²-Ac |
| **112** | α-ara²-α-rha | α-rha³-Ac |
| **116** | α-ara²-α-rha | α-rha²-Ac (3: Ac) |
| **117** | α-ara²-6-deoxy-α-tal (3: β-glc) | H |
| **118** | α-ara²-α-rha (3: β-glc) | H |
| **119** | α-ara²-α-rha (3: β-glc²-β-xyl) | H |
| **120** | α-ara²-α-rha | α-rha |

**113** R = β-gal²-α-rha (3: β-glc)

**114** R = β-glc²-α-rha (3: β-glc)

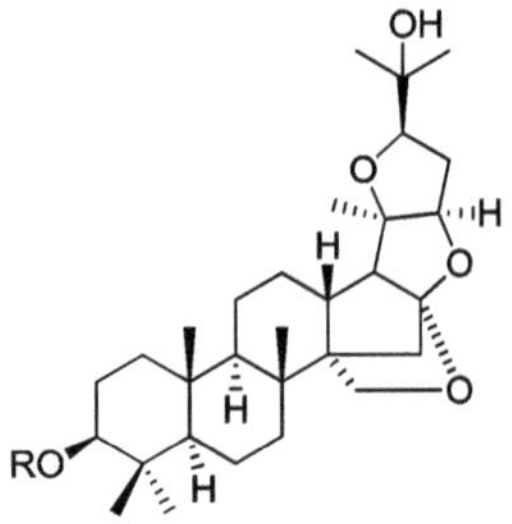


**115** R = β-gal²-α-rha

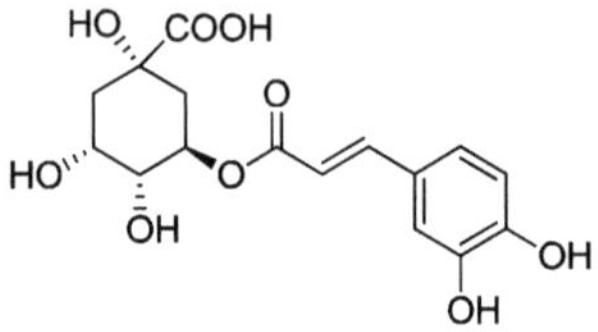


**121**

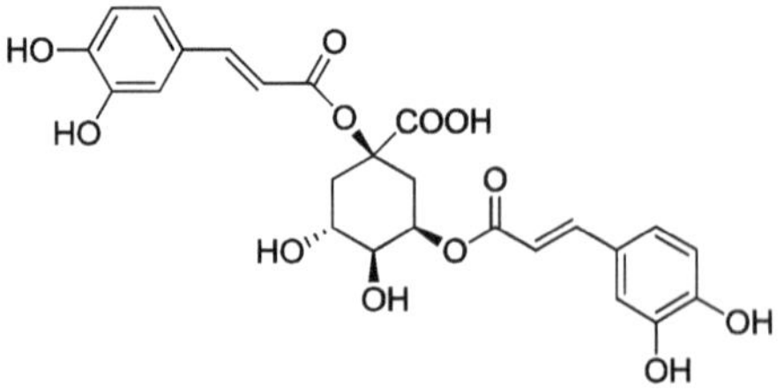


**122**

Unexpectedly, the taste-modifying proteins miraculin and neoculin exhibit a sweet taste-inhibitory activity at nonacidic pH. Neoculin, isolated from *Curculigo latifolia* Dryand. ex W.T. Auden (Hypoxidaceae), elicits a sweet taste in addition to changing the perception of sour solutions to sweetness. Through mutation studies, it was proposed that the mechanism of inducing a sweet taste in the presence of acid is mediated by the protonation of histidine residues on neoculin. Further investigation demonstrated that the inactive forms of neoculin and miraculin bind to the sweet taste receptors and inhibit the binding of other sweeteners [132].

## 3.5 Selected Bitter-Tasting Natural Products

As a result of coevolution, Nature has produced several bitter-tasting compounds and several bitter taste receptors. Thus, while there is only one widely accepted sweet taste receptor in humans, there are 25 identified functional bitter taste receptors and several pseudogenes [10, 133]. Additionally, while monomers of the bitterness receptors respond to bitter tastants in in vitro screening, it is possible that in vivo polymerization occurs broadening the potential number of functional bitter taste receptors. It is estimated that there are tens of thousands of naturally occurring bitter tastants rendering a comprehensive review of these compounds untenable [133]. Therefore, in the following paragraphs, focus will be made on a few structures of commercial interest.

Several bitter-tasting compounds occur in beer due to the inclusion of hops, *Humulus lupulus* L. (Cannabaceae). While beer is now regarded as an acceptable type of bitterness in the diet, the initial reaction to adding bitter-tasting hops to beer was negative [134]. However, hops aided in the stability of beer against microorganisms and subsequently became an integral part of the taste of beer. The bitter-tasting principles of beer are the α-acids humulone (**123**), cohumulone (**124**), and adhumulone (**125**), which are isomerized to *cis*- and *trans*-isohumulone (**126**, **127**), isocohumulone (**128**, **129**), and isoadhumulone (**130**, **131**) during beer production [134, 135]. Additional bitter components include the β-acids lupulone (**132**), colupulone (**133**), adlupulone (**134**), and the flavonoids xanthohumol (**135**), isoxanthohumol (**136**), and 8-prenylnaringenin (**137**) [135]. These bitter-tasting compounds have been found to activate the hTAS2R1, hTAS2R14, and hTAS2R40 receptors. While taste potency results did not correlate directly to the cell-based assays, all of these compounds were determined to be bitter in a taste test. It was then put forth that binding to salivary proteins and the oral cavity may functionally reduce the concentration levels of some of these compounds available to the taste buds in vivo.

Citrus juices are a commonly ingested source of bitter compounds from two chemical classes, namely, flavonoids and limonoids [4, 136]. The bitterness of limonin (**138**) and flavonoids have been referred to as "a major problem for the citrus industry" [4]. The flavonoid naringin (**139**) is considered to be the primary bitter constituent of freshly squeezed juices. In contrast, the bitter limonoid limonin is formed from its tasteless precursor, limonate-A-ring lactone (**140**), over a period of hours after juicing [136]. The bitter citrus flavonoids naringin (**139**) and neohesperidin (**141**) have been screened against all known bitter taste receptors; however, none was activated by these compounds. Limonin activates hTAS2R38, which is perhaps the most widely studied bitter taste receptor, since mutations at this receptor dictate one's sensitivity to 6-*n*-propylthiouracil (PROP). A liking for grapefruit juice has been associated with *TAS2R19* and *TAS2R60* polymorphisms [137]. Interestingly, the mutation associated with *TAS2R60* is a silent mutation. It is possible, however, that this mutation suppresses expression or affects protein folding. Research, investigating reducing the bitterness of the Seville orange, has been important for sweetener discovery; chemical modification of the bitter-tasting neohesperidin resulted in the discovery of the sweetener neohesperidin dihydrochalcone (**14**) [43, 44].

Caffeine (**142**) may be the most widely consumed bitter agent. It has been estimated that the daily consumption rate of caffeine in several countries is greater than 300 mg per capita [138]. Caffeine is produced by several plants used in foods and beverages including coffee, tea, cocoa, yerba mate, guarana, and kola nut. Due to its wide use and availability, caffeine is a common standard for bitter taste experiments. Caffeine activates five human bitter taste receptors, namely, hTAS2R5, hTAS2R10, hTAS2R14, hTAS2R43, and hTAS2R46 [133].

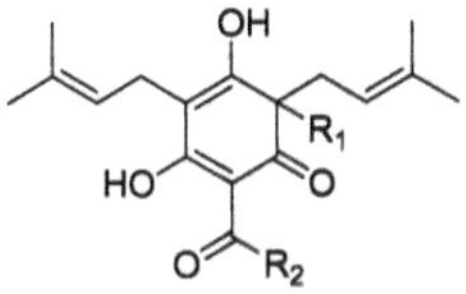

| | $R_1$ | $R_2$ |
|---|---|---|
| **123** | β-OH | isobutyl |
| **124** | β-OH | isopropyl |
| **125** | β-OH | *sec*-butyl |
| **132** | isoprenyl | isobutyl |
| **133** | isoprenyl | isopropyl |
| **134** | isoprenyl | *sec*-butyl |

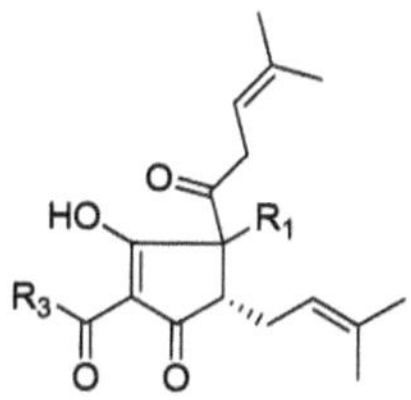

| | $R_1$ | $R_2$ |
|---|---|---|
| **126** | α-OH | isobutyl |
| **127** | β-OH | isobutyl |
| **128** | α-OH | isopropyl |
| **129** | β-OH | isopropyl |
| **130** | α-OH | sec-*butyl* |
| **131** | β-OH | sec-*butyl* |

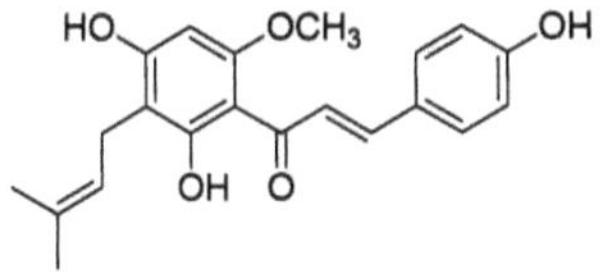

**135**

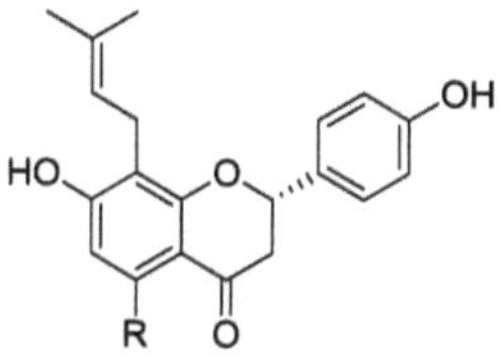

**136** R = $OCH_3$
**137** R = OH

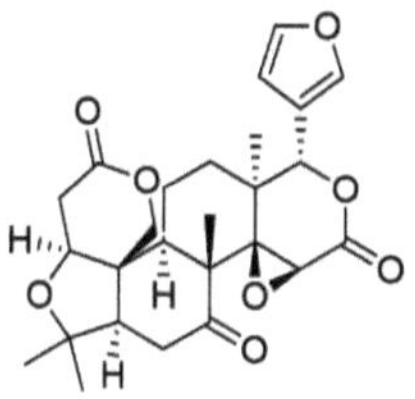

**138**

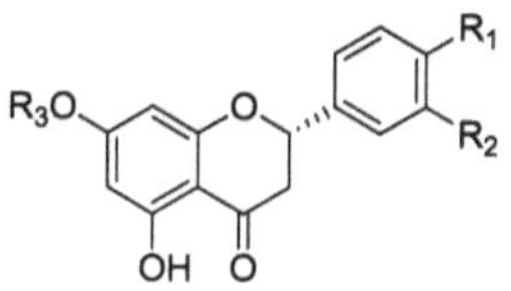

| | $R_1$ | $R_2$ | $R_3$ |
|---|---|---|---|
| **139** | OH | H | β-glc$^2$- α-rha |
| **141** | $OCH_3$ | OH | β-glc$^2$- α-rha |

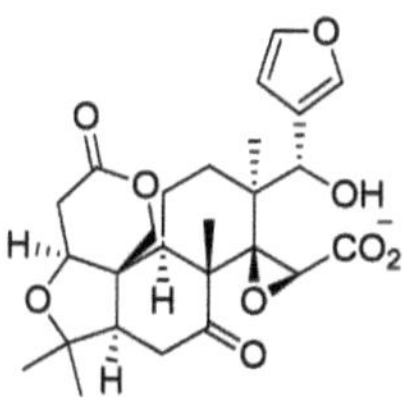

**140**

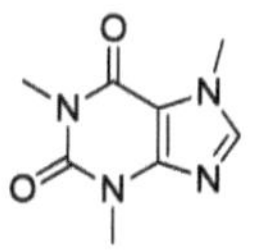

**142**

### 3.6 Bitterness-Masking Natural Products

Since the response to bitter tastants is generally rejection, there exists a commercial impetus to mask bitter tastes. The most common industrial means of masking a bitter taste is by creating a physical barrier to the tastant, although this is not practical for all food applications or some liquid medicines [9]. Physical barriers can include encapsulation (on both the micro and macro levels), forming suspensions, emulsions, coatings, and the use of complexing agents such as ion-exchange resins. Strong flavors can be used to overpower bitter tastants, and congruent flavors such as chocolate, coffee, grapefruit, and mint can be used to place bitterness in a more favorable context [9, 139]. Salts (NaCl and LiCl) have been shown generally to inhibit the bitterness of various bitter tastants, while sour tastants may alter bitterness perception unpredictably, with lower concentrations masking bitterness and higher concentrations showing enhancement [140]. Sweet substances are well known to suppress the intensity of bitter compounds [9]. Umami substances have also been shown to mask bitter taste, but monosodium glutamate also has salty and sour attributes owing to its component ions, which tends to obscure the relationship between umami and bitter taste perception [141]. In order to mask a bitter taste without otherwise changing the flavor of a food, it is necessary to identify compounds that inhibit the ability of the TAS2Rs to detect bitter compounds. Abolishing the bitter taste of bitter-sweet high-potency sweeteners such as acesulfame K, saccharin, and the steviol glycosides [e.g., rebaudioside A (**2**), rubusoside (**9**), and stevioside (**1**)] is of particular commercial interest.

Due to the various methods available to mask bitter taste, it is necessary to give some thought to the mechanism of action. For instance, cyclodextrins may be used to mask bitterness, but the method is much more likely to be a molecular-level encapsulation than direct interaction with the bitter taste receptor. However, investigation of the bitterness-masking properties of phosphatidic acid (**143**) and tannic acid (**144**) has indicated these compounds reduce the bitterness of quinine to below the amount scavenged from solution [142]. Interestingly, tannic acid, while acting as a bitterness-masking agent at lower concentrations, was found to be bitterness enhancing at higher concentrations, with this posited as being due to its astringent properties [142].

Likewise, sweet tastants are often used to mask bitterness; however, this is likely to be due to effects downstream from the taste receptors. This was demonstrated by using bitter-sweet mixtures with and without the sweet taste-inhibiting extract of *Gymnema sylvestre* [143]. Samples containing *G. sylvestre* extract were found to possess increased bitterness; thus, the perception of sweetness mutes the perception of bitterness. However, the existence of high-potency sweeteners with a bitter off-taste reveals that sweeteners may also act on the bitter taste receptor, and therefore, some sweeteners may actually inhibit response at the TAS2Rs. There are examples of natural sweeteners from the patent literature that may have this dual property. Mogroside V (**11**) from *Siraitia grosvenorii* has been patented for use below its taste threshold to block many bitter tastants including coffee, grapefruit, organ meat, and potassium chloride [144]. Likewise, rebaudioside B (**6**) from *Stevia rebaudiana* has been patented to mask the bitterness of other steviol glycosides [145].

Several terpenoids have been found to possess bitterness-masking activity. In addition to the aforementioned mogroside V (**11**), the reportedly tasteless mogroside III (**145**), from the same plant source, also displayed bitterness-masking properties [144]. Hardwickiic acid (**146**), *epi*-hardwickiic acid (**147**), and several derivatives thereof were found to mask the bitterness of high-potency sweeteners [16].

Perhaps the most prolific class of bitterness-masking compounds from Nature are the plant phenolics, especially the flavonoids. Neodiosmin (**148**), a flavone from *Citrus* x *aurantium,* has been demonstrated to have bitterness-masking properties against limonin, caffeine, quinine, and saccharin [146–148]. Another early example of a plant compound with reported bitterness-masking activity is *para*-methoxycinnamaldehyde (**149**), a constituent of cinnamon (*Cinnamomum* spp.). Research, investigating *para*-methoxycinnamaldehyde as a sweetener, led to a report of bitterness-masking capabilities, when used in compositions containing vanillin and saccharin [149]. The sodium salt of another common plant natural product, ferulic acid (**150**), was found to mask the bitterness of acesulfame K, caffeine, quinine, and saccharin [150].

The most intensively studied plant for bitterness-masking activity is the US native plant *Eriodictyon californicum.* As early as 1887, an extract of *E. californicum* leaves was reported to mask the bitterness of quinine with the statement made: "syrup prepared from *Eriodictyon* leaves is extensively used for the administration of quinine in a bitterless form" [151]. Indeed by 1917, it was stated that "its preparations are principally used, however, as vehicles to disguise the taste of disagreeable medicines like quinine" [152]. More recent work by Ley et al. showed eriodictyol (**151**), homoeriodictyol (**152**) and its sodium salt, and sterubin (**153**), all leaf flavonoids of *E. californicum*, to be bitter-masking agents [153]. The sodium salt of homoeriodictyol masks the bitterness of a variety of compounds including caffeine, guaifenesin, paracetamol, and salicin [153]. This work led to the investigation of several structural analogues often including a vanillic moiety. Examples of bitterness-masking compounds identified in this work include the dihydrochalcone phloretin (**17**), a closely related flavan (7,4′-dihydroxy-3′-methoxyflavan) (**154**), various gingerdione derivatives (e.g., [2]-dehydrogingerdione (**155**) and [3]-gingerdione (**156**)), and vanillylamide analogues such as **157** [17, 103, 154, 155]. A compound, only tangentially related to the benzamide derivatives of homoeriodictyol, L-menthane carboxylic acid-*N*-(4-methoxyphenyl)-amide (**158**), was found to mask the bitterness of the similar compound menthol [156].

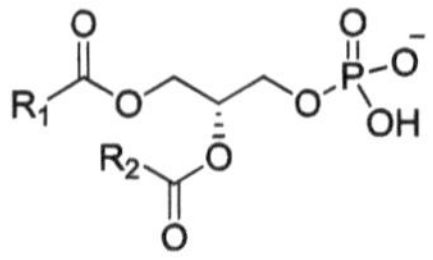

**143** $R_1$, $R_2$ = fatty acid residues

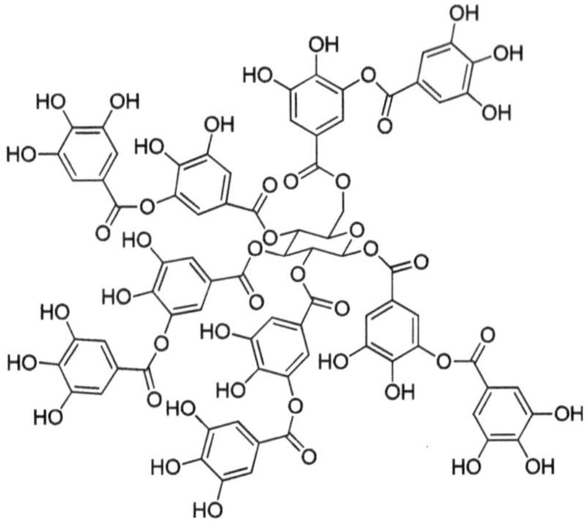

**144**

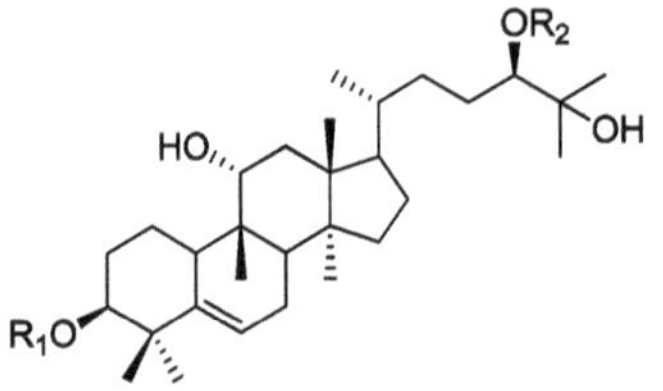

**145** $R_1$ = β-glc, $R_2$ = β-glc$^6$-β-glc

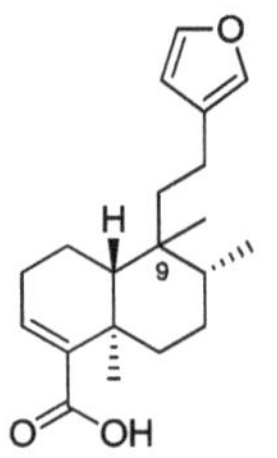

**146** 9*S*
**147** 9*R*

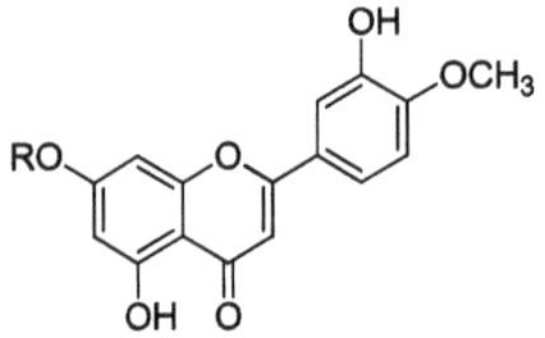

**148** R = β-glc$^2$-α-rha

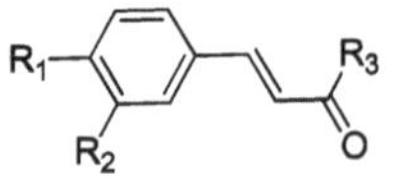

| | $R_1$ | $R_2$ | $R_3$ |
|---|---|---|---|
| **149** | $OCH_3$ | H | H |
| **150** | OH | $OCH_3$ | OH |

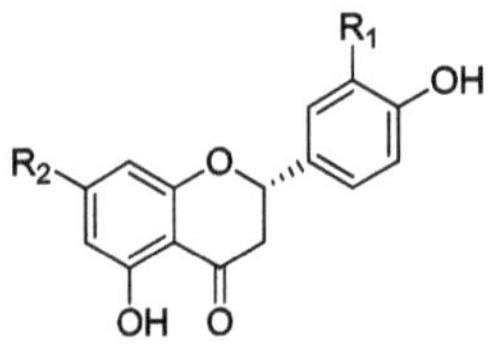

| | $R_1$ | $R_2$ |
|---|---|---|
| **151** | OH | OH |
| **152** | $OCH_3$ | OH |
| **153** | OH | $OCH_3$ |

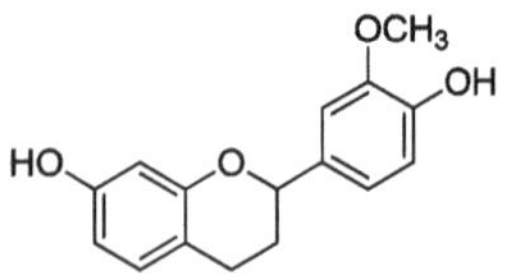

**154**

In an independent research study on the constituents of *E. californicum*, the flavonoids jaceosidin (**159**), sakuranetin (**19**), and 6-methoxysakuranetin (**160**) were found to inhibit the hTAS2R31 bitterness receptor [49]. This receptor is responsible for detecting the bitterness of several tastants including saccharin and acesulfame K. The $IC_{50}$ values for these compounds were roughly equivalent with sakuranetin being the most potent antagonist identified. The activity of sakuranetin and 6-methoxysakuranetin [and the inactivity of naringenin (**21**)] shows that, in addition to an isovanillic substitution pattern on the B ring, A-ring methoxylation at the seven positions is important for bitterness reduction. The hTAS2R31 data obtained also showed that methoxylation at the C6 position did not greatly alter activity at the receptor. Additionally, the activity of jaceosidin and the inactivity of 6-methoxyhomoeriodictyol demonstrates a difference in activity between flavones and flavanones. Homoeriodictyol (**152**) was evaluated in this study but was found to be inactive, indicating that it exerts its bitterness-masking capability through activity at other bitter receptors.

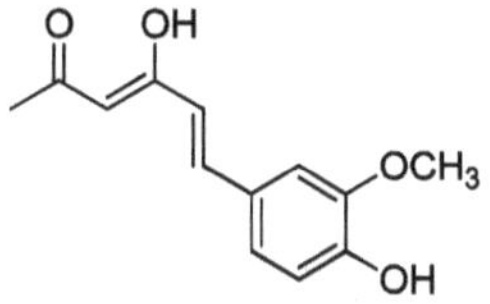


**155**

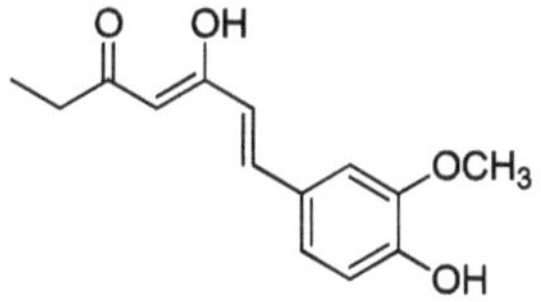


**156**

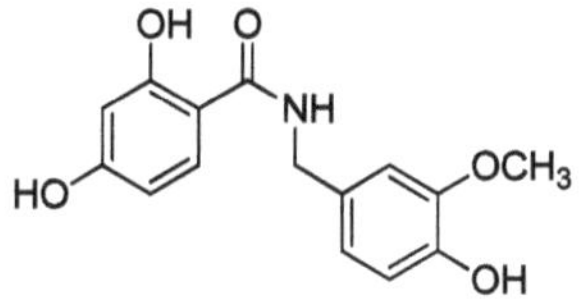


**157**

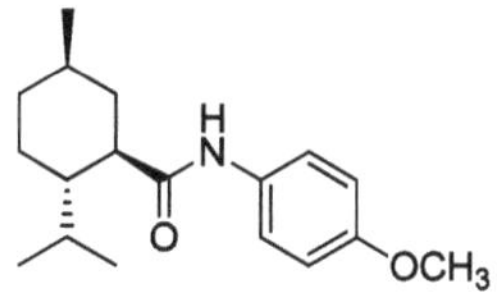


**158**

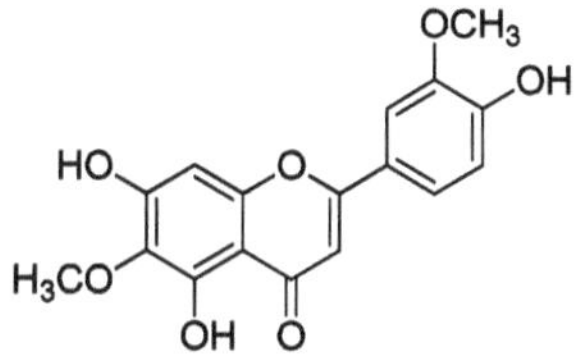


**159**

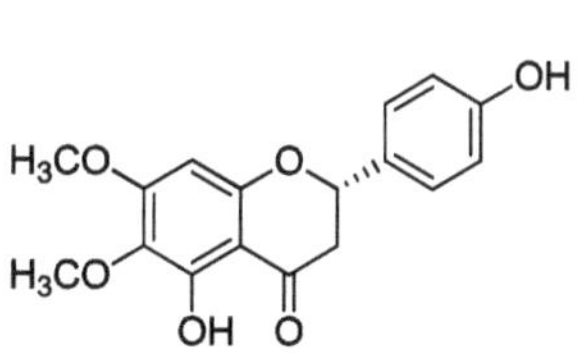


**160**

# 4 Conclusions

As described in this chapter, naturally occurring compounds of various structural classes can activate or inhibit human sweet and bitter taste receptors. Since the discovery of these receptors at the turn of the twenty-first century, two major classes of plant-derived natural products have seen wide commercialization as table top sweeteners, namely, stevia and monk fruit products. Given the recent trend towards "100% natural," it is likely that natural products will continue to be commercialized as sweeteners and for other flavoring purposes in the future.

The development of cell-based assays for sweetness and bitterness can be expected to aid the discovery of novel natural taste-active agents by allowing for the screening of natural products and extract libraries. However, due to their relative novelty and certain restrictions due to intellectual property matters, these in vitro assays have yet to find widespread use.

While many of the plants possessing a sweet taste have been investigated for the presence of high-potency sweeteners already, it is likely that new taste-active compounds will continue to be isolated. For example, only recently has it become appreciated that the chicken egg produces substances that are sweet and others that are sweet taste inhibitors. Furthermore, a comprehensive screening for natural sweet taste enhancers and bitterness-masking agents has not been conducted, and such activity would not necessarily be evident from the historical uses of plants and other organisms, as has been the case for sweet-tasting substances. For instance, the sweet-water effect of the artichoke was not mentioned in the scientific literature until 1935 [157]. Given the diverse chemical space covered by natural products, the increasing availability of taste-related in vitro assays, and the current consumer desire for natural flavors and sweeteners, future continuing discoveries of naturally occurring taste-active molecules should be expected.

# References

1. Lindemann B (2001) Receptors and transduction in taste. Nature 413(6852):219–225
2. Bisognin DA (2002) Origin and evolution of cultivated cucurbits. Ciência Rural 32(4):715–723
3. Binello A, Cravotto G, Nano GM, Spagliardi P (2004) Synthesis of chitosan-cyclodextrin adducts and evaluation of their bitter-masking properties. Flavour Fragrance J 19(5):394–400
4. Drewnowski A, Gomez-Carneros C (2000) Bitter taste, phytonutrients, and the consumer: a review. Am J Clin Nutr 72(6):1424–1435
5. Johns T (1986) Detoxification function of geophagy and domestication of the potato. J Chem Ecol 12(3):635–646
6. Roy GM (1990) The applications and future implications of bitterness reduction and inhibition in food products. Crit Rev Food Sci Nutr 29(2):59–71
7. Wink M (1988) Chemische verteidigung der lupinen: zur biologischen bedeutung der chinolizidinalkaloide. Theor Appl Genet 75(1–2):225–233
8. Haley S (2013) World raw sugar prices – the influence of Brazilian costs of production and world surplus/deficit measures. http://www.ersusda.gov. Accessed 30 Sept 2013

9. Ley JP (2008) Masking bitter taste by molecules. Chemosens Percept 1(1):58–77
10. Go Y, Satta Y, Takenaka O, Takahata N (2005) Lineage-specific loss of function of bitter taste receptor genes in humans and nonhuman primates. Genetics 170(1):313–326
11. Morini G, Bassoli A, Temussi PA (2005) From small sweeteners to sweet proteins: anatomy of the binding sites of the human T1R2_T1R3 receptor. J Med Chem 48(17):5520–5529
12. Winnig M, Bufe B, Kratochwil NA, Slack JP, Meyerhof W (2007) The binding site for neohesperidin dihydrochalcone at the human sweet taste receptor. BMC Struct Biol 7:66
13. Shallenberger RS, Acree TE (1967) Molecular theory of sweet taste. Nature 216(5114):480–482
14. Tancredi T, Temussi PA (1979) Three-dimensional mapping of the bitter taste receptor site. Chem Senses Flavour 4(4):259–265
15. Sheridan C (2004) A taste of the future. Nat Biotechnol 22(10):1203–1205
16. Slack JP, Brockhoff A, Batram C, Menzel S, Sonnabend C, Born S, Galindo MM, Kohl S, Thalmann S, Ostopovici-Halip L, Simons CT, Ungureanu I, Duineveld K, Bologa CG, Behrens M, Furrer S, Oprea TI, Meyerhof W (2010) Modulation of bitter taste perception by a small molecule hTAS2R antagonist. Curr Biol 20(12):1104–1109
17. Ley JP, Paetz S, Blings M, Hoffmann-Lucke P, Bertram HJ, Krammer GE (2008) Structural analogues of homoeriodictyol as flavor modifiers. Part III: short chain gingerdione derivatives. J Agric Food Chem 56(15):6656–6664
18. Bridel M, Lavielle R (1931) Sur le principe sucré des feuilles de Kaâ-hê-é (*Stevia rebaundiana* B). C R Acad Sci Paris 192:1123–1125
19. Kohda H, Kasai R, Yamasaki K, Murakami K, Tanaka O (1976) New sweet diterpene glucosides from *Stevia rebaudiana*. Phytochemistry 15(6):981–983
20. Carakostas MC, Curry LL, Boileau AC, Brusick DJ (2008) Overview: the history, technical function and safety of rebaudioside A, a naturally occurring steviol glycoside, for use in food and beverages. Food Chem Toxicol 46(Suppl 7):S1–S10
21. Cheeseman MA (2011) Agency response letter GRAS notice no GRN 000348. http://www.fda.gov/Food/FoodIngredientsPackaging/GenerallyRecognizedasSafeGRAS/GRASListings/ucm267232htm. Accessed 30 Sept 2013
22. Kruger CL (2009) Agency response letter GRAS notice no GRN 000275. http://www.fda.gov/Food/IngredientsPackagingLabeling/GRAS/NoticeInventory/ucm171539htm. Accessed 30 Sept 2013
23. Roberts A (2011) Agency response letter GRAS notice no GRN 000375. http://www.fda.gov/Food/IngredientsPackagingLabeling/GRAS/NoticeInventory/ucm274761htm. Accessed 30 Sept 2013
24. Fukunaga Y, Miyata T, Nakayasu N, Mizutani K, Kasai R, Tanaka O (1989) Enzymic transglucosylation products of stevioside: separation and sweetness-evaluation. Agric Biol Chem 53(6):1603–1607
25. Tanaka O (1997) Improvement of taste of natural sweeteners. Pure Appl Chem 69(4):675–683
26. Lee C (1975) Intense sweetener from Lo Han Kuo (*Momordica grosvenori*). Experientia 31 (5):533–534
27. Takemoto T, Nakajima T, Arihara S, Okuhira M (1978) Sweet steroid glucosides and related substances from Momordicae. US 4084010
28. Kinghorn AD, Compadre CM (2012) Less common high-potency sweeteners. In: O'Brien Nabors L (ed) Alternative sweeteners, 4th edn. Taylor & Francis, Boca Raton, pp 223–246
29. Cheeseman MA (2010) Agency response letter GRAS notice no GRN 000301. http://www.fda.gov/Food/FoodIngredientsPackaging/GenerallyRecognizedasSafeGRAS/GRASListings/ucm200326htm. Accessed 30 Sept 2013
30. Schwartz TW, Holst B (2007) Allosteric enhancers, allosteric agonists and ago-allosteric modulators: where do they bind and how do they act? Trends Pharmacol Sci 28(8):366–373
31. Marnett LJ, Cohen SM, Fukushima S, Gooderham NJ, Hecht SS, Rietjens IM, Smith RL, Adams TB, Hallagan JB, Harman C, McGowen MM, Taylor SV (2013) GRAS flavoring

substances 26: the 26th publication by the Expert Panel of the Flavor and Extract Manufacturers Association provides an update on recent progress in the consideration of flavoring ingredients generally recognized as safe under the food additive amendment. Food Technol 67(8):38–56
32. Smith RL, Waddell WJ, Cohen SM, Feron VJ, Marnett LJ, Portoghese PS, Rietjens IM, Adams TB, Lucas GC, McGowen MM, Taylor SV, Williams MC (2009) GRAS 24: the 24th publication by the FEMA Expert Panel presents safety and usage data on 236 new generally recognized as safe flavoring ingredients. Food Technol 63(6):46–105
33. Smith RL, Waddell WJ, Cohen SM, Fukushima S, Gooderham NJ, Hecht SS, Marnett LJ, Portoghese PS, Rietjens IMCM, Adams TB, Gavin CL, McGowen MM, Taylor SV (2011) GRAS: flavoring substances 25. Food Technol 65(7):44–75
34. Mizutani K, Tanaka O (2002) The use of *Stevia rebaudiana* sweeteners in Japan. In: Kinghorn AD (ed) stevia: the genus stevia, medicinal and aromatic plants – industrial profiles, vol 19. Taylor & Francis, London, pp 178–195
35. Conn JW, Rovner DR, Cohen EL (1968) Licorice-induced pseudoaldosteronism. Hypertension, hypokalemia, aldosteronopenia, and suppressed plasma renin activity. JAMA 205 (7):492–496
36. Kinghorn AD, Chin Y-W, Pan L, Jia Z (2010) Natural products as sweeteners and sweetness modifiers. In: Verpoorte R (ed) Comprehensive natural products chemistry II: chemistry and biology, vol 6. Elsevier, Oxford, pp 269–315
37. Anonymous (2002) Food and drugs. Fed Reg 67:512–513
38. Mizutani K, Kuramoto T, Tamura Y, Ohtake N, Doi S, Nakaura M, Tanaka O (1994) Sweetness of glycyrrhetic acid 3-*O*-beta-D-monoglucuronide and the related glycosides. Biosci Biotechnol Biochem 58(3):554–555
39. Hall RL, Oser BL (1965) Recent progress in the consideration of flavor ingredients under the food additives amendment 3 GRAS substances. Food Technol 19(2):151–197
40. Van der Wel H, Loeve K (1972) Isolation and characterization of thaumatin I and II, the sweet-tasting proteins from *Thaumatococcus daniellii* Benth. Eur J Biochem 31(2):221–225
41. Crammer B (2008) Recent trends of some natural sweet substances from plants. In: Ikan R (ed) Selected topics in the chemistry of natural products. World Scientific, Hackensack NJ, pp 189–207
42. Oser BL, Ford RA, Bernard BK (1984) Recent progress in the consideration of flavoring ingredients under the food additives amendment. 13. GRAS substances. Food Technol 38 (10):66–89
43. DuBois GE, Crosby GA, Stephenson RA, Wingard RE Jr (1977) Dihydrochalcone sweeteners. Synthesis and sensory evaluation of sulfonate derivatives. J Agric Food Chem 25 (4):763–772
44. Horowitz RM, Gentili B (1963) Flavonoids of citrus-VI: Structure of neohesperidose. Tetrahedron 19(5):773–782
45. Jia Z, Yang X, Hansen CA, Naman CB, Simons CT, Slack JP, Gray K (2010) Consumables. US 20100178389
46. Waddell WJ, Cohen SM, Feron VJ, Goodman JI, Marnett LJ, Portoghese PS, Rietjens IMCM, Smith RL, Adams TB, Gavin CL, McGowen MM, Williams MC (2007) GRAS flavoring substances 23. Food Technol 61(8):22–61
47. Kramer G, Ley J, Riess T, Haug M, Paetz S (2010) Use of 4-Hydroxydihydrochalcons and their salts for enhancing an impression of sweetness. 2007107596 A1. US 20100233102
48. Ley J, Kindel G, Paetz S, Riess T, Haug M, Schmidtmann R, Krammer GE (2008) Use of hesperetin for enhancing the sweet taste.
49. Fletcher JN (2011) Isolation, Identification, and biological evaluation of potential flavor modulatory flavonoids from *Eriodictyon californicum*. Ph.D. dissertation, The Ohio State University
50. Ley JP, Blings M, Paetz S, Kindel G, Freiherr K, Krammer GE, Bertram H-J (2008) Enhancers for sweet taste from the world of non-volatiles: polyphenols as taste modifiers. In: Weerasinghe DK, Dubois GE (eds) Sweetness and sweeteners: biology, chemistry and

psychophysics, vol 979, ACS symposium series. American Chemical Society, Washington, pp 400–409
51. Dick WEJ (1981) Structure-taste correlations for flavans and flavanones conformationally equivalent to phyllodulcin. J Agric Food Chem 29(2):305–312
52. Garcez FR, Garcez WS, Martins M, Lopes FA (2003) Triterpenoids, lignan and flavans from *Terminalia argentea*. Biochem Syst Ecol 31(2):229–232
53. Daniher A, Wang Y (2010) Flavor molecules. US 20100272656
54. Wang Y, Daniher A, De Klerk A, Winkel C (2014) Pyridine derivatives with umami flavour. US 20140295045
55. Asahina Y, Ueno S (1916) Phyllodulcin, a chemical constituent of amacha (*Hydrangea thunbergii* Sieb.). Yakugaku Zasshi (408):146
56. Kinghorn AD, Compadre CM (2001) Less common high-potency sweeteners. In: O'Brien Nabors L (ed) Alternative sweeteners, 3rd edn. Marcel Dekker, New York, pp 209–233
57. Kinghorn AD, Soejarto DD (1986) Sweetening agents of plant origin. CRC Crit Rev Plant Sci 4(2):79–120
58. Oser BL, Ford RA (1978) Recent progress in the consideration of flavoring ingredients under the food additives amendment. 11. GRAS substances. Food Technol 32(2):60–70
59. Kinghorn AD, Soejarto DD (1989) Intensely sweet compounds of natural origin. Med Res Rev 9(1):91–115
60. Ohta M, Sasa S, Inoue A, Tamai T, Fujita I, Morita K, Matsuura F (2010) Characterization of novel steviol glycosides from leaves of *Stevia rebaudiana* Morita. J Appl Glycosci 57 (3):199–209
61. Kobayashi M, Horikawa S, Degrandi IH, Ueno J, Mitsuhashi H (1977) Dulcosides A and B, new diterpene glycosides from *Stevia rebaudiana*. Phytochemistry 16(9):1405–1408
62. Sakamoto I, Yamasaki K, Tanaka O (1977) Application of $^{13}$C NMR spectroscopy to chemistry of natural glycosides: rebaudioside-C, a new sweet diterpene glycoside of *Stevia rebaudiana*. Chem Pharm Bull 25(4):844–846
63. Morita T, Fujita I, Matsuura F, Ota M (2014) New steviol glycoside from stevia for use as sweeteners. US 20140187761
64. Prakash I, Campbell M, Chaturvedula VS (2012) Catalytic hydrogenation of the sweet principles of *Stevia rebaudiana*, rebaudioside B, rebaudioside C, and rebaudioside D and sensory evaluation of their reduced derivatives. Int J Mol Sci 13(11):15126–15136
65. Prakash I, Campbell M, San Miguel RI, Chaturvedula VS (2012) Synthesis and sensory evaluation of *ent*-kaurane diterpene glycosides. Molecules 17(8):8908–8916
66. Kasai R, Kaneda N, Tanaka O, Yamasaki K, Sakamoto I, Morimoto K, Okada S, Kitahata S, Furukawa H (1981) Sweet diterpene-glycosides of leaves of *Stevia rebaudiana* Bertoni – synthesis and structure-sweetness relationship of rebaudiosides-A, -D, -E and their related glycosides. Nippon Kagaku Kaishi 5:726–735
67. Marchal A, Waffo-Teguo P, Genin E, Merillon JM, Dubourdieu D (2011) Identification of new natural sweet compounds in wine using centrifugal partition chromatography-gustatometry and Fourier transform mass spectrometry. Anal Chem 83(24):9629–9637
68. Li YJW, Nakajima J-i, Kimura N, Saito K, Seo S (2007) Oleanane-type triterpene glycosides from *Glycyrrhiza uralensis*. Nat Prod Commun 2(3):243–248
69. Li YJW, Nakajima J, Senoo S, Saito K (2011) Triterpene glycosides, and sweeteners containing them or Glycyrrhiza extracts. JP 4730785
70. Ley J, Reichelt K, Obst K, Wessjohann L, Wessjohann S, Van Sung T, Nguyen TA, Van Trai N (2014) Orally consumed preparations comprising sweet-tasting triterpenes and triterpene glycosides. US 20140170083
71. Jia Z, Yang X (2009) A minor, sweet cucurbitane glycoside from *Siraitia grosvenorii*. Nat Prod Commun 4(6):769–772
72. Li X, Staszewski L, Xu H, Durick K, Zoller M, Adler E (2002) Human receptors for sweet and umami taste. Proc Natl Acad Sci U S A 99(7):4692–4696

73. Nelson G, Hoon MA, Chandrashekar J, Zhang Y, Ryba NJ, Zuker CS (2001) Mammalian sweet taste receptors. Cell 106(3):381–390
74. Sainz E, Korley JN, Battey JF, Sullivan SL (2001) Identification of a novel member of the T1R family of putative taste receptors. J Neurochem 77(3):896–903
75. Zhao GQ, Zhang Y, Hoon MA, Chandrashekar J, Erlenbach I, Ryba NJ, Zuker CS (2003) The receptors for mammalian sweet and umami taste. Cell 115(3):255–266
76. Galvez T, Parmentier ML, Joly C, Malitschek B, Kaupmann K, Kuhn R, Bittiger H, Froestl W, Bettler B, Pin JP (1999) Mutagenesis and modeling of the GABAB receptor extracellular domain support a venus flytrap mechanism for ligand binding. J Biol Chem 274 (19):13362–13369
77. Pin JP, Galvez T, Prezeau L (2003) Evolution, structure, and activation mechanism of family 3/C G-protein-coupled receptors. Pharmacol Ther 98(3):325–354
78. Cui M, Jiang P, Maillet E, Max M, Margolskee RF, Osman R (2006) The heterodimeric sweet taste receptor has multiple potential ligand binding sites. Curr Pharm Des 12(35):4591–4600
79. Gal J (2012) The discovery of stereoselectivity at biological receptors: Arnaldo Piutti and the taste of the asparagine enantiomers – history and analysis on the 125th anniversary. Chirality 24(12):959–976
80. Li X, Servant G (2008) Functional characterization of human sweet taste receptor: high-throughput screening assay development and structure function relation. In: Weerasinghe DK, DuBois GE (eds) Sweetness and sweeteners: biology, chemistry and psychophysics. American Chemical Society, Washington, pp 368–385
81. Kunishima N, Shimada Y, Tsuji Y, Sato T, Yamamoto M, Kumasaka T, Nakanishi S, Jingami H, Morikawa K (2000) Structural basis of glutamate recognition by a dimeric metabotropic glutamate receptor. Nature 407(6807):971–977
82. Xu H, Staszewski L, Tang H, Adler E, Zoller M, Li X (2004) Different functional roles of T1R subunits in the heteromeric taste receptors. Proc Natl Acad Sci U S A 101(39):14258–14263
83. Nie Y, Vigues S, Hobbs JR, Conn GL, Munger SD (2005) Distinct contributions of T1R2 and T1R3 taste receptor subunits to the detection of sweet stimuli. Curr Biol 15(21):1948–1952
84. Temussi PA (2002) Why are sweet proteins sweet? Interaction of brazzein, monellin and thaumatin with the T1R2-T1R3 receptor. FEBS Lett 526(1–3):1–4
85. Jiang P, Ji Q, Liu Z, Snyder LA, Benard LM, Margolskee RF, Max M (2004) The cysteine-rich region of T1R3 determines responses to intensely sweet proteins. J Biol Chem 279 (43):45068–45075
86. Koizumi A, Nakajima K, Asakura T, Morita Y, Ito K, Shmizu-Ibuka A, Misaka T, Abe K (2007) Taste-modifying sweet protein, neoculin, is received at human T1R3 amino terminal domain. Biochem Biophys Res Commun 358(2):585–589
87. Assadi-Porter FM, Maillet EL, Radek JT, Quijada J, Markley JL, Max M (2010) Key amino acid residues involved in multi-point binding interactions between brazzein, a sweet protein, and the T1R2-T1R3 human sweet receptor. J Mol Biol 398(4):584–599
88. Urwyler S (2011) Allosteric modulation of family C G-protein-coupled receptors: from molecular insights to therapeutic perspectives. Pharmacol Rev 63(1):59–126
89. Hellfritsch C, Brockhoff A, Stahler F, Meyerhof W, Hofmann T (2012) Human psychometric and taste receptor responses to steviol glycosides. J Agric Food Chem 60(27):6782–6793
90. Krohn M, Seibert S, Kleber A, Wonschik J (2012) Sweetener and/or sweetness enhancer, sweetener composition, methods of making the same and consumables containing the same. WO 2012107203
91. Krohn M, Seibert S, Kleber A, Wonschik J (2012) Sweetener and/or sweetness enhancer, sweetener composition, methods of making the same and consumables containing the same. WO 2012107205
92. Krohn M, Seibert S, Kleber A, Wonschik J (2012) Sweetener, sweetener compositions, methods of making the same and consumables containing the same. WO 2012107202

93. Krohn M, Seibert S, Kleber A, Wonschik J (2012) Sweetener and/or sweetness enhancer, sweetener composition, methods of making the same and consumables containing the same. WO 2012107207
94. Krohn M, Seibert S, Kleber A, Wonschik J (2012) Sweetness enhancer, sweetener compositions, methods of making the same and consumables containing the same. WO 2012107204
95. Krohn M, Seibert S, Kleber A, Wonschik J (2012) Sweetener and/or sweetness enhancer, sweetener composition, methods of making the same and consumables containing the same. WO 2012107206
96. Gaydou EM, Bianchini JP, Randriamiharisoa RP (1987) Orange and mandarin peel oils differentiation using polymethoxylated flavone composition. J Agric Food Chem 35 (4):525–529
97. Moharram FA, Marzouk MS, Ibrahim MT, Mabry TJ (2006) Antioxidant galloylated flavonol glycosides from *Calliandra haematocephala*. Nat Prod Res 20(10):927–934
98. Bokesch HR, Wamiru A, Le Grice SF, Beutler JA, McKee TC, McMahon JB (2008) HIV-1 ribonuclease H inhibitory phenolic glycosides from *Eugenia hyemalis*. J Nat Prod 71 (9):1634–1636
99. Potier P, Das B, Bui AM, Janot MM, Pourrat A, Pourrat H (1966) Structure of tormentic acid, a pentacyclic triterpene acid isolated from the roots of *Potentilla tormentosa*. Bull Soc Chim Fr 11:3458–3465
100. Otsuka H, Takeda Y, Yamasaki K (1990) Xyloglucosides of benzyl and phenethyl alcohols and Z-hex-3-en-1-ol from leaves of *Alangium platanifolium* var. *trilobum*. Phytochemistry 29 (11):3681–3683
101. Kleinhofs A, Haskins FA, Gorz HJ (1967) *trans*-o-Hydroxycinnamic acid glucosylation in cell-free extracts of *Melilotis alba*. Phytochemistry 6(10):1313–1318
102. Ratnayake AS, Yoshida WY, Mooberry SL, Hemscheidt T (2001) The structure of microcarpalide, a microfilament disrupting agent from an endophytic fungus. Org Lett 3 (22):3479–3481
103. Wessjohann L, Backes M, Ley J, Paetz S, Reichelt K (2010) Use of hydroxyflavan derivatives for taste modification. US 20100292175
104. Fraser AW, Lewis JR (1974) Rutaceae constituents. IV. Flavonoids from *Merrillia caloxylon*. Phytochemistry 13(8):1561–1564
105. Krohn M, Seibert S, Kleber A (2012) Sweetness enhancer, sweetener compositions, methods of making the same and consumables containing the same. US 20120201935
106. Kashket S (1990) Sweetness and flavor enhancement with flavanols. US 4906480
107. DuBois GE, Prakash I (2012) Non-caloric sweeteners, sweetness modulators, and sweetener enhancers. Ann Rev Food Sci Technol 3:353–380
108. Krohn M, Zinke H (2009) Cyclic lipopeptides for use as taste modulators. WO 2009156112
109. Krohn M, Zinke H (2009) Oligopeptides for use as taste modulators. US 20090202698
110. Sigoillot M, Brockhoff A, Meyerhof W, Briand L (2012) Sweet-taste-suppressing compounds: current knowledge and perspectives of application. Appl Microbiol Biotechnol 96 (3):619–630
111. Acree TE, Lindley M (2008) Structure-activity relationship and AH-B after 40 years. In: Weerasinghe DK, DuBois GE (eds) Sweetness and sweeteners: biology, chemistry and psychophysics, vol 979, ACS symposium series. American Chemical Society, Washington, pp 96–108
112. Jiang P, Cui M, Zhao B, Liu Z, Snyder LA, Benard LM, Osman R, Margolskee RF, Max M (2005) Lactisole interacts with the transmembrane domains of human T1R3 to inhibit sweet taste. J Biol Chem 280(15):15238–15246
113. Burdock GA, Wagner BM, Smith RL, Munro IC, Newberne PM (1990) Recent progress in the consideration of flavoring ingredients under the food additives amendment. 15. GRAS substances. Food Technol 44(2):78–86
114. Hooper D (1887) An examination of the leaves of *Gymnema sylvestre*. Pharm J Trans 17:867–868

115. Liu HM, Kiuchi F, Tsuda Y (1992) Isolation and structure elucidation of gymnemic acids, antisweet principles of *Gymnema sylvestre*. Chem Pharm Bull 40(6):1366–1375
116. Maeda M, Iwashita T, Kurihara Y (1989) Studies on taste modifiers. II. Purification and structure determination of gymnemic acids, antisweet active principle from *Gymnema sylvestre* leaves. Tetrahedron Lett 30(12):1547–1550
117. Yoshikawa K, Amimoto K, Arihara S, Matsuura K (1989) Structure studies of new antisweet constituents from *Gymnema sylvestre*. Tetrahedron Lett 30(9):1103–1106
118. Yoshikawa K, Amimoto K, Arihara S, Matsuura K (1989) Gymnemic acid V, VI and VII from Gur-Ma, the leaves of *Gymnema sylvestre* R. Br. Chem Pharm Bull 37(3):852–854
119. Yoshikawa K, Kondo Y, Arihara S, Matsuura K (1993) Antisweet natural products. IX. Structures of gymnemic acids XV–XVIII from *Gymnema sylvestre* R. Br. V. Chem Pharm Bull 41(10):1730–1732
120. Yoshikawa K, Nakagawa M, Yamamoto R, Arihara S, Matsuura K (1992) Antisweet natural products. V. Structures of gymnemic acids VIII–XII from *Gymnema sylvestre* R. Br. Chem Pharm Bull 40(7):1779–1782
121. Yoshikawa K, Arihara S, Matsuura K (1991) A new type of antisweet principles occurring in *Gymnema sylvestre*. Tetrahedron Lett 32(6):789–792
122. Ye W, Liu X, Zhang Q, Che CT, Zhao S (2001) Antisweet saponins from *Gymnema sylvestre*. J Nat Prod 64(2):232–235
123. Imoto T, Miyasaka A, Ishima R, Akasaka K (1991) A novel peptide isolated from the leaves of *Gymnema sylvestre*–I. Characterization and its suppressive effect on the neural responses to sweet taste stimuli in the rat. Comp Biochem Physiol A Comp Physiol 100(2):309–314
124. Yoshikawa K, Nagai Y, Yoshida M, Arihara S (1993) Antisweet natural products. VIII. Structures of hodulosides VI–X from *Hovenia dulcis* Thunb. var. *tomentella* Makino. Chem Pharm Bull 41(10):1722–1725
125. Yoshikawa K, Tumura S, Yamada K, Arihara S (1992) Antisweet natural products. VII. Hodulosides I, II, III, IV, and V from the leaves of *Hovenia dulcis* Thunb. Chem Pharm Bull 40(9):2287–2291
126. Yoshikawa K, Shimono N, Arihara S (1991) Antisweet substances, jujubasaponins I-III from *Zizyphus jujuba*. Revised structure of ziziphin. Tetrahedron Lett 32(48):7059–7062
127. Yoshikawa K, Shimono N, Arihara S (1992) Antisweet natural products. VI. Jujubasaponins IV, V and VI from *Zizyphus jujuba* Mill. Chem Pharm Bull 40(9):2275–2278
128. Maehashi K, Matano M, Kondo A, Yamamoto Y, Udaka S (2007) Riboflavin-binding protein exhibits selective sweet suppression toward protein sweeteners. Chem Senses 32(2):183–190
129. Masuda T, Ueno Y, Kitabatake N (2001) Sweetness and enzymatic activity of lysozyme. J Agric Food Chem 49(10):4937–4941
130. Galindo-Cuspinera V, Winnig M, Bufe B, Meyerhof W, Breslin PA (2006) A TAS1R receptor-based explanation of sweet 'water-taste'. Nature 441(7091):354–357
131. Bartoshuk LM, Lee CH, Scarpellino R (1972) Sweet taste of water induced by artichoke (*Cynara scolymus*). Science 178(4064):988–989
132. Nakajima K, Morita Y, Koizumi A, Asakura T, Terada T, Ito K, Shimizu-Ibuka A, Maruyama J, Kitamoto K, Misaka T, Abe K (2008) Acid-induced sweetness of neoculin is ascribed to its pH-dependent agonistic-antagonistic interaction with human sweet taste receptor. FASEB J 22(7):2323–2330
133. Meyerhof W, Batram C, Kuhn C, Brockhoff A, Chudoba E, Bufe B, Appendino G, Behrens M (2010) The molecular receptive ranges of human TAS2R bitter taste receptors. Chem Senses 35(2):157–170
134. Verzele M (1986) 100 Years of hop chemistry and its relevance to brewing. J Inst Brew 92 (1):32–48
135. Intelmann D, Batram C, Kuhn C, Haseleu G, Meyerhof W, Hofmann T (2009) Three TAS2R bitter taste receptors mediate the psychophysical responses to bitter compounds of hops (*Humulus lupulus* L) and beer. Chemosens Percept 2:118–132

136. Puri M, Marwaha SS, Kothari RM, Kennedy JF (1996) Biochemical basis of bitterness in citrus fruit juices and biotech approaches for debittering. Crit Rev Biotechnol 16(2):145–155
137. Hayes JE, Wallace MR, Knopik VS, Herbstman DM, Bartoshuk LM, Duffy VB (2011) Allelic variation in TAS2R bitter receptor genes associates with variation in sensations from and ingestive behaviors toward common bitter beverages in adults. Chem Senses 36 (3):311–319
138. Heckman MA, Weil J, Gonzalez de Mejia E (2010) Caffeine (1,3,7-trimethylxanthine) in foods: a comprehensive review on consumption, functionality, safety, and regulatory matters. J Food Sci 75(3):R77–87
139. Sharma S, Lewis S (2010) Taste masking technologies: a review. Int J Pharm Pharm Sci 2 (2):6–13
140. Breslin PAS (1996) Interactions among salty, sour and bitter compounds. Trends Food Sci Technol 7(12):390–399
141. Fuke S, Ueda Y (1996) Interactions between umami and other flavor characteristics. Trends Food Sci Technol 7(12):407–411
142. Nakamura T, Tanigake A, Miyanaga Y, Ogawa T, Akiyoshi T, Matsuyama K, Uchida T (2002) The effect of various substances on the suppression of the bitterness of quinine-human gustatory sensation, binding, and taste sensor studies. Chem Pharm Bull 50(12):1589–1593
143. Lawless HT (1979) Evidence for neural inhibition in bittersweet taste mixtures. J Comp Physiol Psychol 93(3):538–547
144. Finley JW, Holliday D, King JM, Prudente ADJ (2011) Masking bitter flavors by using mogrosides. WO 2011046768
145. Bridges JR, Carlson A, Patton PA (2012) Stevia blends containing rebaudioside B. WO 2012102769
146. Del Rio JA, Benavente O, Castillo J, Borrego F (1992) Neodiosmin, a flavone glycoside of *Citrus aurantium*. Phytochemistry 31(2):723–724
147. Guadagni DG, Horowitz RM, Gentili B, Maier VP (1977) Method of reducing bitterness in citrus juices. US 4031265
148. Guadagni DG, Maier VP, Turnbaugh JG (1976) Effect of neodiosmin on threshold and bitterness of limonin in water and orange juice. J Food Sci 41(3):681–684
149. Neely JS, Thompson JA (1975) Culinary composition containing paramethoxycinnamaldehyde as a flavoring agent and sweetener. US 3867557
150. Riemer JA (1994) Bitterness inhibitors for foods and artificial sweeteners. US 5336513
151. Rother R (1887) Some constituents of Yerba Santa. Am J Pharm 59:225
152. Sayre LEA (1917) A manual of organic materia medica and pharmacognosy. P Blakiston's Son and Co, Philadelphia.
153. Ley JP, Krammer G, Reinders G, Gatfield IL, Bertram HJ (2005) Evaluation of bitter masking flavanones from Herba Santa (*Eriodictyon californicum* (H. and A.) Torr., Hydrophyllaceae). J Agric Food Chem 53(15):6061–6066
154. Ley JP, Blings M, Paetz S, Krammer GE, Bertram HJ (2006) New bitter-masking compounds: hydroxylated benzoic acid amides of aromatic amines as structural analogues of homoeriodictyol. J Agric Food Chem 54(22):8574–8579
155. Ley JP, Paetz S, Blings M, Riess T, Krammer GE (2010) Structural analogues of hispolon as flavor modifiers. In: Blank I, Wuest M, Yeretzian C (eds) 12th Weurman Symposium, Interlaken, Switzerland, 1–4 July 2008. Expression of multidisciplinary flavour science. Zuercher Hochschule fuer Angewandte Wissenschaften, Institut fuer Chemie und Biologische Chemie Waedenswil, Switzerland, pp 402–405
156. Oertling H, Loges H, Machinek A, Simchen U, Surburg H (2011) Teeth cleaning compound containing menthol with reduced bitter sensation. US 20110081303
157. Blakeslee AF (1935) A dinner demonstration of threshold differences in taste and smell. Science 81(2108):504–507

Top Med Chem (2017) 23: 73–114
DOI: 10.1007/7355_2015_98

Published online: 14 February 2014

# Chemical Activation of Sensory TRP Channels

**Brett Boonen, Justyna B. Startek, and Karel Talavera**

**Abstract** The overall perception of flavor results from the integration of taste, smell, and somatosensory information streaming out of specialized receptor cells located in the oronasal cavities. Several members of the transient receptor potential family of cation channels contribute to the signal transduction of chemical stimuli. All bona fide TRP channel chemosensors contribute to flavor detection by acting on epithelial cells and/or sensory nerve endings in the mucosa of the nose, mouth, and throat. Chemical activation of these channels results in a very obvious, but yet obscure, sensory modality called trigeminality or chemesthesis, which is related to the perception of texture, temperature, and pungency. These sensations arise when chemical compounds activate receptor cells associated with other senses that mediate touch, thermal perception, and pain. In this chapter we illustrate the huge diversity of chemical agonists of TRP channels and underscore the need of more basic research on this amazing family of molecular sensors, which are very likely to hold the key for better understanding of human sensory pathophysiology.

**Keywords** TRPV1, Allyl isothiocyanate, Capsaicin, Menthol, TRPA1, TRPM8

## Contents



---

Brett Boonen and Justyna B. Startek are equally contributed to this work.

B. Boonen, J.B. Startek, and K. Talavera (✉)
Department of Cellular and Molecular Medicine, Laboratory of Ion Channel Research and TRP Research Platform Leuven (TRPLe), KU Leuven, Herestraat 49, Campus Gasthuisberg O&N1 bus 802, 3000 Leuven, Belgium
e-mail: Karel.talavera@med.kuleuven.be

# 1 Introduction

The survival of organisms depends critically on their capacity to sense environmental cues and to respond with appropriate adaptive behaviors. Smell and taste confer the animals' ability to recognize and discriminate different chemicals, airborne or dissolved in the saliva, respectively. Chemicals can be also detected by nerve endings present in nearly all tissues, particularly the mucosa (nose, mouth, throat, eyes) and skin. This type of response is called trigeminality or chemesthesis and is related to sensing the texture and temperature [1]. Activation of free nerve endings can trigger soothing feelings and freshness but can be also associated with unpleasant and even painful sensations. The study of the molecular mechanisms underlying these perceptions is a major task in modern biology.

In recent years, several members of the transient receptor potential (TRP) protein superfamily have been identified as key players in multiple chemosensory processes. TRP proteins are cation channels with great diversity in activation mechanisms and are expressed in numerous tissues and cell types [2, 3]. The mammalian TRP superfamily consists of 28 proteins essential in sensory physiology including contributions to taste, olfaction, vision, hearing, touch, and thermo- and osmosensation [4, 5]. Furthermore, they mediate responses to many endogenous molecules such as hormones, growth factors, and metabolic stress modulators.

Interestingly, none of the TRP channels that have been unequivocally involved in the mechanisms of taste or olfaction have been shown to function themselves as chemoreceptors. Indeed, the only TRP channels, shown so far to be directly involved in the transduction of taste and/or olfaction (TRPM5 and TRPC2) [82, 83], remain orphans of exogenous chemical agonists. In contrast, several TRP channels, expressed in free sensory nerve endings and epithelial cells (i.e., TRPA1, TRPV1, TRPV2, TRPV3, and TRPM8), are activated by a myriad of chemicals contained in food spices, odorants, and cosmetics. This chapter provides for an update on the sensitivity of such TRP channels to compounds relevant for human chemosensation.

## 2 TRP Channel Classification and General Structure

The mammalian TRP superfamily can be divided into seven subfamilies: TRPC (canonical), TRPV (vanilloid), TRPM (melastatin), TRPA (ankyrin), TRPN (NOMPC-like), TRPP (polycystin), and TRPML (mucolipin) [3]. TRP channels are classified as nonselective cationic channels, although they vary in the relative permeability for different ions [3]. With exception of TRPM4 and TRPM5 [84, 85], all are $Ca^{2+}$ permeable and thus constitute cellular $Ca^{2+}$ entry gateways and play crucial roles in different signaling cascades. TRP channels are assumed to assemble into homo- or heteromers, where four subunits form a cationic pore. The carboxyl-(C)- and amino-(N)-terminals are located intracellularly, and their length varies between different members of the TRP family. The N- and C-terminals contain many motifs and domains that are essential for channel function, including groups with enzymatic activity as, for example, the Nudix hydrolase domain of TRPM2 having an ADP-ribose pyrophosphatase activity [86, 87]. Other domains include ankyrin repeats, TRP domains, phosphorylation sites, EF hands, calmodulin-binding motifs, and lipid-interaction domains, which are not fully conserved between different members, but play crucial roles in channel activity.

The major homology region encompasses the six transmembrane domains with the pore loop located between the fifth and sixth transmembrane segments [2–4]. Apart from TRPM, TRPP, and TRPML channels, all TRPs contain multiple N-terminal ankyrin (ANK) repeats, with TRPA1 having the longest stretch of 18 ankyrin repeats [2, 3, 88]. Ankyrins are composed of a 33 amino acid-long repeated motif, which forms short (seven to eight residues) and long (nine to ten residues) alpha-helixes interconnected by a short loop [89, 90]. ANK repeats are well-described protein-interaction motifs mediating various cellular functions including ion transport, cell signaling, cytoskeleton interactions, and inflammatory processes [91–93]. It is not yet clear whether specific sequences and/or number of repeats are required to form specialized interactions or complete function in the tetramerization of TRP channels. Nevertheless, numerous studies demonstrate interactions between ANK repeats and various proteins and their role in channel multimerization and gating [93–96]. Another weakly conserved sequence of approximately 25 amino acids is referred to as the TRP domain, which is located in the proximal C-termini of all TRP channels with exception of TRPA1 and TRPP [97–99]. It contains the TRP box 1 (EWKFAR) motif and proline-rich box 2, which are variable between different channels [99]. In the TRPC, TRPM, and TRPV subfamilies, channel tetramerization is also mediated by a coiled-coil motif [100–102]. The structural feature of the coiled-coil domain comprises a heptad repeat of amino acids located roughly 88–120 amino acid from S6 transmembrane domain [100]. It was shown that mutations or deletion of the coiled-coil domain result in severe disruption of the subunit interactions and can cause loss of the channel function [100, 102].

## 3 TRPA1: A Broadly Tuned Chemosensor

Human TRPA1 was first identified in a screen for genes downregulated upon oncogenic transformation of fibroblasts [2]. Previously called ANKTM1 (ankyrin repeat-containing ion channel 1), TRPA1 is the only TRPA protein reported in mammals [2, 3] and is a target for noxious chemical irritants in peripheral sensory neurons implicating a functional role in pain and neurogenic inflammation [1]. This channel also participates in additional sensory processes such as cold and mechanosensation. TRPA1 is expressed in the dorsal root ganglion (DRG), vagal ganglion (VG), and trigeminal ganglion (TG) neurons [4], specifically in small diameter Aδ and C-sensory fibers [103, 104]. TRPA1 is mostly expressed in neurons expressing the capsaicin receptor TRPV1. For example, in rat trigeminal ganglion, TRPV1 is expressed in approximately 44% of the neurons, of which over 83% expresses TRPA1 [103]. It has been reported that TRPA1 is also expressed in non-neural cells that are also relevant for chemosensation such as skin keratinocytes [105, 106] and airway epithelial cells [107].

TRPA1 is activated by a large number of noxious chemicals found in many plants, food, cosmetics, and pollutants. Many TRPA1 agonists are highly reactive electrophiles with shared ability to modify thiol groups as cysteine or lysine residues in the N-terminus of TRPA1. Numerous mutagenesis experiments revealed that modification of only three cysteine residues C619, C639, and C663 and to some extent lysine K708 in the cytoplasmic N-terminus region of the channel leads to its activation [6, 108]. Electrophilic, pungent compounds such as unsaturated aldehydes, ketones, isothiocyanates, and thiosulfinates can be found in many plants, for example, in Brassicaceae (the cabbage family), Allioideae (onion and garlic, leek, chives, and shallots subfamily), or *Cinnamomum* genus. Some of these compounds serve as plant defensive traits against herbivores.

The plants from the Brassicaceae family (cabbage, broccoli, cauliflower, mustard, horseradish, radish, wasabi, and watercress) are important crops for human and animal diets. Allyl isothiocyanate (AITC, aka mustard oil) and derivatives such as benzyl isothiocyanate, phenylethyl isothiocyanate, isopropyl isothiocyanate, and methyl isothiocyanate are the main pungent ingredients inducing TRPA1 activation (Fig. 1; [6–10]). Topical mustard oil application induces activation of underlying nerve endings and a strong burning sensation and inflammation. Electrophysiology and $Ca^{2+}$ imaging experimental data established that hTRPA1 is activated by all compounds mentioned above and the underlying gating mechanism involves covalent modification of the cytoplasmic N-terminal tail [9, 108].

Allicin (S-allyl 2-propene-1-sulfinothioate) is another organosulfur TRPA1 activator found in extracts of raw garlic with a pungent taste and odor [13, 14]. Fresh cloves of garlic contain alliin, which is a natural plant protector against pathogens. If the structural integrity of the clove is compromised, for instance, by fungi attack or by just crushing the clove, an enzymatic reaction occurs, transforming alliin into allicin within seconds [109]. Produced this way, allicin is short lived and converted into more stable, sulfide compounds such as diallyl sulfide (DAS), diallyl disulfide

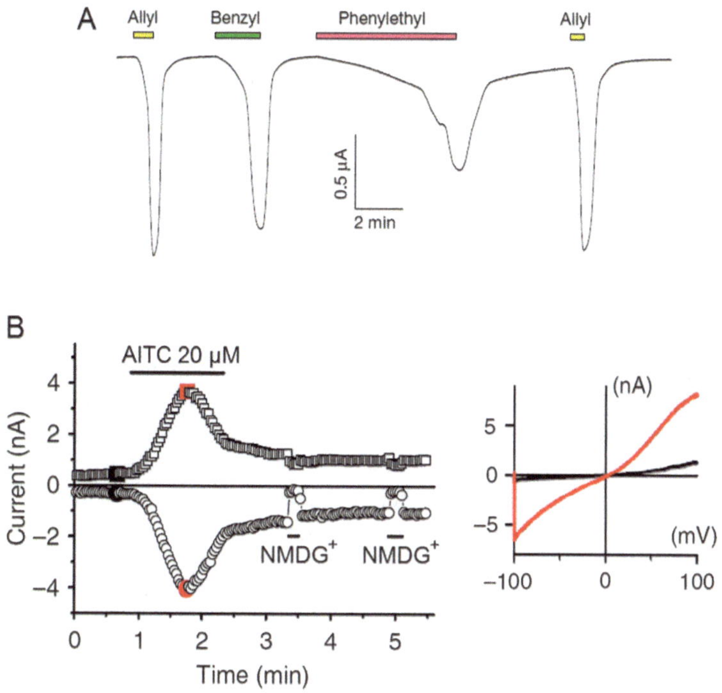


**Fig. 1** Activation of TRPA1 by isocyanates. (**a**) Effects of allyl, benzyl, or phenylethyl isothiocyanate (10 mmol/L) on currents recorded at −60 mV in a *Xenopus laevis* oocyte expressing human TRPA1. Reproduced with permission from Jordt et al. [9]. (**b**) Time course of the effects of allyl isothiocyanate (AITC) on the amplitude of whole-cell currents recorded at −75 mV (*circles*) and +50 mV (*squares*) in a Chinese hamster ovary cell stably transfected with mouse TRPA1. The *colored filled symbols* correspond to the current traces shown on the *right panel*. Reproduced with permission from Talavera et al. [67]. µM (µmol/L)

(DADS), and diallyl trisulfide (DATS) [15]. Ratiometric $Ca^{2+}$ imaging in dissociated neurons from rodent trigeminal ganglia confirmed that approximately 30% of neurons respond to the garlic extract, and this response could be eliminated with ruthenium red, a nonselective TRP channel blocker [15]. Further, using HEK and CHO cells transfected with TRPA1 or TRPV1, it was confirmed that both channels are activated by allicin and its derivatives, but the response to allicin by-products was less intense and much slower [13–15]. Additional studies revealed that DAS, DADS, and DATS activate both channels but show higher affinity for TRPA1 [13, 15, 16]. It has also been proposed that these compounds induce vasodilatation through the release of CGRP following the activation of TRPA1 on sensory nerve endings [15].

Ligustilide is a lipophilic and electrophilic dihydrophthalide extracted from Dang Gui (Chinese Angelica root) commonly used in traditional Chinese medicine due to its analgesic and anti-inflammatory properties [17, 18]. This compound can be also found in celery and lovage and is responsible for the plants' specific aroma and taste [110]. Ligustilide was demonstrated to activate TRPA1, with an effective concentration ($EC_{50}$) of 44 μmol/L [18]. Mutation of TRPA1 cysteines revealed covalent modification of the channel by the compound as the action mechanism [18].

Cinnamon oil and its main component cinnamaldehyde (CA) are obtained from the plants of the *Cinnamomum* genus and are widely used as flavoring agent. It is known for its fragrance and when administrated orally, it induces a sweet, lightly pungent sensation. Cinnamon oil was reported to have many beneficial pharmaceutical effects as antioxidant activity, antimicrobial effect, and even antidiabetic action [111, 112]. The burning and tingling sensation produced by cinnamaldehyde is attributed to its action on TRPA1. CA has in fact a bimodal effect on TRPA1 inducing activation in the micromolar range but also inhibition in the millimolar range, with an effective inhibitory concentration ($IC_{50}$) of 3.5 mmol/L [19]. This bimodal action was confirmed in mouse DRG neurons. It is worth noticing that at 5 mmol/L, CA also induced $Ca^{2+}$ increase in neurons isolated from TRPA1 knockout mice, suggesting that other receptors may be also activated by this compound [19]. The inhibitory action of cinnamaldehyde on TRPA1 may also explain, at least in part, its weak ability to induce painful sensations. Interestingly, the non-electrophilic cinnamaldehyde derivatives cinnamyl alcohol and cinnamic acid are much less potent, with cinnamic acid being unable to induce TRPA1 activation [20, 21]. These observations further support a mechanism of channel activation by modification of cysteine residues.

The psychoactive component of plants of the *Cannabis* genus, $\Delta^9$-tetrahydrocannabinol (THC), is well known for its analgesic properties and for inducing other effects such as anxiety, memory and concentration problems, paranoia, hallucinations, and increased heart rate [23, 113]. The main action of THC is related to activation of cannabinoid receptors 1 and 2 (CB1 & CB2), but TRPA1 is another target of this compound. This was demonstrated using electrophysiology and $Ca^{2+}$ imaging experiments in rat trigeminal sensory neurons as well as in TRPA1-transfected HEK cells and oocytes [9]. Other cannabinoids, structurally related to THC, including cannabidiol, cannabidiolic acid, cannabichromene, cannabigerol, and $\Delta^9$-tetrahydrocannabinol acid, were also able to activate TRPA1 [24], with $EC_{50}$ values of 0.096 μmol/L, 12 μmol/L, 0.06 μmol/L, 3.4 μmol/L, and 0.24 μmol/L, respectively, compared to 0.24 μmol/L for THC. These values, obtained using rat TRPA1-transfected HEK cells with the use of ratiometric $Ca^{2+}$ imaging [24], correlate to the predicted electrophilicity of the compounds, which supports the proposed mechanism of TRPA1 activation by covalent modification. It is also important to mention that cannabinoids activate TRPV2 and inhibit TRPM8. Additionally, they are potent TRP channel desensitizers, with activity dependent on several factors, for instance, the compound lipophilicity. Previously, it has been demonstrated that bulkiness and lipophilicity of the activating molecule could facilitate interaction with the binding moieties of the channel as well as penetration of the compound into the cell [114, 115]. Of note, the differential desensitization and activation of different targets by the cannabinoids may be a potential mechanism of their analgesic therapeutical action.

Curcumin, an electrophilic compound found in the turmeric root of the plant from the ginger family, was also shown to activate TRPA1 [31, 32]. This compound has pungent, bitter, and sharp taste and is used as spice and known to have anti-inflammatory and antiseptic properties [32]. Electrophysiological measurements in TRPA1-transfected cells and native mouse vagal neurons demonstrated an increase in TRPA1 currents in a concentration-dependent manner at concentrations up to 30 μmol/L [31]. TRPA1-deficient mice did not respond to curcumin and, accordingly, it failed to activate TRPM8 and TRPV1 [31]. Miogadial (MD), miogatrial (MT), polygodial (PG), and 1′-acetoxychavicol acetate (ACA) are α,β-unsaturated 1,4-dialdehydes found in extracts from ginger plants [33]. They have a strong, pungent taste attributed to TRPA1 activation in sensory nerve endings. These terpenoids have pain-relieving and anti-inflammatory properties. Although the initial sensation is very painful, sensory neurons are quickly desensitized bringing lasting analgesic effect [33]. Activation of TRPA1 occurs with $EC_{50}$ of 0.2 μmol/L for MD, 0.13 μmol/L for MT, 0.059 μmol/L for PG, and 0.16 μmol/L for ACA [33, 34]. Except ACA, they were also able to activate TRPV1, but the potency of these terpenoids is among the highest of TRPA1 agonists, which points to a role of TRPA1 in the pungency of these compounds [33, 34].

Isovelleral, a pungent terpenoid naturally present in the fungus *Lactifluus vellereus*, has some unusual properties [38]. This sesquiterpene dialdehyde is rapidly produced upon damage of the basidioma and released as a white pungent milk [38, 116]. These fungi have very hot and peppery taste, which serves to deter foragers. In fact, skin contact with this milk induces strong inflammatory responses such as blistering [38]. Isovelleral inhibits TRPV1 function with comparable potency to ruthenium red and strongly activates TRPA1 channels [38]. Unexpectedly, the gating mechanism does not involve covalent modification of channel cysteine or histidine residues, even though the chemical moieties present in isovelleral would suggest this activation mechanism [38]. Experiments conducted in TRPA1 KO mice demonstrated some residual activity of the compound suggesting involvement of another yet to be discovered target [38]. Previous reports demonstrate cytotoxic effects induced by this sesquiterpene dialdehyde including disturbance of cell integrity [117]. A large group of pungent sesquiterpene dialdehydes and its derivatives with unknown properties is present in many plants and animals. They include compounds such as muzigadial, drimanial, ugandensidial or warburganal (pepper-bark tree), sacculatal, and isosacculatal (*Trichocoleopsis sacculata* worm) or ancistrodial (*Ancistrotermes cavithorax* termite) [38, 118, 119].

Another terpenoid, umbellulone present in the tree *Umbellularia californica*, is well known for its headache- and migraine-inducing properties [39, 120]. It also causes sneezing, coughing, and airway irritation [39]. This potent monoterpene ketone is reported to affect respiration, heartbeat, and even induce death [39]. Perception of the compound is described as cold and painful and is associated with TRPA1 and TRPM8 activation in trigemino-vascular nerve endings leading to CGRP release and nociceptive responses [39, 40]. Umbellulone was reported to have bimodal action against TRPA1, with an $IC_{50}$ of 408 μmol/L. The action mechanism probably involves an alternative non-electrophilic pathway [40]. However, it was reported that this compound also weakly reacts with thiol groups [40]. The headache-causing effect may be explained by possible diffusion of this highly lipophilic compound from the nasal mucosa into the circulation and activating perivascular sensory nerve endings in the meningeal vessels [39]. Then again, stimulation of TRPA1 in trigeminal nerve endings in the nasal mucosa and signal propagation through reflex pathways could also lead to meningeal vasodilatation [39].

Extracted from a plant belonging to the mint family (*Perilla frutescens*), perillaldehyde (PA) and perilla ketone (PK) produce an unusual cinnamon-mint flavor with tingling and cooling mouth sensations [46]. The perilla herb is widely used in Asian cuisine, and it is well known in traditional Chinese medicine for its anti-allergic and anti-inflammatory properties [121]. Both compounds activate and quickly desensitize TRPA1-containing neurons, which may be a reason for their unusual sensation. In fact, due to their interesting properties and simple structure, they are interesting targets for drug design. Recently, a series of active chemicals were developed based on the structure of perilla ketone including heteroaryl ketone, which has a very distinctive structure compared to other TRPA1 agonists [47]. All compounds presented in that study had high biological activity and potency toward TRPA1 [47]. Furthermore, crucial chemical groups in their structures were identified that are needed for TRPA1 gating via channel covalent modification.

Other interesting plants evoking pungent sensation are the Sichuan and Melegueta peppers. In contrast to the hot and burning sensations caused by chili peppers, Sichuan peppers induce a slight lemony, mildly hot, and tingly sensation in the mouth and are used as spices in Chinese and Japanese cuisine. This sensation is attributed to unsaturated hydroxyalkamides especially α-,β-,γ-,δ-hydroxysanshools and their by-products [36, 55]. It was shown that HαSS but not HβSS is able to excite sensory neurons [55–57]. Furthermore, from 17 TRP channels tested, only TRPA1 and TRPV1 could be activated. These results were further confirmed by electrophysiological and ratiometric $Ca^{2+}$ imaging in transfected HEK cells [36, 57]. Also, only HαSS was able to produce the specific pungent and simultaneous cooling sensation once applied to the tongue [56, 57], which supports the role of TRPA1 in compound perception. The proposed TRPA1 activation mechanism involves covalent modification of the channel, and activation is highly decreased, once three crucial cysteine residues in the N-terminal are mutated [36]. Linalool is a monoterpenoid present in extracts of Sichuan pepper that has also been shown to activate TRPA1 in DRG neurons as well as transfected in HEK cells [36]. Surprisingly, linalool has a strong floral aroma but not a pungent taste. Instead, it has a very unpleasant flavor. The TRPA1 triple cysteine mutant channel was equally sensitive to linalool as the wild-type channel suggesting for another TRPA1 gating mechanism. Linalool can be included into the group of non-electrophilic compounds together with thymol, menthol, or camphor, for which the activation mechanisms are still not understood.

6-Shogaol and 6-paradol, two pungent compounds from Melegueta peppers or plants from *Zingiber* genus (ginger) that contain a vanilloid moiety, are also reported to activate TRPA1 and TRPV1 [36, 37]. Surprisingly, they bear structural similarity to capsaicin present in chili peppers, which is specific for TRPV1 but has no effect on TRPA1. On the other hand, it was shown that capsaicin-like capsinoids such as capsiate, dihydrocapsiate, and nordihydrocapsiate, naturally present in CH-19 sweet pepper and having Scoville heat rating 1,000 times less than that of capsaicin, are activators of both TRPA1 and TRPV1 [62, 63]. Capsinoids are nonpungent, as they cause no burning sensation when applied to the tongue, but a behavioral study reported nociceptive reactions after injection in the mouse hind paw [62]. Due to the fact that the capsinoids 6-shogaol and 6-paradol are highly lipophilic, the lack of response in the oral cavity and eyes could be explained by retention of the compounds in the epithelium or cornea before reaching the sensory nerve endings. It is possible that capsiate and its derivatives are confined in the membranes of epithelium or cornea similarly to olvanil (a nonpungent TRPV1 agonist), which has a very low rate of penetration in the epidermis [62, 64, 65]. Alternatively, it is possible that compounds are degraded by esterases and lipases found in the oral cavity and cornea before reaching the sensory nerve endings [122]. So far, the mechanisms of TRPA1 activation by these non-electrophilic compounds are unknown, since they are unable to modify thiol groups in channel structure. Since many of the non-electrophilic compounds are lipophilic and spontaneously insert into lipid membranes, thus changing its properties, it is possible that TRPA1 activation is induced by mechanical perturbations rather than channel modification. This is the case of amphipathic trinitrophenol, which accumulates in the outer leaflet of the plasma membrane and changes membrane curvature. As a result, the mechanical alterations of the membrane may induce TRPA1 activation [123].

Menthol ((1*R*,2*S*,5*R*)-2-isopropyl-5-methylcyclohexanol), obtained from *Mentha* genus plants, is another compound inducing complex sensations. Once applied at room temperature to the tongue or mucus membranes of airways, it gives a cooling sensation, but at temperature above 37°C, it potentiates the feeling of warmth [48]. Differences in compound perception are related to its ability to activate multiple TRP channels including TRPM8, TRPA1, and TRPV3 [41]. The cooling feeling is primarily attributed to the activation of TRPM8 [51, 164]. Other menthol effects such as pain, skin irritation, or burning could not be explained by TRPM8 activity. It has been shown that menthol has a bimodal effect acting as a TRPA1 activator in low concentrations (between 1 and 30 μmol/L) and leading to channel block at high concentrations (above 1 mmol/L) (Fig. 2a; [50]). Mechanisms of channel activation and blocking seem to be distinctive from each other, since menthol was unable to activate TRPA1 during or after mustard oil application; however, the blocking effect was preserved [50]. Previous studies show that menthol can also alter plasma membrane fluidity [124], which may in fact be sensed by TRPA1 and thus causing its activation. It is also worth to note that not all species of TRPA1 are sensitive to menthol. Nonmammalian TRPA1 channels were reported to be menthol insensitive [49].

Several other monoterpenes such as camphor and thymol also exhibit bimodal action with TRPA1 activation at low concentrations and inhibition in the millimolar range (Fig. 2b; [19]). Camphor and its derivatives (e.g., linalool, 1,8-cineole, or

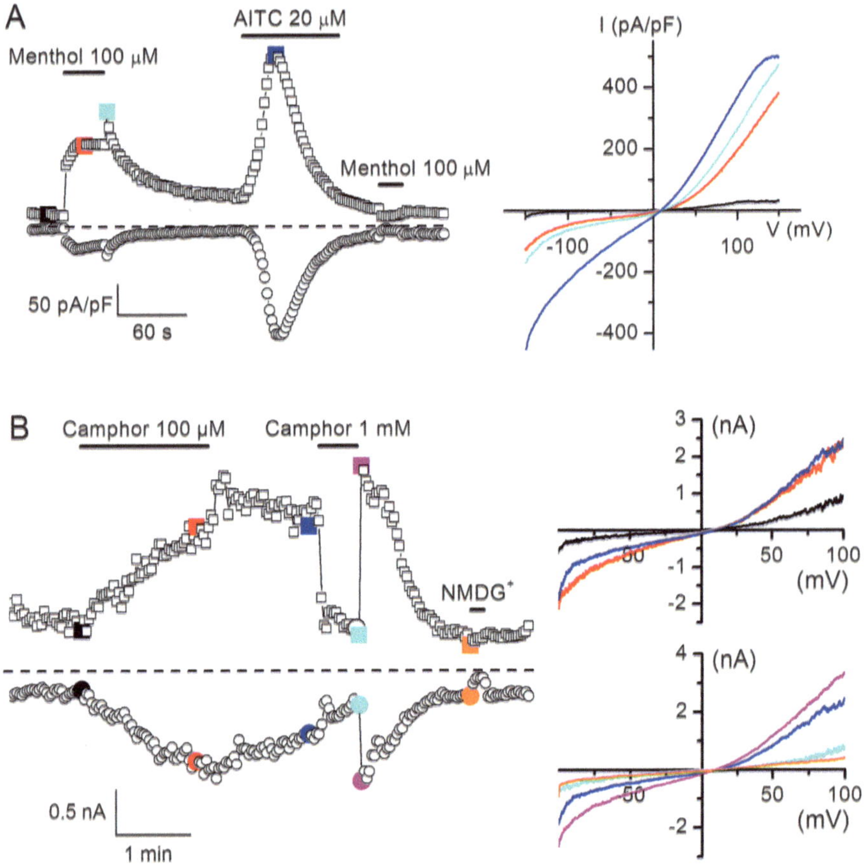


**Fig. 2** Bimodal action of menthol and camphor on mouse TRPA1. (**a**) Effects of menthol on the amplitude of whole-cell currents recorded at −80 mV (*circles*) and +80 mV (*squares*) in a Chinese hamster ovary cell stably transfected with mouse TRPA1. The filled symbols correspond to the colored current traces shown on the *right panel*. Note the sudden increase in current amplitudes upon washout of menthol and that application of this compound after AITC induces current inhibition. These effects are reminiscent of the bimodal action of menthol on TRPA1 channels. Reproduced with permission from Karashima et al. [50]. (**b**) Effects of camphor on the amplitude of whole-cell currents recorded at −75 mV (*circles*) and +75 mV (*squares*) in a Chinese hamster ovary cell stably transfected with mouse TRPA1. The *colored filled symbols* correspond to the current traces shown on the *right panel*. Note that camphor 100 μM induces current increase, but camphor 1 mM inhibits the current. Washout of 1 mM camphor elicits a strong rebound of the current. Reproduced with permission from Alpizar et al. [19]. μM (μmol/L), mM (mmol/L)

borneol) are found in the wood of camphor tree *Cinnamomum camphora*. Camphor is widely used for its medical properties as, for example, decongestive and calming actions. Due to its analgesic and antiseptic properties, it is used as a component of ointments. It can also be used as a spice and has a strong, fragrant smell and a bitter, pungent taste [125]. The pungency of camphor is associated with activation of TRPV1 and TRPV3 channels [42], and its soothing action could be attributed to TRPA1 inhibition with an $IC_{50}$ of 0.66 mmol/L [19, 42, 43]. However, it was reported that camphor activates mouse TRPA1 at concentrations below 300 μmol/L (Fig. 2b; [19]). Application of 100 μmol/L of camphor in TRPA1-transfected CHO cells induced a small, approximately threefold increase of TRPA1 currents. Also, over one third of primary mouse DRG neurons responded to 300 μmol/L of camphor with an increase in intracellular $Ca^{2+}$ [19]. Consequently, TRPA1 activation could explain the burning sensation caused by camphor at low temperatures [125], which may be difficult to explain by activity of heat-activated TRPV channels.

Thymol is one of the main components of the oils extracted from plants of *Thymus* and *Origanum* genus [66]. It is commonly used as a spice for its characteristics, very strong aromatic odor, bitter taste, and pungency [126]. Thymol is widely used in the food industry as well as in cosmetics and pharmaceutical production due to its antimicrobial, antifungal, and antispasmodic properties [66]. Thymol is known to activate TRPM8, ionotropic GABA receptors, and TRPV3 channels, but these effects are not responsible for its pungent taste [53, 66, 127]. In fact, the bimodal action of the compound on TRPA1 could explain many of its properties. Similarly to the monoterpenoids mentioned above, thymol activates TRPA1 at concentrations in micromolar range and inhibits at high concentrations [50]. The mechanism of thymol action on TRPA1 remains unknown.

Interestingly, similar to thymol and camphor, many other alkyl phenols share the ability to activate TRPA1, for instance, thyme oil-derived carvacrol [53], 2,5-dimethylphenol, 2,6-diethylphenol, and cresols mainly found in air pollutions [66]. So far, it is not known whether all of them have bimodal actions or how they gate TRPA1. Due to their non-electrophilic properties, it is unlikely that activation involves the Michael addition mechanism. There is also a large group of plant-derived phenolic compounds that are likely to activate TRP channels, for example, alkyl phenols in cashew anacardic acids, cardanols, and cardols. The screening and identification of all these chemicals as well as the elucidation of their properties and action mechanisms will take many years.

Nicotine is a potent pyrrolidine alkaloid produced by plants from the Solanaceae family, especially in high concentrations by the tobacco plant (*Nicotiana tabacum*). In lower concentrations, it can be also found in tomato, eggplant, potato, and bell and chili peppers. Nicotine can act on nicotinic acetylcholine receptors (nAChRs) expressed in nociceptive neurons [128], but this may not fully explain the painful burning and stinging sensation it causes. Unfortunately, many of the commonly used nicotine replacement therapies produce strong side effects. For instance, the nicotine gum can cause tingling, burning sensation, and irritation in the mouth, which are potentiated with increasing nicotine doses [129]. Recently, it was shown that nicotine activates TRPA1, which may explain all abovementioned unpleasant feelings. Nicotine has a bimodal effect on TRPA1 inducing activation at low concentrations ($EC_{50} = 17$ μmol/L) and inhibition in high concentrations [67]. Nicotine was found to inhibit TRPV1 suggesting for a lack of contribution of this channel to nicotine's pungency [67, 130]. Two distinct responses to nicotine were identified in mouse trigeminal ganglion neurons, one mediated by TRPA1 activation and other by nAChRs [67]. The use of the nAChR inhibitor hexamethonium served to confirm the specific action of nicotine on TRPA1 currents, both in transfected CHO cells and in mouse TG neurons [67]. Due to the fact that nicotine belongs to a group of non-electrophilic TRPA1 activators, it is very unlikely that this compound would induce channel opening via covalent modification. Consequently, the mechanism of its action remains unknown.

TRPA1 is also known to be activated by many different highly reactive compounds found in cigarette smoke, for example, crotonaldehyde and acrolein [68]. Both compounds induced strong intracellular $Ca^{2+}$ influx in guinea pig juxtaglomerular (JG) neurons. Based on the use of the TRPV1 inhibitor capsazepine and the TRPA1 blocker HC030031, it was shown that TRPA1 is responsible for the neuronal excitation produced by these compounds [68]. Tobacco has also considerable amounts of other alkaloids structurally similar to nicotine such as nornicotine, anabasine, and anatabine [131]. It was reported that anabasine, a potent herbivore repellant, was able to activate TRPA1-transfected CHO cells with similar potency and concentration dependence as nicotine [67]. The activities of nornicotine and anatabine on TRPA1 and other TRP channels as well as of thousands of other chemicals, present in cigarette smoke [132], on TRPA1, and other TRP channels, remain unknown.

Piperine and at least 20 other chemicals, including piperolein A and piperolein B, pipernonaline, dehydropipernonaline, or isochavicine, are the main compounds responsible for the pungent, burning effect of pepper (*Piper* genus) [69]. Many of these compounds, especially the ones containing the piperidine ring, are able to activate both TRPA1 and TRPV1 channels indicating importance of this moiety for channel activation. Other piperine-derived compounds result in stronger activation of TRPV1 but are also able to activate TRPA1. On the other hand, only one compound, *N*-isobutyl-(2*E*,4*E*)-tetradeca-2,4-diamide (*N*-tetra), was able to specifically activate TRPA1-transfected HEK cells [69]. The flavor of freshly ground black pepper can induce reflexes such as sneezing, coughing, or eye tearing related to TRP channel activation in sensory nerve endings. The floral, nutty, or citrusy notes, present in different types of pepper, are related to terpenes, for example, linalool, sabinene, limonene, α- and β-pinene, myrcene, and phellandrene found in the outer layer of the peppercorn [70]. Differences in compound compositions and concentrations between black and white peppercorns are attributed to distinctive flavor sensations. In white pepper, the presence of the sesquiterpene rotundone (3,4,5,6,7,8-Hexahydro-3α,8α-dimethyl-5α-(1-methylethenyl)azulene-1 (2*H*)-one), also found in other spices such as oregano and rosemary, as well as *p*-cresol, gives corns its distinctive aroma [133, 134]. Many of the compounds mentioned above are known activators or inhibitors of TRPA1. For example, limonene, which induces strong irritation and odor of oranges, can be also found in Rutaceae family (citrus plants) and Apiaceae plants like celery along with species from *Eucalyptus* genus. Recently, it has been shown that limonene is able to activate TRPA1 [66]. Nonetheless, many of other pungent chemicals present in essential oils derived from these plants remain unknown.

The Brazilian green propolis extract, used as dietary supplement, has an uncommon herbal flavor and induces a distinctive unpleasant, bitter taste and pungency at the back of the throat [72]. Propolis, also produced by honeybees, has been used for many decades in traditional medicine because of its countless healing properties. It has antimicrobial, antifungal, and anti-inflammatory activity and in addition was reported to induce apoptosis of cancer cells [72, 135–139]. Pungency of propolis was recently related to activation of TRPA1 channels in sensory nerve endings by the main propolis component artepillin C [72]. Surprisingly, other types of propolis do not have a pungent taste. Brazilian green propolis extract has been also shown to contain some other TRPA1 activators such as cinnamic acids. It was reported that prenyl-containing artepillin C is a highly specific and potent activator of TRPA1 expressed in HEK cells with a reported $EC_{50}$ of 1.8 μmol/L [72]. The compound is also very lipophilic, thus it has the potential to insert into cellular membranes. Nevertheless, the mechanism of action on TRPA1 remains unknown.

Finally, it was recently described that the pungent component of extra-virgin olive oil, oleocanthal, was able to activate TRPA1 channels [73]. This phenylethanoid triggers a distinctive burning sensation in the back of the throat, inducing cough, and was reported to have anti-inflammatory, anticarcinogenic, and antioxidant activities [140–142]. Similarly to the cases of artepillin C or ibuprofen, the exclusive effect of oleocanthal at the back of the throat may be related to the uneven TRPA1 distribution in the human oral mucosa, with poor expression in sensory fibers innervating the anterior tongue and high expression on pharyngeal fibers and nasal epithelia [73]. It was reported that oleocanthal cannot modify TRPA1 cysteine residues, although it contains two aldehyde moieties in its structure that are necessary to interact with TRPA1 [73]. Yet, the mechanism of action still remains unknown.

## 4 TRPV1: The Capsaicin Receptor

TRPV1 is well known as a transducer of heat and the receptor of capsaicin [58]. The in vivo activation of TPRV1 receptors by natural agonists like capsaicin is associated with a sharp and burning pain, perceived as pungency. However, pungency of TRPV1 agonists is not depending on potency but rather critically dependent on lipophilicity. Highly lipophilic agonists are less pungent, since they induce slow TRPV1 activation, delaying or even suppressing its ability to trigger action potentials in sensory neurons [114]. This peculiar quality of TRPV1 activation has set off the quest for finding or developing TRPV1 agonists that have a high potency/pungency ratio, which facilitates oral administration.

TRPV1 can be found in different tissues throughout the body. The highest expression levels could be found in dorsal root ganglion, nodose ganglion, and trigeminal ganglion neurons [143]. In the population of sensory neurons, expression of TRPV1 is predominantly found in small- and medium diameter-peptidergic and non-peptidergic neurons. Moreover, TRPV1 expression has been reported also in different parts of the brain, the afferent innervation of the urinary bladder, the liver, granulocytes, and mast cells [2–5, 86, 144]. TRPV1, as other TRP channels, has six transmembrane domains with a pore-forming hydrophobic stretch between the fifth and the sixth transmembrane segments. The pore of TRPV1 shows a greater selectivity for $Ca^{2+}$ over $Na^{+}$ ($P_{Ca}/P_{Na} = 9.6{:}1$) [58].

TRPV1 can be activated by several stimuli: capsaicin, the most specific chemical activator [58, 145], protons [146, 147], and a plethora of other chemical compounds [145, 148]. Because of the extensive amount of chemical and physical stimuli, TRPV1 channels, as other members of the TRP superfamily, are able to integrate several chemical and physical stimuli. This speaks for the important role of TRPV1 as a pain sensor and its role in inflammatory hyperalgesia [59, 149]. Numerous synthetic or endogenous TRPV1 agonists used for medicinal purposes are described in the literature. However, here we will focus only on channel activators relevant for flavor, namely, tastants and odorants.

As mentioned above, the first ligand for TRPV1 to be found was the pungent compound *capsaicin* (Fig. 3; [58]). Capsaicin (trans-8methyl-*N*-vanillyl-6-nonenamide) is a main component of the fruit of chili peppers (genus *Capsicum*). Several structurally related compounds, called capsaicinoids, are found in the capsicum spices. Other examples of capsaicinoids are dihydrocapsaicin, nordihydrocapsaicin, homodihydrocapsaicin, and homocapsaicin [151]. Capsaicin is a potent alkaloid of the vanilloid family containing an aromatic ring, an amide bond, and a hydrophobic side chain. Chemical groups in the position three as well as hydroxyl group in the position four of the ring were shown to be highly important

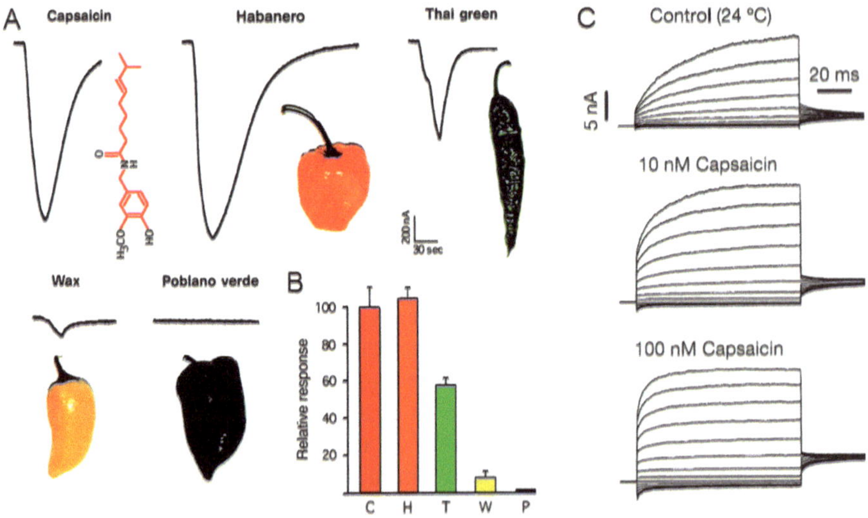


**Fig. 3** Activation of TRPV1 by capsaicin and pepper extracts. (**a**) Effect of application of capsaicin or extracts from different peppers on currents recorded in a *Xenopus laevis* oocyte expressing TRPV1 channels. Reproduced with permission from Caterina et al. [58]. (**b**) Comparison of the amplitude of the responses to these agents, normalized to the average effect of capsaicin. Reproduced with permission from Caterina et al. [58]. (**c**) Currents elicited by voltage steps in the range from −100 to +160 mV in a HEK293 cell transfected with TRPV1 in the absence and presence of capsaicin. Note that capsaicin increases the rate of channel activation. Reproduced with permission from Voets et al. [150]. nM (nmol/L)

for activity. This compound is a hydrophobic, odorless, off-white solid with a melting point of 62–65°C and a molecular mass of 305.4 g/mol [151]. As it is not water soluble, alcohols and other organic solvents are used to solubilize capsaicin [152]. Capsaicin is highly specific for TRPV1, as the disruption of TRPV1 gene in vivo abrogates responses to capsaicin [60]. Capsaicin acts as a gating modifier of TRPV1 by shifting the voltage dependence of activation to more negative potentials, therefore increasing the open probability of the channel at the resting potential of sensory neurons [168]. TRPV1 activation by capsaicin induces rapid influx of $Ca^{2+}$ and causes desensitization. This effect made capsaicin an attractive compound for medicinal use, as it renders the channel unable to open and in such a way causes an analgesic effect [59].

A related family of compounds is the capsinoids. These are chemicals found in the CH-19 nonpungent red peppers (*Capsicum annuum* or *Capsicum frutescens*). They are structurally similar to capsaicinoids with the alteration caused by a different central linker between the aliphatic hydroxyl group of the vanillyl moiety and the fatty acids resulting in an ether instead of an amide [152]. Capsinoids are of particular interest as they are nonpungent and easily administrable, yet they are still active on the TRPV1 channel [62]. For instance, capsiate has been an interesting tool in research on weight regulation by activation of TRPV1 [60].

TRPV1 is activated by strong extracellular acidosis [146, 153, 154] and has been suggested that the modulation of this channel by changes in pH may contribute to the interactions between salty and sour stimuli [155]. Furthermore, it has been reported that intracellular alkalosis activates TRPV1 [147], whereas intracellular acidosis has the opposite effect [156] suggesting for a complex regulation. However, evidence for functional expression of TRPV1 in taste bud cells is inconclusive, and it is likely that the contribution of this channel to chemosensation is via its expression in trigeminal nerve endings [82].

Another interesting modulator of TRPV1 is ethanol. It was shown that at concentrations ranging from 0.1 to 3%, this compound increases the intracellular $Ca^{2+}$ concentration in TRPV1-expressing cells at physiological temperatures. Furthermore, ethanol sensitized the response of this channel to capsaicin, heat, and extracellular acidosis [157]. However, it is noticeable that no convincing electrophysiological characterization of the effects of ethanol on TRPV1 has been reported thus far, and therefore, the underlying mechanisms remain largely unknown.

AITC was initially described as a specific agonist of TRPA1 [9]. However, it has been found that TRPA1 knockout mice show avoidance behavior toward AITC [10]. This behavior could be explained by the activation of TRPV1 (Fig. 4). AITC interacts in a direct and reversible manner with this channel through a mechanism depending on an amino acid residue (S513) that is also required for capsaicin-induced activation [11]. The interaction is unlike that described for TRPA1 and AITC, as it is not depending on cysteine modification. However, the mechanism of activation is similar to capsaicin, as AITC also induces a shift of the activation curve to more negative membrane voltages [11]. Most interestingly, AITC cross-sensitizes with the effect of low pH. This is relevant, as AITC is mostly presented in

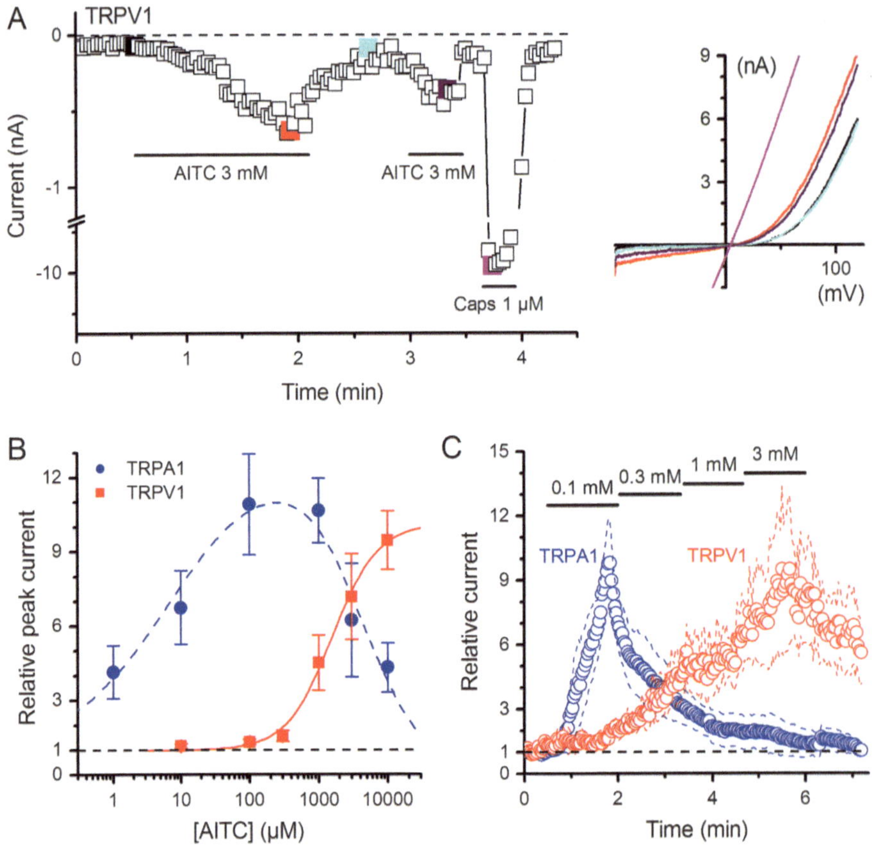


**Fig. 4** Activation of TRPV1 channels by AITC. (**a**) Application of AITC (3 mmol/L) induces a reversible increase in the amplitude of the current recorded at −75 mV in a HEK293 cell transfected with mouse TRPV1 ($T = 25$ C). The *colored filled symbols* correspond to the current traces shown on the *right panel*. (**b**) Concentration dependence of the maximal effect of AITC on mouse TRPA1 and mouse TRPV1 current amplitudes measured at −75 mV. In each cell, currents were normalized to the amplitude obtained in control solution. The *dashed blue line* represents the fit, with a *bell-shaped function* accounting for stimulatory and inhibitory effects of AITC on TRPA1. In contrast, the amplitude of the TRPV1 current increases with the concentration of AITC following a classical Hill-type behavior (*solid red line*). (**c**) Average time course of the effects of cumulative application of AITC on mouse TRPA1 and mouse TRPV1 currents at 35°C. The *dashed lines* represent the means ± the corresponding standard error of the mean (SEM). In each cell, current amplitudes were normalized to the value measured in control solution. Note that AITC induces a strong desensitization of TRPA1 currents, but a steady increase of TRPV1 currents. Reproduced with permission from Everaerts et al. [10]. μM (μmol/L), mM (mmol/L)

acidic preparations (mustard and vinaigrettes) [11]. Moreover, it has been described that heat also cross-sensitizes with AITC for TRPV1 activation [10, 12]. As mentioned above, allicin, the pungent compound in garlic and onion, has been described as an activator of both TRPA1 and TPRV1 [13, 14]. However, unlike for AITC, allicin does activate these channels via cysteine modification.

Piperine has been reported to activate TRPV1 but with less potency and greater efficacy than capsaicin [69]. The same study reported a stronger desensitizing effect of piperine, which has been proposed be due to a dephosphorylation of the channel [61]. Screening studies identified several compounds in pepper with activity on TRPV1. These include, in addition to piperine, isopiperine, isochavicine, piperanine, pipernonaline, dehydropipernonaline, retro C, piperolein A, and piperolein B [69]. They were reported to be more effective in causing an intracellular $Ca^{2+}$ increase in a TRPV1 heterologous expression system than capsaicin. These compounds were however not specific for TRPV1, as they also activate TRPA1.

Camphor is another described TRPV1 agonist [42]. This compound induces a rise of intracellular $Ca^{2+}$ concentration in capsaicin-responding neurons or heterologous expression systems [19, 41]. The exact mechanism for TRPV1 activation is not clear, but it is known that this effect requires sites different from those important for capsaicin and other vanilloid compounds [42]. Camphor desensitizes TRPV1 in a distinct manner than capsaicin. The desensitization is independent from extracellular $Ca^{2+}$ and might be partially explained through a block of the channel by the compound itself [42].

Several alkylamides in the pericarp of the dried Sichuan pepper have been attributed to the pungent nature of the spice [55]. Alpha-sanshool is one of these compounds described to induce a tingling and burning sensation, which was attributed to the activation of TPRV1 in sensory neurons and confirmed to activate heterologously expressed TRPV1 [36, 57]. The essential oil of Melegueta pepper (*Aframomum Melegueta*) contains the hydroxyarylalkanones 6-shogaol and 6-paradol in approximately equal concentrations. Using $Ca^{2+}$ imaging, it was confirmed that both compounds were able to activate capsaicin-responding DRG neurons [36]. In the same study, it was confirmed that transfection of TRPV1 channels was sufficient to produce a $Ca^{2+}$ influx upon challenge with 6-shogaol and 6-paradol, thus confirming their agonist properties for this channel. Using the TRPV1-C158A mutant, it was shown that activation by these compounds does not depend on the covalent binding of the cysteine known to be critical for activation by other electrophilic compounds such as allicin [14, 36].

Commonly found in ginger family plants, terpenoids with an $\alpha,\beta$-unsaturated 1,4-dialdehyde moiety, such as miogadial, miogatrial, and polygadial, activate rat TRPV1 in heterologous expression systems, and their effects are reversed by the TRPV1 blocker capsazepine [33]. The order of efficacy for increasing the intracellular $Ca^{2+}$ concentration in capsaicin-responding neurons was reported as follows: miogatrial > miogadial > polygadial [33].

[6]-Gingerol and its derivatives gingerone and olvanil are able to activate rTPRV1 and hTRPV1 [35]. When dehydrated, gingerol becomes a shogaol, also found in Melegueta peppers, and produces a stronger activation of TRPV1.

Eugenol is a chemical that can be found in clove and cinnamon leafs. This compound has been ascribed many properties, among which is the ability to act as an analgesic [74]. This can be explained by its activating effect on TRPV1 [75] and by its blocking effect on voltage-gated sodium channels [158]. The inward currents evoked by eugenol could be prevented by the TRPV1 antagonist capsazepine [75].

It has been proposed that a constitutively open channel variant of TRPV1 underlies the amiloride-insensitive component of salty taste [159, 160]. Behavioral experiments, comparing wild-type and TRPV1 KO mice, suggested that TRPV1 is involved in the avoidance of NaCl but also showed that TRPV1-independent mechanisms contribute to the amiloride-insensitive NaCl responses [161]. A more recent study concluded that TRPV1 may contribute to the perception of $Na^+$ but not via activation of chorda tympani nerves [162].

Interestingly, TRPV1 was shown to be activated by the noncaloric sweeteners aspartame, saccharin, acesulfame-K, and sodium cyclamate and by $Cu^{2+}$, $Mg^{2+}$, $Fe^{2+}$, and $Zn^{2+}$ sulfate salts. This leads to the hypothesis that TRPV1 may contribute to the mechanisms underlying the bitter/metallic taste of noncaloric sweeteners and to the metallic taste sensation triggered by sulfate salts [164]. However, the functional expression of TRPV1 in taste receptor cells remains a matter of debate [82], and the most plausible contribution of TRPV1 to flavor is through its activation in trigeminal nerve endings innervating the oral mucosa [164].

## 5 TRPM8: A Minty-Flavored Channel

TRPM8 channels were found to be expressed in fraction of neurons in trigeminal ganglia as well as in the dorsal root ganglia [163]. TRPM8 expression was associated with the Aδ or C fibers, known to convey the fast and slower pain responses [103, 165]. Moreover, this channel was reported to be expressed in lingual nerve fibers of the tongue [166]. Remarkably, TRPM8-expressing fibers were found adjacent to the taste buds but not to innervate these structures [166]. TRPM8 is considered as one of cold sensors [167, 168] and is activated by several compounds inducing cooling sensations and analgesia. These compounds include, as mentioned before, menthol (Fig. 5), thymol, camphor, or eucalyptol [150, 163].

Menthol, well-known for inducing a cooling sensation, is widely used as food additive (gums, candy), pharmaceuticals (creams, toothpastes), and household products. TRPM8 was reported to be directly activated by menthol with a reported $EC_{50}$ value of 66.7 μmol/L [51]. Single-channel currents induced by menthol are described to have a slope conductance of 83 pS and a strong outward rectification pattern [51]. Repeated TRPM8 stimulation with menthol induces channel desensitization and downregulation [169–173]. It was shown that $PIP_2$ maintains a high functional state of TRPM8 in sensory neurons [169, 171, 174]. After TRPM8 activation, influx of $Ca^{2+}$ via the channel causes depletion of $PIP_2$ that induces its desensitization [171, 172]. Subsequently, another protein, calmodulin, was suggested to play a role in TRPM8 acute desensitization by mediating a gate switch [171]. The binding sites for calmodulin are present in the N-terminus of TRPM8, which further confirms the $Ca^{2+}$-dependent regulation of this channel. It is also possible that in high-$Ca^{2+}$ conditions, direct binding of calmodulin induces

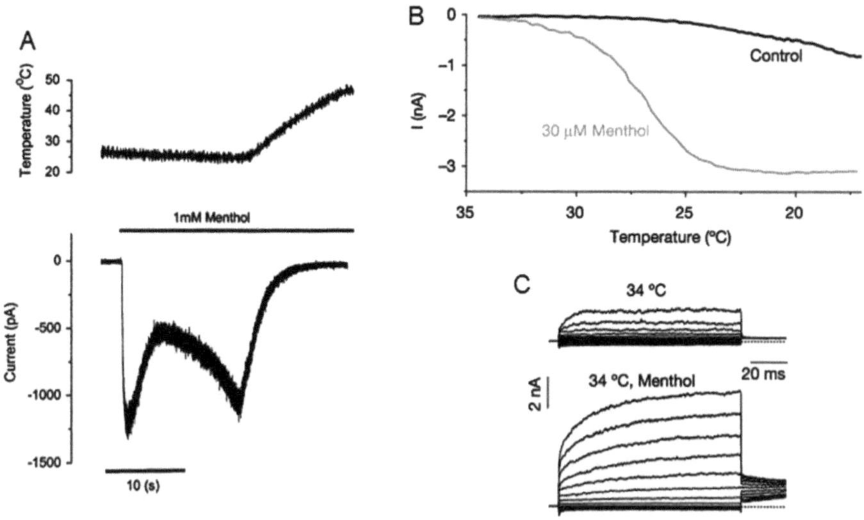


**Fig. 5** Activation of TRPM8 channels by menthol. (**a**) Effects of application of menthol on the amplitude of currents recorded at −60 mV in a Chinese hamster ovary cell transfected with TRPM8 channels. Note that heating inhibits the current. Reproduced with permission from Peier et al. [163]. (**b**) Temperature dependence of the amplitude of currents recorded in HEK293 cells transfected with TRPM8 in control and in the presence of 30 μmol/L (μM) menthol. Reproduced with permission from Voets et al. [150]. (**c**) TRPM8 currents elicited at 34°C in response to voltage steps in the range from −120 to +160 mV in the absence and presence of menthol. Reproduced with permission from Voets et al. [150]

channels' low functional state and acute desensitization. Unlike acute desensitization, tachyphylaxis was reported to be significantly decreased, when $PIP_2$ levels were drastically lowered; thus $PIP_2$ is necessary to keep TRPM8 at the high functional state [171]. In high-$Ca^{2+}$ conditions, $PIP_2$ is hydrolyzed via $Ca^{2+}$-dependent PLC, which in turn reduces the channel functional state inducing tachyphylaxis [171]. Summarizing, $PIP_2$ and calmodulin are important in the channel gating and induction of acute desensitization and tachyphylaxis. Also other proteins are essential in channel activity and include PLC, PKC, and protein phosphatases [171]. Therefore, TRPM8 activation and desensitization in sensory nerve endings is controlled by multiple intracellular signaling pathways.

Eucalyptol, also known as 1,8-cineole, is one of the components of the eucalyptus oil obtained from plants of Myrtaceae family such as *Eucalyptus polybractea* [76]. The eucalyptus oil is used as a pharmaceutical, flavoring agent, antiseptic, and repellent. The taste of the compound is described as initially spicy and pungent, then cooling with camphor-like fragrance [175]. Due to its properties, eucalyptol is used as a medication relieving symptoms of rhinosinusitis and bronchial asthma cough [76, 176]. Unusual sensations induced by this compound could be attributed to its action on several channels from TRP superfamily activating TRPM8 and TRPV3 and inhibiting TRPA1 [77]. Unexpectedly, the structural analog of eucalyptol 1,4-cinole, also present in eucalyptus oil, was able to activate both TRPM8 and TRPA1 channels [77]. Thus, the initial spicy and pungent taste of eucalyptus oil could be explained by the activating and inhibiting actions of these compounds on different TRP channels.

Referred to above as TRPV1 and TRPA1 agonist, eugenol was also reported to activate TRPM8 channels [20]. Similarly, another compound from the monoterpenoid family, menthone, present in oils of *Mentha arvensis* or *Pelargonium geraniums*, was described to activate TRPM8 [77]. But then again, activation induced by menthone was unspecific, as the compound could also activate TRPA1 [77]. Other compounds such as geraniol, linalool, thymol, and umbellulone activate TRPM8, but their potency is much lower than that of menthol [40, 44, 52, 54, 177, 178]. Finally, a study reporting the effects of 70 different odorants in HEK293 cells expressing mouse TRPM8 [52] showed that some commercially used chemicals such as *N*-ethyl-5-methyl-2-(1-methylethyl)cyclohexanecarboxamide, isopulegol, menthoxypropanediol, or *p*-menthane-3,8-diol have similar or higher potency than menthol for TRPM8 activation and induce even stronger cooling sensations [52].

# 6 Role of Other TRP Channels in Chemosensation

## *6.1 TRPC Channels*

Until now, not much is known about the involvement of channels from the canonical TRP subfamily in chemesthesis. What is uncommon for TRPC channels is their participation in slow, sustained elevations of intracellular $Ca^{2+}$ on the top of conventional ligand-based gating [179]. TRPC activation is associated with activity of store-operated $Ca^{2+}$ channels, while ligand-based activation involves phospholipase C signaling pathway engaging G-protein-coupled receptors [179, 180]. It was reported that several members of the TRPC family, TRPC1, TRPC3, and TRPC6, are expressed in DRG neurons [110, 181–184].

Hyperforin is an acylphloroglucinol derivative isolated from the *Hypricum* genus plants proposed as a TRPC6 activator [81]. Extracts from *H. perforatum* have slightly warming, bitter, sweet, and astringent taste. They are commonly used due to its anti-inflammatory, antimicrobial, and antidepressive properties [185]. Sensation caused by the plant extracts might not be attributed to hyperforin only, since numerous other compounds were found in *Hypericum* plants including naphthodianthrones such as hypericin or pseudohypericin, biflavones such as I3 or amentoflavone, and flavonoid glycosides such as rutin, hyperoside, quercitrin, and its derivatives [185]. Antidepressive properties of hyperforin and its derivative adhyperforin were shown to inhibit neuronal uptake of few neurotransmitters as

GABA or serotonin followed by the increase of their concentrations in the brain [186, 187]. This effect was correlated to the intracellular $Na^+$ increase induced by TRPC6 activation [188]. Roughly 5% of olfactory cells from the main olfactory epithelium are reported to express TRPC6 channels [189], but specific mechanisms of activation and the role of this channel in smell remain unknown.

TRPC2 is a pseudogene in humans [190], but it has been shown to be expressed in the vomeronasal organ (VNO) of rodents with its essential role in sensing pheromones as benzaldehyde, 2-tridecanone, dehydro-exo-brevicomin, and (*Z*)-7-dodecen-1-yl acetate. The main channel population is located in the sensory villi of VNO cells, which are highly specialized in recognition of a variety of chemicals [191, 192]. TRPC2 was shown to be activated by depletion of $Ca^{2+}$ stores in different cell types, but it is rather questionable if this mechanism would be responsible for channel activation, since it is located far from $Ca^{2+}$ stores [191, 192]. The activation mechanism for this channel remains unknown, but may include diacylglycerol-induced channel activation. Lack of TRPC2 was reported to dramatically reduce response to pheromones followed by behavioral changes. For instance TRPC2 knockout mice were not able to distinguish the sexual identity of their mates, and the aggression levels in males were significantly reduced [193, 194].

## 6.2 TRPV Channels

TRPV2 is a member of the TRPV family with a high sequence similarity to TRPV1 [195]. So far, only three plant-derived compounds have been shown to strongly activate TRPV2, namely, the cannabinoids obtained from *Cannabis sativa* Δ9-tetrahydrocannabinol, cannabidiol, and tetrahydrocannabivarin [25]. Nevertheless, cannabinoids are not selective agonists for TRPV2, as they also modulate other previously mentioned members of the TRP family [26].

TRPV3 is expressed in sensory neurons and keratinocytes and is typically known for its involvement in skin pathologies [196]. This channel is activated by multiple plant-derived compounds including monoterpenes, carvacrol, thymol, carveol, 6-*tert*-butyl-*m*-cresol, dihydrocarveol, menthol, camphor, borneol, eugenol, cresol, cinnamon, and thujone, as well as the acyclic monoterpenes linalool, geraniol, and propofol [22, 41, 42, 45, 53, 61].

TRPV4 has been shown to be expressed in skin keratinocytes, where it is proposed to play a role in barrier function [197]. Only one natural exogenous activator of TRPV4 has been described so far, bisandrographolide A (BAA), a compound extracted from *Andrographis paniculata* [79]. BAA was described as a potent and selective agonist of the TRPV4 channel, with an $EC_{50}$ of 800 nmol/L [79, 80].

## 6.3 TRPP Channels

The TRPP (PKD) subfamily of TRP channels includes two interesting protein groups, PKD1 and PKD2, which were indicated as possible chemosensors or/and mechanosensors [198, 199]. The discovery of PKD channels was related to search of gene mutations inducing autosomal dominant polycystic kidney disease [200–202]. Mutation of the genes encoding for PKD1 or PKD2 results in severe loss of $Ca^{2+}$ signaling in kidney epithelial cells leading to abnormal cell proliferation [203]. The PKD2-like subfamily includes channels such as PKD2 (TRPP2), TRPP3, and PKD2L2 (TRPP5) [199]. All of them share similar structural homology to other TRP channels [204, 205]. TRPP2 and TRPP3 are nonselective cation channels with permeability to $K^+$, $Na^+$, and $Ca^{2+}$ ions [206–208].

It has been also demonstrated that PKD2L1 and PKD1L3 are expressed in subset of taste receptor cell (TCR) distinct from once responsible for sensing sweet, bitter, and umami flavors [209]. PKD2L1 was mainly located in the tongue and palate taste buds, and PKD1L3 expression was restricted to the posterior tongue taste buds [209, 210]. They were proposed to play direct role in detection of acidic and salty compounds [210]. HEK cells co-transfected with PKD2L1 and PKD1L3 responded with an increase in intracellular $Ca^{2+}$ levels to different low-pH chemicals including citric, malic, acetic, succinic, hydrochloric, phosphoric, or sulfuric acids [210–213]. It must be noticed, however, that the role of PKD2L1/PKD1L3 channels as transducers of sour taste has been challenged by genetic evidence showing that mouse devoid of these channels show only mildly reduced or unchanged responses to acid [214, 215]. Furthermore, it has been argued that the effects of low pH on PKD2L1/PKD1L3 channels heterologously expressed in HEK293T cells do not correspond to the responses of taste cells responding to acid stimuli [82].

## 6.4 TRPM3 Channels

TRPM3 is expressed in the kidney, liver, ovaries, testes, vascular smooth muscle as well as in the nervous system including a subset of sensory neurons from trigeminal and dorsal root ganglia [182, 216, 217]. The *Trpm3* gene encodes several splice variants of the channel, which are characterized by different ligand sensitivities [218, 219]. The best described TRPM3 isoforms are TRPM3$\alpha$1 and TRPM3$\alpha$2 [219].

TRPM3 channels can be activated by several molecules such as nifedipine or pregnenolone sulfate (PS) [218, 220]. PS belongs to the class of neurosteroids and activates the channel in highly selective manner [220]. It was demonstrated that even small modifications in the compound structure significantly diminish or abolish interaction with the channel [220]. Conversely, other members of the TRP cation superfamily are not activated by PS [220]. Another group of compounds, including citrus fruit flavonoids such as naringen, hesperetin, and isosakuranetin, was demonstrated to interact with TRPM3 [221, 222]. The glycosylated flavonoid hesperidin is predominantly found in oranges and mandarins, and upon ingestion, this flavonoid is transformed into hesperetin [223]. Naringen, most commonly found in grapefruits and oranges, could be also found in low concentrations in tomatoes and tomato-based products [223]. Isosakuranetin is mostly found in blood oranges and grapefruits [221]. All these flavonoids were reported to have slight or strong bitter taste, and TRPM3 was suspected as molecular target for their action. However, it was shown that the taste related to these compounds is limited to the particular glycoside forms of the chemicals but absent in the related TRPM3-interacting aglycones forms, thus excluding TRPM3 involved in bitter taste recognition [222]. Furthermore these flavonoids were able to inhibit TRPM3 currents elicited by well-established TRPM3 activators in a heterologous expression system and in rat DRG neurons [222]. The reported $IC_{50}$ values were 0.5 μmol/L for naringen, 2 μmol/L for hesperetin, and 50 nmol/L for isosakuranetin after a current induction by 35 μmol/L of PS [222]. The TRPM3 function still in chemosensation remains very poorly defined.

## 7 Concluding Remarks

It is clear that the relevance of the TRP channel family for human chemosensation has been well established (Table 1). However, it is important to keep in mind that the description of the pharmacological properties of sensory TRP channels is far from being completed, as the effects of multiple relevant compounds remain unknown. It is also notable that the actions of TRP channel modulators are species dependent. Thus, ahead of us is the painstaking, but compulsory task of determining how much of what has been found in heterologous expression systems and in mouse trigeminal neurons can be translated to human chemosensation. Even more troubling is the fact that we do not know yet the patterns of functional expression of sensory TRP channels in the human oronasal mucosa. A number of laudable attempts have been made to determine the expression of TRP channels in human tissues using chemical agonists, but in the absence of a thorough pharmacological characterization of these compounds, inside and outside the TRP family, we are at the risk of extracting wrong or incomplete conclusions. Furthermore, the understanding of the mechanisms underlying the chemical activation of TRP channels and the relationship with other properties, such as voltage-dependent gating and thermo- and mechano-sensitivity, are yet to be fully elucidated. Also important properties such as sensitization, desensitization, pore dilation, modulation of ionic selectivity, bimodal effects, and channel interactions need to be considered when trying to understand how TRP channels influence the excitability and intracellular signaling in epithelial cells and sensory nerve endings. Thus, despite the massive advances obtained in recent years, the field of sensory TRP channels remains quite unexplored and open to many tasteful surprises.

**Table 1** Summary of the discussed compound activity on the TRP channels

| Origin | Compound name | Chemical structure | TRPA1 | TRPV1 | TRPM8 | Other TRP channels | Reference |
|---|---|---|---|---|---|---|---|
| Brassicaceae | Allyl isothiocyanate |  | ✓ | ✓ | | | [6–12] |
| | Benzyl isothiocyanate | 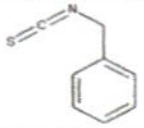 | ✓ | | | | |
| | Phenylethyl isothiocyanate | 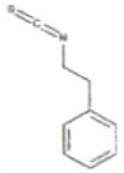 | ✓ | | | | |
| | Isopropyl isothiocyanate | 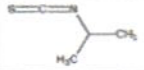 | ✓ | | | | |
| | Methyl isothiocyanate |  | ✓ | | | | |
| Allioideae | Allicin |  | ✓ | ✓ | | | [13–16] |
| | Diallyl sulfide |  | ✓ | | | | |
| | Diallyl disulfide |  | ✓ | | | | |
| | Diallyl trisulfide |  | ✓ | | | | |
| Apiaceae | Ligustilide | 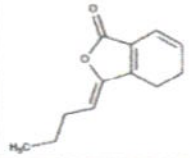 | ✓ | | | | [17, 18] |
| *Cinnamomum* | Cinnamaldehyde | 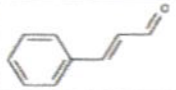 | ✓ | ✓ | | TRPV3 | [19–22] |
| | Cinnamyl alcohol | 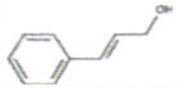 | ✓ | ✓ | | | |
| | Cinnamic acid | 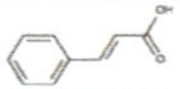 | ✓ | | | | |
| | Borneol | 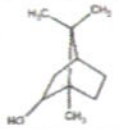 | ✓ | | | TRPV3 | |

| | | | | | | | |
|---|---|---|---|---|---|---|---|
| *Cannabis* | $\Delta^9$-Tetrahydrocannabinol | 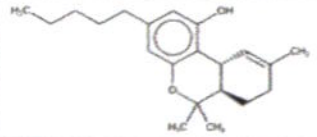 | ✓ | ✓ | | TRPV2 | [9, 23–30] |
| Zingiberaceae | Curcumin | 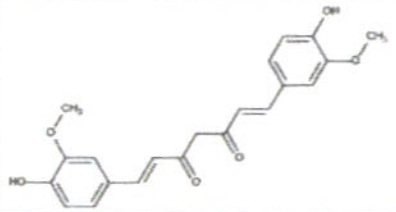 | ✓ | ✓ | | | [20, 31–35] |
| | [6]-Gingerol | 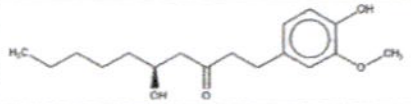 | ✓ | ✓ | | | |
| | Gingerone | 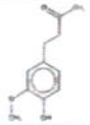 | ✓ | ✓ | | | |
| | Miogadial | 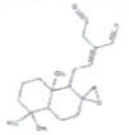 | ✓ | ✓ | | | |
| | Miogatrial | 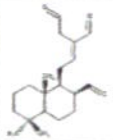 | ✓ | ✓ | | | |
| | Polygodial | 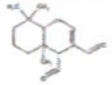 | ✓ | ✓ | | | |
| | 1’-Acetoxychavicol acetate | 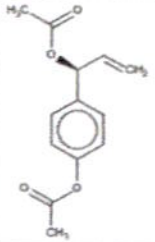 | ✓ | | | | |
| | 6-Shogaol | 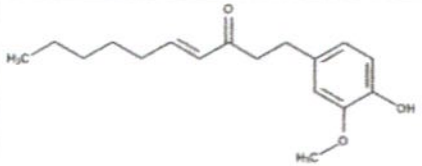 | ✓ | ✓ | | | [36, 37] |
| | 6-Paradol | 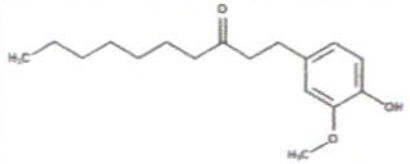 | ✓ | ✓ | | | |

(continued)

**Table 1** (continued)

| Origin | Compound name | Chemical structure | TRPA1 | TRPV1 | TRPM8 | Other TRP channels | Reference |
|---|---|---|---|---|---|---|---|
| Russulaceae | Isovelleral | 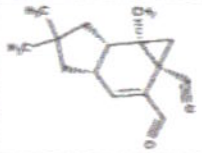 | ✓ | | | | [38] |
| Lauraceae | Umbellulone | 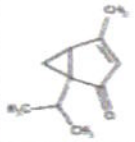 | ✓ | | ✓ | | [39, 40] |
| | Camphor |  | ✓ | ✓ | ✓ | TRPV3<br>TRPC5 | [19, 41–45] |
| Lamiaceae | Perillaldehyde | 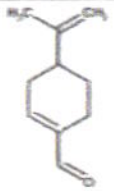 | ✓ | | | | [46, 47] |
| | Perilla ketone | 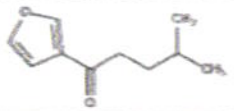 | ✓ | | | | |
| | Menthol | 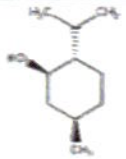 | ✓ | | ✓ | TRPV3 | [41, 48–52] |
| | Thymol | 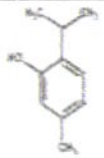 | ✓ | | | TRPV3 | [22, 41, 50, 53, 54] |
| | Carvacrol | 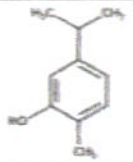 | ✓ | | | TRPV3 | [22, 53, 54] |
| Rutaceae | α-,β-,γ-,δ-Hydroxy-sanshools | 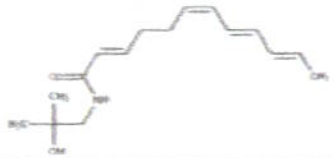 | ✓ | ✓ | | | [36, 52, 55–57] |
| | Linalool |  | ✓ | | ✓ | TRPV3 | |

| | | | | | | | |
|---|---|---|---|---|---|---|---|
| Solanaceae | Capsaicin | 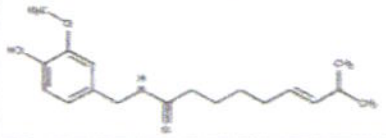 | | ✓ | | | [58–61] |
| | Capsiate | 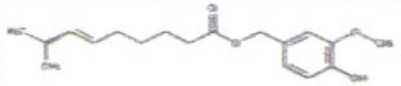 | ✓ | ✓ | | | [60, 62–65] |
| | Dihydrocapsiate | 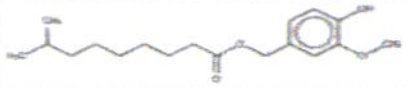 | ✓ | | | | |
| | Nordihydrocapsiate | 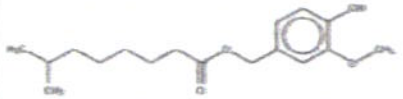 | ✓ | | | | |
| | Olvanil | 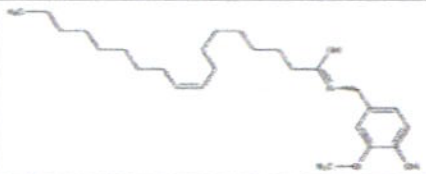 | ✓ | | | | [66] |
| | Nicotine | 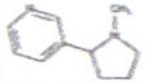 | ✓ | | | | [67] |
| Piperaceae | Piperine | 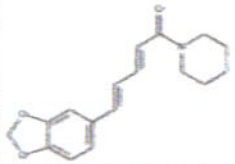 | ✓ | ✓ | | | [69–71] |
| | Piperolein A |  | ✓ | ✓ | | | |
| | Piperolein B |  | ✓ | ✓ | | | |
| | Pipernonaline |  | ✓ | ✓ | | | |
| | Isopiperine | 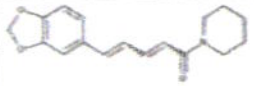 | ✓ | ✓ | | | |
| | Isochavicine | 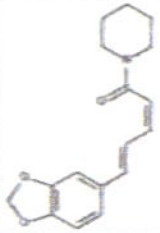 | ✓ | ✓ | | | |

(continued)

**Table 1** (continued)

| Origin | Compound name | Chemical structure | TRPA1 | TRPV1 | TRPM8 | Other TRP channels | Reference |
|---|---|---|---|---|---|---|---|
| | Dehydropipernonaline | 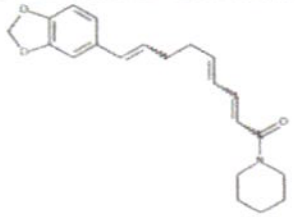 | ✓ | ✓ | | | |
| | Piperanine | 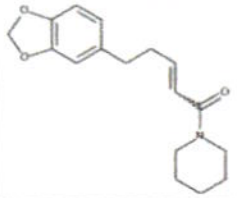 | ✓ | ✓ | | | |
| | Retrofractamide C | 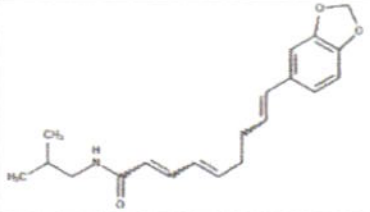 | | ✓ | | | |
| | *N*-Isobutyl-(2*E*,4*E*)-tetradeca-2,4-diamide (*N*-tetra) |  | ✓ | | | | |
| | Dehydropipernonaline | 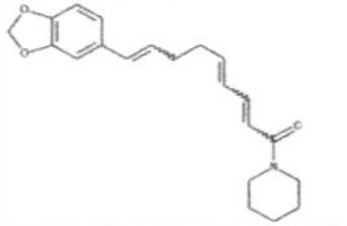 | ✓ | ✓ | | | |
| | Sabinene | 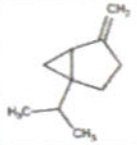 | ✓ | | | | |
| | Limonene | 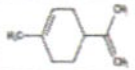 | ✓ | | | | |
| | α- and β-Pinene | 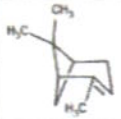 | ✓ | | | | |
| | Myrcene | 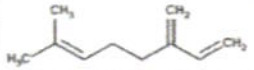 | ✓ | | | | |
| | Phellandrene | 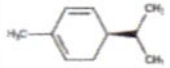 | ✓ | | | | |

| | | | | | | | |
|---|---|---|---|---|---|---|---|
| Asteraceae | Artepillin C (propolis) | 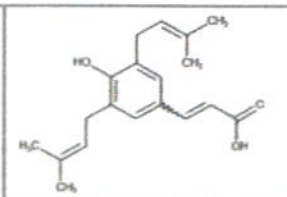 | ✓ | | | | [72] |
| Oleaceae | Oleocanthal | 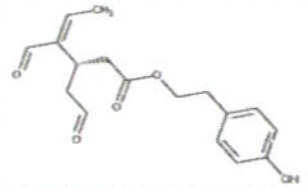 | ✓ | | | | [73] |
| Myrtaceae | Eugenol | 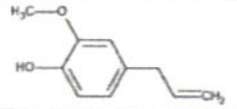 | ✓ | ✓ | | TRPV3 | [20, 74, 75] |
| | 1,8-Cineole | 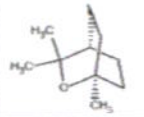 | | | ✓ | TRPV3 | [52, 76, 77] |
| Euphorbiaceae | Resiniferatoxin | 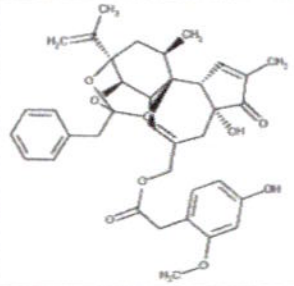 | | ✓ | | | [58, 78] |
| Acanthaceae | Bisandrographolide A | 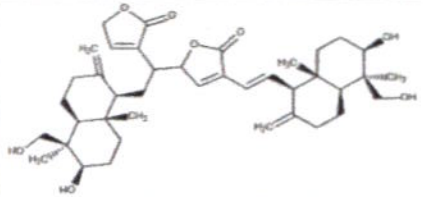 | | | | TRPV4 | [79, 80] |
| Hypericaceae | Hyperforin | 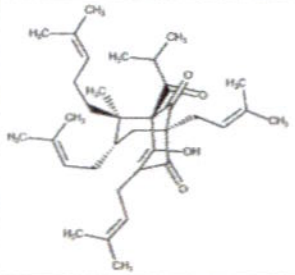 | | | | TRPC6 | [81] |
| Poaceae | Geraniol | 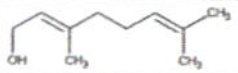 | | ✓ | ✓ | TRPV3 | n [52] |

Compounds activating TRPA1-blue tick, TRPV1-red tick, TRPM8-green tick

**Acknowledgments** This work was supported by grants from the Research Council of the KU Leuven (OT/12/091, GOA/14/011, EF/95/010, and PF-TRPLe) and the Flemish Research Foundation (FWO, G.0702.12, G.0896.12, G.0765.13, and G.0C77.15 N).

# References

1. Nilius B, Appendino G (2013) Spices: the savory and beneficial science of pungency. Rev Physiol Biochem Pharmacol 164:1–76
2. Venkatachalam K, Montell C (2007) TRP channels. Annu Rev Biochem 76:387–417
3. Nilius B, Owsianik G (2011) The transient receptor potential family of ion channels. Genome Biol 12(3):218
4. Pedersen SF, Owsianik G, Nilius B (2005) TRP channels: an overview. Cell Calcium 38 (3–4):233–252
5. Nilius B, Owsianik G, Voets T, Peters JA (2007) Transient receptor potential cation channels in disease. Physiol Rev 87(1):165–217
6. Hinman A, Chuang HH, Bautista DM, Julius D (2006) TRP channel activation by reversible covalent modification. Proc Natl Acad Sci U S A 103(51):19564–19568
7. Vriens J, Nilius B, Vennekens R (2008) Herbal compounds and toxins modulating TRP channels. Curr Neuropharmacol 6(1):79–96
8. Nagata K (2007) TRP channels as target sites for insecticides: physiology, pharmacology and toxicology. Invert Neurosci 7(1):31–37
9. Jordt SE, Bautista DM, Chuang HH, McKemy DD, Zygmunt PM, Högestätt ED, Meng ID, Julius D (2004) Mustard oils and cannabinoids excite sensory nerve fibres through the TRP channel ANKTM1. Nature 427(6971):260–265
10. Everaerts W, Gees M, Alpizar YA, Farre R, Leten C, Apetrei A, Dewachter I, van Leuven F, Vennekens R, De Ridder D, Nilius B, Voets T, Talavera K (2011) The capsaicin receptor TRPV1 is a crucial mediator of the noxious effects of mustard oil. Curr Biol 21(4):316–321
11. Gees M, Alpizar YA, Boonen B, Sanchez A, Everaerts W, Segal A, Xue F, Janssens A, Owsianik G, Nilius B, Voets T, Talavera K (2013) Mechanisms of transient receptor potential vanilloid 1 activation and sensitization by allyl isothiocyanate. Mol Pharmacol 84 (3):325–334
12. Alpizar YA, Boonen B, Gees M, Sanchez A, Nilius B, Voets T, Talavera K (2014) Allyl isothiocyanate sensitizes TRPV1 to heat stimulation. Pflugers Arch 466(3):507–515
13. Macpherson LJ, Geierstanger BH, Viswanath V, Bandell M, Eid SR, Hwang S, Patapoutian A (2005) The pungency of garlic: activation of TRPA1 and TRPV1 in response to allicin. Curr Biol 15(10):929–934
14. Salazar H, Llorente I, Jara-Oseguera A, García-Villegas R, Munari M, Gordon SE, Islas LD, Rosenbaum T (2008) A single N-terminal cysteine in TRPV1 determines activation by pungent compounds from onion and garlic. Nat Neurosci 11(3):255–261
15. Bautista DM, Movahed P, Hinman A, Axelsson HE, Sterner O, Högestätt ED, Julius D, Jordt SE, Zygmunt PM (2005) Pungent products from garlic activate the sensory ion channel TRPA1. Proc Natl Acad Sci U S A 102(34):12248–12252
16. Koizumi K, Iwasaki Y, Narukawa M, Iitsuka Y, Fukao T, Seki T, Ariga T, Watanabe T (2009) Diallyl sulfides in garlic activate both TRPA1 and TRPV1. Biochem Biophys Res Commun 382(3):545–548
17. Feng Z, Lu Y, Wu X, Zhao P, Li J, Peng B, Qian Z, Zhu L (2012) Ligustilide alleviates brain damage and improves cognitive function in rats of chronic cerebral hypoperfusion. J Ethnopharmacol 144(2):313–321
18. Zhong J, Pollastro F, Prenen J, Zhu Z, Appendino G, Nilius B (2011) Ligustilide: a novel TRPA1 modulator. Pflugers Arch 462(6):841–849

19. Alpizar YA, Gees M, Sanchez A, Apetrei A, Voets T, Nilius B, Talavera K (2013) Bimodal effects of cinnamaldehyde and camphor on mouse TRPA1. Pflugers Arch 465(6):853–864
20. Bandell M, Story GM, Hwang SW, Viswanath V, Eid SR, Petrus MJ, Earley TJ, Patapoutian A (2004) Noxious cold ion channel TRPA1 is activated by pungent compounds and bradykinin. Neuron 41(6):849–857
21. Sadofsky LR, Boa AN, Maher SA, Birrell MA, Belvisi MG, Morice AH (2011) TRPA1 is activated by direct addition of cysteine residues to the N-hydroxysuccinyl esters of acrylic and cinnamic acids. Pharmacol Res 63(1):30–36
22. Vogt-Eisele AK, Weber K, Sherkheli MA, Vielhaber G, Panten J, Gisselmann G, Hatt H (2007) Monoterpenoid agonists of TRPV3. Br J Pharmacol 151(4):530–540
23. Malik Z, Baik D, Schey R (2015) The role of cannabinoids in regulation of nausea and vomiting, and visceral pain. Curr Gastroenterol Rep 17(2):429
24. De Petrocellis L, Vellani V, Schiano-Moriello A, Marini P, Magherini PC, Orlando P, Di Marzo V (2008) Plant-derived cannabinoids modulate the activity of transient receptor potential channels of ankyrin type-1 and melastatin type-8. J Pharmacol Exp Ther 325 (3):1007–1015
25. Qin N, Neeper MP, Liu Y, Hutchinson TL, Lubin ML, Flores CM (2008) TRPV2 is activated by cannabidiol and mediates CGRP release in cultured rat dorsal root ganglion neurons. J Neurosci 28(24):6231–6238
26. De Petrocellis L, Orlando P, Moriello AS, Aviello G, Stott C, Izzo AA, Di Marzo V (2012) Cannabinoid actions at TRPV channels: effects on TRPV3 and TRPV4 and their potential relevance to gastrointestinal inflammation. Acta Physiol (Oxf) 204(2):255–266
27. Ahluwalia J, Urban L, Capogna M, Bevan S, Nagy I (2000) Cannabinoid 1 receptors are expressed in nociceptive primary sensory neurons. Neuroscience 100(4):685–688
28. Izzo AA, Sharkey KA (2010) Cannabinoids and the gut: new developments and emerging concepts. Pharmacol Ther 126(1):21–38
29. Munro S, Thomas KL, Abu-Shaar M (1993) Molecular characterization of a peripheral receptor for cannabinoids. Nature 365(6441):61–65
30. Schatz AR, Lee M, Condie RB, Pulaski JT, Kaminski NE (1997) Cannabinoid receptors CB1 and CB2: a characterization of expression and adenylate cyclase modulation within the immune system. Toxicol Appl Pharmacol 142(2):278–287
31. Leamy AW, Shukla P, McAlexander MA, Carr MJ, Ghatta S (2011) Curcumin ((*E*,*E*)-1,7-bis (4-hydroxy-3-methoxyphenyl)-1,6-heptadiene-3,5-dione) activates and desensitizes the nociceptor ion channel TRPA1. Neurosci Lett 503(3):157–162
32. Premkumar LS (2014) Transient receptor potential channels as targets for phytochemicals. ACS Chem Neurosci 5(11):1117–1130
33. Iwasaki Y, Tanabe M, Kayama Y, Abe M, Kashio M, Koizumi K, Okumura Y, Morimitsu Y, Tominaga M, Ozawa Y, Watanabe T (2009) Miogadial and miogatrial with alpha, beta-unsaturated 1,4-dialdehyde moieties – novel and potent TRPA1 agonists. Life Sci 85 (1–2):60–69
34. Narukawa M, Koizumi K, Iwasaki Y, Kubota K, Watanabe T (2010) Galangal pungent component, 1'-acetoxychavicol acetate, activates TRPA1. Biosci Biotechnol Biochem 74 (8):1694–1696
35. Witte DG, Cassar SC, Masters JN, Esbenshade T, Hancock AA (2002) Use of a fluorescent imaging plate reader-based calcium assay to assess pharmacological differences between the human and rat vanilloid receptor. J Biomol Screen 7(5):466–475
36. Riera CE, Menozzi-Smarrito C, Affolter M, Michlig S, Munari C, Robert F, Vogel H, Simon SA, le Coutre J (2009) Compounds from Sichuan and Melegueta peppers activate, covalently and non-covalently, TRPA1 and TRPV1 channels. Br J Pharmacol 157(8):1398–1409
37. Weng CJ, Wu CF, Huang HW, Ho CT, Yen GC (2010) Anti-invasion effects of 6-shogaol and 6-gingerol, two active components in ginger, on human hepatocarcinoma cells. Mol Nutr Food Res 54(11):1618–1627

38. Escalera J, von Hehn CA, Bessac BF, Sivula M, Jordt SE (2008) TRPA1 mediates the noxious effects of natural sesquiterpene deterrents. J Biol Chem 283(35):24136–24144
39. Nassini R, Materazzi S, Vriens J, Prenen J, Benemei S, De Siena G, la Marca G, Andrè E, Preti D, Avonto C, Sadofsky L, Di Marzo V, De Petrocellis L, Dussor G, Porreca F, Taglialatela-Scafati O, Appendino G, Nilius B, Geppetti P (2012) The 'headache tree' via umbellulone and TRPA1 activates the trigeminovascular system. Brain 135(Pt 2):376–390
40. Zhong J, Minassi A, Prenen J, Taglialatela-Scafati O, Appendino G, Nilius B (2011) Umbellulone modulates TRP channels. Pflugers Arch 462(6):861–870
41. Macpherson LJ, Hwang SW, Miyamoto T, Dubin AE, Patapoutian A, Story GM (2006) More than cool: promiscuous relationships of menthol and other sensory compounds. Mol Cell Neurosci 32(4):335–343
42. Xu H, Blair NT, Clapham DE (2005) Camphor activates and strongly desensitizes the transient receptor potential vanilloid subtype 1 channel in a vanilloid-independent mechanism. J Neurosci 25(39):8924–8937
43. Takaishi M, Uchida K, Fujita F, Tominaga M (2014) Inhibitory effects of monoterpenes on human TRPA1 and the structural basis of their activity. J Physiol Sci 64(1):47–57
44. Selescu T, Ciobanu AC, Dobre C, Reid G, Babes A (2013) Camphor activates and sensitizes transient receptor potential melastatin 8 (TRPM8) to cooling and icilin. Chem Senses 38 (7):563–575
45. Moqrich A, Hwang SW, Earley TJ, Petrus MJ, Murray AN, Spencer KS, Andahazy M, Story GM, Patapoutian A (2005) Impaired thermosensation in mice lacking TRPV3, a heat and camphor sensor in the skin. Science 307(5714):1468–1472
46. Bassoli A, Borgonovo G, Caimi S, Scaglioni L, Morini G, Moriello AS, Di Marzo V, De Petrocellis L (2009) Taste-guided identification of high potency TRPA1 agonists from Perilla frutescens. Bioorg Med Chem 17(4):1636–1639
47. Bassoli A, Borgonovo G, Morini G, De Petrocellis L, Schiano Moriello A, Di Marzo V (2013) Analogues of perillaketone as highly potent agonists of TRPA1 channel. Food Chem 141 (3):2044–2051
48. Green BG (1985) Menthol modulates oral sensations of warmth and cold. Physiol Behav 35 (3):427–434
49. Xiao B, Dubin AE, Bursulaya B, Viswanath V, Jegla TJ, Patapoutian A (2008) Identification of transmembrane domain 5 as a critical molecular determinant of menthol sensitivity in mammalian TRPA1 channels. J Neurosci 28(39):9640–9651
50. Karashima Y, Damann N, Prenen J, Talavera K, Segal A, Voets T, Nilius B (2007) Bimodal action of menthol on the transient receptor potential channel TRPA1. J Neurosci 27 (37):9874–9884
51. McKemy DD, Neuhausser WM, Julius D (2002) Identification of a cold receptor reveals a general role for TRP channels in thermosensation. Nature 416(6876):52–58
52. Behrendt HJ, Germann T, Gillen C, Hatt H, Jostock R (2004) Characterization of the mouse cold-menthol receptor TRPM8 and vanilloid receptor type-1 VR1 using a fluorometric imaging plate reader (FLIPR) assay. Br J Pharmacol 141(4):737–745
53. Xu H, Delling M, Jun JC, Clapham DE (2006) Oregano, thyme and clove-derived flavors and skin sensitizers activate specific TRP channels. Nat Neurosci 9(5):628–635
54. Lee SP, Buber MT, Yang Q, Cerne R, Cortés RY, Sprous DG, Bryant RW (2008) Thymol and related alkyl phenols activate the hTRPA1 channel. Br J Pharmacol 153(8):1739–1749
55. Sugai E, Morimitsu Y, Kubota K (2005) Quantitative analysis of sanshool compounds in Japanese pepper (Xanthoxylum piperitum DC.) and their pungent characteristics. Biosci Biotechnol Biochem 69(10):1958–1962
56. Bryant BP, Mezine I (1999) Alkylamides that produce tingling paresthesia activate tactile and thermal trigeminal neurons. Brain Res 842(2):452–460
57. Koo JY, Jang Y, Cho H, Lee CH, Jang KH, Chang YH, Shin J, Oh U (2007) Hydroxy-alpha-sanshool activates TRPV1 and TRPA1 in sensory neurons. Eur J Neurosci 26(5):1139–1147

58. Caterina MJ, Schumacher MA, Tominaga M, Rosen TA, Levine JD, Julius D (1997) The capsaicin receptor: a heat-activated ion channel in the pain pathway. Nature 389 (6653):816–824
59. Watanabe M, Ueda T, Shibata Y, Kumamoto N, Ugawa S (2015) The role of TRPV1 channels in carrageenan-induced mechanical hyperalgesia in mice. Neuroreport 26 (3):173–178
60. Caterina MJ, Leffler A, Malmberg AB, Martin WJ, Trafton J, Petersen-Zeitz KR, Koltzenburg M, Basbaum AI, Julius D (2000) Impaired nociception and pain sensation in mice lacking the capsaicin receptor. Science 288(5464):306–313
61. Vriens J, Appendino G, Nilius B (2009) Pharmacology of vanilloid transient receptor potential cation channels. Mol Pharmacol 75(6):1262–1279
62. Iida T, Moriyama T, Kobata K, Morita A, Murayama N, Hashizume S, Fushiki T, Yazawa S, Watanabe T, Tominaga M (2003) TRPV1 activation and induction of nociceptive response by a non-pungent capsaicin-like compound, capsiate. Neuropharmacology 44(7):958–967
63. Shintaku K, Uchida K, Suzuki Y, Zhou Y, Fushiki T, Watanabe T, Yazawa S, Tominaga M (2012) Activation of transient receptor potential A1 by a non-pungent capsaicin-like compound, capsiate. Br J Pharmacol 165(5):1476–1486
64. Yang L, Fujita T, Jiang CY, Piao LH, Yue HY, Mizuta K, Kumamoto E (2011) TRPV1 agonist piperine but not olvanil enhances glutamatergic spontaneous excitatory transmission in rat spinal substantia gelatinosa neurons. Biochem Biophys Res Commun 410(4):841–845
65. Del Prete D, Caprioglio D, Appendino G, Minassi A, Schiano-Moriello A, Di Marzo V, De Petrocellis L (2015) Discovery of non-electrophilic capsaicinoid-type TRPA1 ligands. Bioorg Med Chem Lett 25(5):1015–1017
66. Lee SP, Buber MT, Yang Q, Cerne R, Cortés RY, Sprous DG, Bryant RW (2008) Thymol and related alkyl phenols activate the hTRPA1 channel. Br J Pharmacol 153(8):1739–1749
67. Talavera K, Gees M, Karashima Y, Meseguer VM, Vanoirbeek JA, Damann N, Everaerts W, Benoit M, Janssens A, Vennekens R, Viana F, Nemery B, Nilius B, Voets T (2009) Nicotine activates the chemosensory cation channel TRPA1. Nat Neurosci 12(10):1293–1299
68. Andrè E, Campi B, Materazzi S, Trevisani M, Amadesi S, Massi D, Creminon C, Vaksman N, Nassini R, Civelli M, Baraldi PG, Poole DP, Bunnett NW, Geppetti P, Patacchini R (2008) Cigarette smoke-induced neurogenic inflammation is mediated by alpha, beta-unsaturated aldehydes and the TRPA1 receptor in rodents. J Clin Invest 118(7):2574–2582
69. Okumura Y, Narukawa M, Iwasaki Y, Ishikawa A, Matsuda H, Yoshikawa M, Watanabe T (2010) Activation of TRPV1 and TRPA1 by black pepper components. Biosci Biotechnol Biochem 74(5):1068–1072
70. Jirovetz L, Buchbauer G, Ngassoum MB, Geissler M (2002) Aroma compound analysis of piper nigrum and piper guineense essential oils from Cameroon using solid-phase microextraction-gas chromatography, solid-phase microextraction-gas chromatography–mass spectrometry and olfactometry. J Chromatogr A 976(1–2):265–275
71. McNamara FN, Randall A, Gunthorpe MJ (2005) Effects of piperine, the pungent component of black pepper, at the human vanilloid receptor (TRPV1). Br J Pharmacol 144(6):781–790
72. Hata T, Tazawa S, Ohta S, Rhyu MR, Misaka T, Ichihara K (2012) Artepillin C, a major ingredient of Brazilian propolis, induces a pungent taste by activating TRPA1 channels. PLoS One 7(11):e48072
73. Peyrot des Gachons C, Uchida K, Bryant B, Shima A, Sperry JB, Dankulich-Nagrudny L, Tominaga M, Smith AB, Beauchamp GK, Breslin PA (2011) Unusual pungency from extra-virgin olive oil is attributable to restricted spatial expression of the receptor of oleocanthal. J Neurosci 31(3):999–1009
74. Ohkubo T, Shibata M (1997) The selective capsaicin antagonist capsazepine abolishes the antinociceptive action of eugenol and guaiacol. J Dent Res 76(4):848–851
75. Yang BH, Piao ZG, Kim YB, Lee CH, Lee JK, Park K, Kim JS, Oh SB (2003) Activation of vanilloid receptor 1 (VR1) by eugenol. J Dent Res 82(10):781–785

76. Juergens UR, Dethlefsen U, Steinkamp G, Gillissen A, Repges R, Vetter H (2003) Anti-inflammatory activity of 1.8-cineol (eucalyptol) in bronchial asthma: a double-blind placebo-controlled trial. Respir Med 97(3):250–256
77. Takaishi M, Fujita F, Uchida K, Yamamoto S, Ms C, Shimizu M, Tominaga M (2012) 1,8-cineole, a TRPM8 agonist, is a novel natural antagonist of human TRPA1. Mol Pain 8:86
78. Maggi CA, Patacchini R, Tramontana M, Amann R, Giuliani S, Santicioli P (1990) Similarities and differences in the action of resiniferatoxin and capsaicin on central and peripheral endings of primary sensory neurons. Neuroscience 37(2):531–539
79. Smith PL, Maloney KN, Pothen RG, Clardy J, Clapham DE (2006) Bisandrographolide from *Andrographis paniculata* activates TRPV4 channels. J Biol Chem 281(40):29897–29904
80. Vriens J, Owsianik G, Janssens A, Voets T, Nilius B (2007) Determinants of 4 alpha-phorbol sensitivity in transmembrane domains 3 and 4 of the cation channel TRPV4. J Biol Chem 282 (17):12796–12803
81. Leuner K, Heiser JH, Derksen S, Mladenov MI, Fehske CJ, Schubert R, Gollasch M, Schneider G, Harteneck C, Chatterjee SS, Müller WE (2010) Simple 2,4-diacylphloroglucinols as classic transient receptor potential-6 activators – identification of a novel pharmacophore. Mol Pharmacol 77(3):368–377
82. Roper SD (2014) TRPs in taste and chemesthesis. Handb Exp Pharmacol 223:827–871
83. Zufall F (2014) TRPs in olfaction. Handb Exp Pharmacol 223:917–933
84. Nilius B, Prenen J, Tang J, Wang C, Owsianik G, Janssens A, Voets T, Zhu MX (2005) Regulation of the $Ca^{2+}$ sensitivity of the nonselective cation channel TRPM4. J Biol Chem 280(8):6423–6433
85. Nilius B, Prenen J, Janssens A, Owsianik G, Wang C, Zhu MX, Voets T (2005) The selectivity filter of the cation channel TRPM4. J Biol Chem 280(24):22899–22906
86. Montell C, Birnbaumer L, Flockerzi V (2002) The TRP channels, a remarkably functional family. Cell 108(5):595–598
87. Tóth B, Iordanov I, Csanády L (2014) Putative chanzyme activity of TRPM2 cation channel is unrelated to pore gating. Proc Natl Acad Sci U S A 111(47):16949–16954
88. Jabba S, Goyal R, Sosa-Pagán JO, Moldenhauer H, Wu J, Kalmeta B, Bandell M, Latorre R, Patapoutian A, Grandl J (2014) Directionality of temperature activation in mouse TRPA1 ion channel can be inverted by single-point mutations in ankyrin repeat six. Neuron 82 (5):1017–1031
89. Li J, Mahajan A, Tsai MD (2006) Ankyrin repeat: a unique motif mediating protein-protein interactions. Biochemistry 45(51):15168–15178
90. Sedgwick SG, Smerdon SJ (1999) The ankyrin repeat: a diversity of interactions on a common structural framework. Trends Biochem Sci 24(8):311–316
91. Mosavi LK, Cammett TJ, Desrosiers DC, Peng ZY (2004) The ankyrin repeat as molecular architecture for protein recognition. Protein Sci 13(6):1435–1448
92. Jernigan KK, Bordenstein SR (2014) Ankyrin domains across the tree of life. PeerJ 2:e264
93. Gaudet R (2008) A primer on ankyrin repeat function in TRP channels and beyond. Mol Biosyst 4(5):372–379
94. Gaudet R (2009) Divide and conquer: high resolution structural information on TRP channel fragments. J Gen Physiol 133(3):231–237
95. Huynh KW, Cohen MR, Chakrapani S, Holdaway HA, Stewart PL, Moiseenkova-Bell VY (2014) Structural insight into the assembly of TRPV channels. Structure 22(2):260–268
96. Jin X, Touhey J, Gaudet R (2006) Structure of the N-terminal ankyrin repeat domain of the TRPV2 ion channel. J Biol Chem 281(35):25006–25010
97. Clapham DE (2003) TRP channels as cellular sensors. Nature 426(6966):517–524
98. Ramsey IS, Delling M, Clapham DE (2006) An introduction to TRP channels. Annu Rev Physiol 68:619–647
99. Montell C (2005) The TRP superfamily of cation channels. Sci STKE 2005(272):re3
100. Tsuruda PR, Julius D, Minor DL (2006) Coiled coils direct assembly of a cold-activated TRP channel. Neuron 51(2):201–212

101. Fujiwara Y, Minor DL (2008) X-ray crystal structure of a TRPM assembly domain reveals an antiparallel four-stranded coiled-coil. J Mol Biol 383(4):854–870
102. García-Sanz N, Fernández-Carvajal A, Morenilla-Palao C, Planells-Cases R, Fajardo-Sánchez E, Fernández-Ballester G, Ferrer-Montiel A (2004) Identification of a tetramerization domain in the C terminus of the vanilloid receptor. J Neurosci 24 (23):5307–5314
103. Kobayashi K, Fukuoka T, Obata K, Yamanaka H, Dai Y, Tokunaga A, Noguchi K (2005) Distinct expression of TRPM8, TRPA1, and TRPV1 mRNAs in rat primary afferent neurons with adelta/c-fibers and colocalization with trk receptors. J Comp Neurol 493(4):596–606
104. Ji G, Zhou S, Carlton SM (2008) Intact Adelta-fibers up-regulate transient receptor potential A1 and contribute to cold hypersensitivity in neuropathic rats. Neuroscience 154 (3):1054–1066
105. Anand U, Otto WR, Facer P, Zebda N, Selmer I, Gunthorpe MJ, Chessell IP, Sinisi M, Birch R, Anand P (2008) TRPA1 receptor localisation in the human peripheral nervous system and functional studies in cultured human and rat sensory neurons. Neurosci Lett 438 (2):221–227
106. Atoyan R, Shander D, Botchkareva NV (2009) Non-neuronal expression of transient receptor potential type A1 (TRPA1) in human skin. J Invest Dermatol 129(9):2312–2315
107. Nassini R, Pedretti P, Moretto N, Fusi C, Carnini C, Facchinetti F, Viscomi AR, Pisano AR, Stokesberry S, Brunmark C, Svitacheva N, McGarvey L, Patacchini R, Damholt AB, Geppetti P, Materazzi S (2012) Transient receptor potential ankyrin 1 channel localized to non-neuronal airway cells promotes non-neurogenic inflammation. PLoS One 7(8):e42454
108. Macpherson LJ, Dubin AE, Evans MJ, Marr F, Schultz PG, Cravatt BF, Patapoutian A (2007) Noxious compounds activate TRPA1 ion channels through covalent modification of cysteines. Nature 445(7127):541–545
109. Wallock-Richards D, Doherty CJ, Doherty L, Clarke DJ, Place M, Govan JR, Campopiano DJ (2014) Garlic revisited: antimicrobial activity of Allicin-containing garlic extracts against *Burkholderia cepacia* complex. PLoS One 9(12):e112726
110. Meotti FC, Lemos de Andrade E, Calixto JB (2014) TRP modulation by natural compounds. Handb Exp Pharmacol 223:1177–1238
111. Ranasinghe P, Pigera S, Premakumara GA, Galappaththy P, Constantine GR, Katulanda P (2013) Medicinal properties of 'true' cinnamon (Cinnamomum zeylanicum): a systematic review. BMC Complement Altern Med 13:275
112. Rao PV, Gan SH (2014) Cinnamon: a multifaceted medicinal plant. Evid Based Complement Alternat Med 2014:642942
113. Hirst RA, Lambert DG, Notcutt WG (1998) Pharmacology and potential therapeutic uses of cannabis. Br J Anaesth 81(1):77–84
114. Ursu D, Knopp K, Beattie RE, Liu B, Sher E (2010) Pungency of TRPV1 agonists is directly correlated with kinetics of receptor activation and lipophilicity. Eur J Pharmacol 641 (2–3):114–122
115. Morita A, Iwasaki Y, Kobata K, Iida T, Higashi T, Oda K, Suzuki A, Narukawa M, Sasakuma S, Yokogoshi H, Yazawa S, Tominaga M, Watanabe T (2006) Lipophilicity of capsaicinoids and capsinoids influences the multiple activation process of rat TRPV1. Life Sci 79(24):2303–2310
116. Camazine SM, Resch JF, Eisner T, Meinwald J (1983) Mushroom chemical defense: Pungent sesquiterpenoid dialdehyde antifeedant to opossum. J Chem Ecol 9(10):1439–1447
117. Aujard I, Röme D, Arzel E, Johansson M, de Vos D, Sterner O (2005) Tridemethylisovelleral, a potent cytotoxic agent. Bioorg Med Chem 13(22):6145–6150
118. Jonassohn M, Anke H, Morales P, Sterner O (1995) Structure-activity relationships for unsaturated dialdehydes. 10. The generation of bioactive products by autoxidation of isovelleral and merulidial. Acta Chem Scand 49(7):530–535
119. Feld H, Hertewich UM, Zapp J, Becker H (2005) Sacculatane diterpenoids from axenic cultures of the liverwort *Fossombronia wondraczekii*. Phytochemistry 66(10):1094–1099

120. Starkenmann C, Cayeux I, Birkbeck AA (2011) Exploring natural products for new taste sensations. Chimia (Aarau) 65(6):407–410
121. Baraldi PG, Preti D, Materazzi S, Geppetti P (2010) Transient receptor potential ankyrin 1 (TRPA1) channel as emerging target for novel analgesics and anti-inflammatory agents. J Med Chem 53(14):5085–5107
122. Kasting GB, Francis WR, Bowman LA, Kinnett GO (1997) Percutaneous absorption of vanilloids: in vivo and in vitro studies. J Pharm Sci 86(1):142–146
123. Hill K, Schaefer M (2007) TRPA1 is differentially modulated by the amphipathic molecules trinitrophenol and chlorpromazine. J Biol Chem 282(10):7145–7153
124. Kashiwayanagi M, Suenaga A, Enomoto S, Kurihara K (1990) Membrane fluidity changes of liposomes in response to various odorants. Complexity of membrane composition and variety of adsorption sites for odorants. Biophys J 58(4):887–895
125. Green BG (1990) Sensory characteristics of camphor. J Invest Dermatol 94(5):662–666
126. Gavliakova S, Biringerova Z, Buday T, Brozmanova M, Calkovsky V, Poliacek I, Plevkova J (2013) Antitussive effects of nasal thymol challenges in healthy volunteers. Respir Physiol Neurobiol 187(1):104–107
127. Mohammadi B, Haeseler G, Leuwer M, Dengler R, Krampfl K, Bufler J (2001) Structural requirements of phenol derivatives for direct activation of chloride currents via GABA (A) receptors. Eur J Pharmacol 421(2):85–91
128. Alimohammadi H, Silver WL (2000) Evidence for nicotinic acetylcholine receptors on nasal trigeminal nerve endings of the rat. Chem Senses 25(1):61–66
129. Hatsukami DK, Stead LF, Gupta PC (2008) Tobacco addiction. Lancet 371 (9629):2027–2038
130. Fucile S, Sucapane A, Eusebi F (2005) $Ca^{2+}$ permeability of nicotinic acetylcholine receptors from rat dorsal root ganglion neurones. J Physiol 565(Pt 1):219–228
131. Jacob P, Hatsukami D, Severson H, Hall S, Yu L, Benowitz NL (2002) Anabasine and anatabine as biomarkers for tobacco use during nicotine replacement therapy. Cancer Epidemiol Biomarkers Prev 11(12):1668–1673
132. Talhout R, Schulz T, Florek E, van Benthem J, Wester P, Opperhuizen A (2011) Hazardous compounds in tobacco smoke. Int J Environ Res Public Health 8(2):613–628
133. Wood C, Siebert TE, Parker M, Capone DL, Elsey GM, Pollnitz AP, Eggers M, Meier M, Vössing T, Widder S, Krammer G, Sefton MA, Herderich MJ (2008) From wine to pepper: rotundone, an obscure sesquiterpene, is a potent spicy aroma compound. J Agric Food Chem 56(10):3738–3744
134. Jagella T, Grosch W (1999) Flavour and off-flavour compounds of black and white pepper (piper nigrum L.) III. Desirable and undesirable odorants of white pepper. Eur Food Res Technol 209(1):27–31
135. Nakamura R, Watanabe K, Oka K, Ohta S, Mishima S, Teshima R (2010) Effects of propolis from different areas on mast cell degranulation and identification of the effective components in propolis. Int Immunopharmacol 10(9):1107–1112
136. Inui S, Hatano A, Yoshino M, Hosoya T, Shimamura Y, Masuda S, Ahn MR, Tazawa S, Araki Y, Kumazawa S (2014) Identification of the phenolic compounds contributing to antibacterial activity in ethanol extracts of Brazilian red propolis. Nat Prod Res 28 (16):1293–1296
137. Hattori H, Okuda K, Murase T, Shigetsura Y, Narise K, Semenza GL, Nagasawa H (2011) Isolation, identification, and biological evaluation of HIF-1-modulating compounds from Brazilian green propolis. Bioorg Med Chem 19(18):5392–5401
138. Patel S (2016) Emerging adjuvant therapy for cancer: propolis and its constituents. J Diet Suppl. 13(3):245–268
139. Begnini KR, Moura de Leon PM, Thurow H, Schultze E, Campos VF, Martins Rodrigues F, Borsuk S, Dellagostin OA, Savegnago L, Roesch-Ely M, Moura S, Padilha FF, Collares T, Pêgas Henriques JA, Seixas FK (2014) Brazilian red propolis induces apoptosis-like cell

death and decreases migration potential in bladder cancer cells. Evid Based Complement Alternat Med 2014:639856
140. Lucas L, Russell A, Keast R (2011) Molecular mechanisms of inflammation. Anti-inflammatory benefits of virgin olive oil and the phenolic compound oleocanthal. Curr Pharm Des 17(8):754–768
141. Elnagar AY, Sylvester PW, El Sayed KA (2011) (−)-Oleocanthal as a c-Met inhibitor for the control of metastatic breast and prostate cancers. Planta Med 77(10):1013–1019
142. Rosignoli P, Fuccelli R, Fabiani R, Servili M, Morozzi G (2013) Effect of olive oil phenols on the production of inflammatory mediators in freshly isolated human monocytes. J Nutr Biochem 24(8):1513–1519
143. Pingle SC, Matta JA, Ahern GP (2007) Capsaicin receptor: TRPV1 a promiscuous TRP channel. Handb Exp Pharmacol 179:155–171
144. Alawi K, Keeble J (2010) The paradoxical role of the transient receptor potential vanilloid 1 receptor in inflammation. Pharmacol Ther 125(2):181–195
145. Caterina MJ, Julius D (2001) The vanilloid receptor: a molecular gateway to the pain pathway. Annu Rev Neurosci 24:487–517
146. Aneiros E, Cao L, Papakosta M, Stevens EB, Phillips S, Grimm C (2011) The biophysical and molecular basis of TRPV1 proton gating. EMBO J 30(6):994–1002
147. Dhaka A, Uzzell V, Dubin AE, Mathur J, Petrus M, Bandell M, Patapoutian A (2009) TRPV1 is activated by both acidic and basic pH. J Neurosci 29(1):153–158
148. Brito R, Sheth S, Mukherjea D, Rybak LP, Ramkumar V (2014) TRPV1: a potential drug target for treating various diseases. Cells 3(2):517–545
149. Raoux M, Rodat-Despoix L, Azorin N, Giamarchi A, Hao J, Maingret F, Crest M, Coste B, Delmas P (2007) Mechanosensor channels in mammalian somatosensory neurons. Sensors 7 (9):1667–1682
150. Voets T, Droogmans G, Wissenbach U, Janssens A, Flockerzi V, Nilius B (2004) The principle of temperature-dependent gating in cold- and heat-sensitive TRP channels. Nature 430(7001):748–754
151. Mueller-Seitz E, Hiepler C, Petz M (2008) Chili pepper fruits: content and pattern of capsaicinoids in single fruits of different ages. J Agric Food Chem 56(24):12114–12121
152. Luo XJ, Peng J, Li YJ (2011) Recent advances in the study on capsaicinoids and capsinoids. Eur J Pharmacol 650(1):1–7
153. Tominaga M, Caterina MJ, Malmberg AB, Rosen TA, Gilbert H, Skinner K, Raumann BE, Basbaum AI, Julius D (1998) The cloned capsaicin receptor integrates multiple pain-producing stimuli. Neuron 21(3):531–543
154. Jordt SE, Tominaga M, Julius D (2000) Acid potentiation of the capsaicin receptor determined by a key extracellular site. Proc Natl Acad Sci U S A 97(14):8134–8139
155. DeSimone JA, Lyall V (2006) Taste receptors in the gastrointestinal tract III. Salty and sour taste: sensing of sodium and protons by the tongue. Am J Physiol Gastrointest Liver Physiol 291(6):G1005–G1010
156. Chung S, Kim YH, Koh JY, Nam TS, Ahn DS (2011) Intracellular acidification evoked by moderate extracellular acidosis attenuates transient receptor potential V1 (TRPV1) channel activity in rat dorsal root ganglion neurons. Exp Physiol 96(12):1270–1281
157. Trevisani M, Smart D, Gunthorpe MJ, Tognetto M, Barbieri M, Campi B, Amadesi S, Gray J, Jerman JC, Brough SJ, Owen D, Smith GD, Randall AD, Harrison S, Bianchi A, Davis JB, Geppetti P (2002) Ethanol elicits and potentiates nociceptor responses via the vanilloid receptor-1. Nat Neurosci 5(6):546–551
158. Park CK, Kim K, Jung SJ, Kim MJ, Ahn DK, Hong SD, Kim JS, Oh SB (2009) Molecular mechanism for local anesthetic action of eugenol in the rat trigeminal system. Pain 144 (1–2):84–94
159. Lyall V, Heck GL, Vinnikova AK, Ghosh S, Phan TH, Alam RI, Russell OF, Malik SA, Bigbee JW, DeSimone JA (2004) The mammalian amiloride-insensitive non-specific salt taste receptor is a vanilloid receptor-1 variant. J Physiol 558(Pt 1):147–159

160. Lyall V, Heck GL, Vinnikova AK, Ghosh S, Phan TH, Desimone JA (2005) A novel vanilloid receptor-1 (VR-1) variant mammalian salt taste receptor. Chem Senses 30(Suppl 1):i42–i43
161. Ruiz C, Gutknecht S, Delay E, Kinnamon S (2006) Detection of NaCl and KCl in TRPV1 knockout mice. Chem Senses 31(9):813–820
162. Smith KR, Treesukosol Y, Paedae AB, Contreras RJ, Spector AC (2012) Contribution of the TRPV1 channel to salt taste quality in mice as assessed by conditioned taste aversion generalization and chorda tympani nerve responses. Am J Physiol Regul Integr Comp Physiol 303(11):R1195–R1205
163. Peier AM, Moqrich A, Hergarden AC, Reeve AJ, Andersson DA, Story GM, Earley TJ, Dragoni I, McIntyre P, Bevan S, Patapoutian A (2002) A TRP channel that senses cold stimuli and menthol. Cell 108(5):705–715
164. Riera CE, Vogel H, Simon SA, le Coutre J (2007) Artificial sweeteners and salts producing a metallic taste sensation activate TRPV1 receptors. Am J Physiol Regul Integr Comp Physiol 293(2):R626–R634
165. Takashima Y, Daniels RL, Knowlton W, Teng J, Liman ER, McKemy DD (2007) Diversity in the neural circuitry of cold sensing revealed by genetic axonal labeling of transient receptor potential melastatin 8 neurons. J Neurosci 27(51):14147–14157
166. Abe J, Hosokawa H, Okazawa M, Kandachi M, Sawada Y, Yamanaka K, Matsumura K, Kobayashi S (2005) TRPM8 protein localization in trigeminal ganglion and taste papillae. Brain Res Mol Brain Res 136(1–2):91–98
167. Colburn RW, Lubin ML, Stone DJ Jr, Wang Y, Lawrence D, D'Andrea MR, Brandt MR, Liu Y, Flores CM, Qin N (2007) Attenuated cold sensitivity in TRPM8 null mice. Neuron 54 (3):379–386
168. Bautista DM, Siemens J, Glazer JM, Tsuruda PR, Basbaum AI, Stucky CL, Jordt SE, Julius D (2007) The menthol receptor TRPM8 is the principal detector of environmental cold. Nature 448(7150):204–208
169. Kühn FJ, Kühn C, Lückhoff A (2009) Inhibition of TRPM8 by icilin distinct from desensitization induced by menthol and menthol derivatives. J Biol Chem 284(7):4102–4111
170. Rohács T, Lopes CM, Michailidis I, Logothetis DE (2005) PI(4,5)P2 regulates the activation and desensitization of TRPM8 channels through the TRP domain. Nat Neurosci 8(5):626–634
171. Sarria I, Ling J, Zhu MX, Gu JG (2011) TRPM8 acute desensitization is mediated by calmodulin and requires $PIP_2$: distinction from tachyphylaxis. J Neurophysiol 106 (6):3056–3066
172. Yudin Y, Lukacs V, Cao C, Rohacs T (2011) Decrease in phosphatidylinositol 4,5-bisphosphate levels mediates desensitization of the cold sensor TRPM8 channels. J Physiol 589(Pt 24):6007–6027
173. Rohács T, Lopes CM, Jin T, Ramdya PP, Molnár Z, Logothetis DE (2003) Specificity of activation by phosphoinositides determines lipid regulation of Kir channels. Proc Natl Acad Sci U S A 100(2):745–750
174. Voets T, Nilius B (2007) Modulation of TRPs by PIPs. J Physiol 582(Pt 3):939–944
175. Laska M, Distel H, Hudson R (1997) Trigeminal perception of odorant quality in congenitally anosmic subjects. Chem Senses 22(4):447–456
176. Kehrl W, Sonnemann U, Dethlefsen U (2004) Therapy for acute nonpurulent rhinosinusitis with cineole: results of a double-blind, randomized, placebo-controlled trial. Laryngoscope 114(4):738–742
177. Mahieu F, Owsianik G, Verbert L, Janssens A, De Smedt H, Nilius B, Voets T (2007) TRPM8-independent menthol-induced $Ca^{2+}$ release from endoplasmic reticulum and Golgi. J Biol Chem 282(5):3325–3336
178. Lübbert M, Kyereme J, Schöbel N, Beltrán L, Wetzel CH, Hatt H (2013) Transient receptor potential channels encode volatile chemicals sensed by rat trigeminal ganglion neurons. PLoS One 8(10):e77998

179. Birnbaumer L (2009) The TRPC class of ion channels: a critical review of their roles in slow, sustained increases in intracellular $Ca^{2+}$ concentrations. Annu Rev Pharmacol Toxicol 49:395–426
180. Boulay G, Zhu X, Peyton M, Jiang M, Hurst R, Stefani E, Birnbaumer L (1997) Cloning and expression of a novel mammalian homolog of *Drosophila* transient receptor potential (Trp) involved in calcium entry secondary to activation of receptors coupled by the $G_q$ class of G protein. J Biol Chem 272(47):29672–29680
181. Alkhani H, Ase AR, Grant R, O'Donnell D, Groschner K, Séguéla P (2014) Contribution of TRPC3 to store-operated calcium entry and inflammatory transductions in primary nociceptors. Mol Pain 10:43
182. Vandewauw I, Owsianik G, Voets T (2013) Systematic and quantitative mRNA expression analysis of TRP channel genes at the single trigeminal and dorsal root ganglion level in mouse. BMC Neurosci 14:21
183. Kress M, Karasek J, Ferrer-Montiel AV, Scherbakov N, Haberberger RV (2008) TRPC channels and diacylglycerol dependent calcium signaling in rat sensory neurons. Histochem Cell Biol 130(4):655–667
184. Elg S, Marmigere F, Mattsson JP, Ernfors P (2007) Cellular subtype distribution and developmental regulation of TRPC channel members in the mouse dorsal root ganglion. J Comp Neurol 503(1):35–46
185. Crockett SL, Robson NK (2011) Taxonomy and Chemotaxonomy of the Genus Hypericum. Med Aromat Plant Sci Biotechnol 5 (Special Issue 1):1–13
186. Philippu A (2001) In vivo neurotransmitter release in the locus coeruleus – effects of hyperforin, inescapable shock and fear. Pharmacopsychiatry 34(Suppl 1):S111–S115
187. Kaehler ST, Sinner C, Chatterjee SS, Philippu A (1999) Hyperforin enhances the extracellular concentrations of catecholamines, serotonin and glutamate in the rat locus coeruleus. Neurosci Lett 262(3):199–202
188. Treiber K, Singer A, Henke B, Müller WE (2005) Hyperforin activates nonselective cation channels (NSCCs). Br J Pharmacol 145(1):75–83
189. Elsaesser R, Montani G, Tirindelli R, Paysan J (2005) Phosphatidyl-inositide signalling proteins in a novel class of sensory cells in the mammalian olfactory epithelium.Eur. J Neurosci 21(10):2692–2700
190. Vannier B, Peyton M, Boulay G, Brown D, Qin N, Jiang M, Zhu X, Birnbaumer L (1999) Mouse *Trp2*, the homologue of the human *Trpc2* pseudogene, encodes mTrp2, a store depletion-activated capacitative $Ca^{2+}$ entry channel. Proc Natl Acad Sci U S A 96 (5):2060–2064
191. Menco BP, Carr VM, Ezeh PI, Liman ER, Yankova MP (2001) Ultrastructural localization of G-proteins and the channel protein TRP2 to microvilli of rat vomeronasal receptor cells. J Comp Neurol 438(4):468–489
192. Liman ER, Corey DP, Dulac C (1999) TRP2: a candidate transduction channel for mammalian pheromone sensory signaling. Proc Natl Acad Sci U S A 96(10):5791–5796
193. Stowers L, Holy TE, Meister M, Dulac C, Koentges G (2002) Loss of sex discrimination and male-male aggression in mice deficient for TRP2. Science 295(5559):1493–1500
194. Leypold BG, Yu CR, Leinders-Zufall T, Kim MM, Zufall F, Axel R (2002) Altered sexual and social behaviors in *trp2* mutant mice. Proc Natl Acad Sci U S A 99(9):6376–6381
195. Caterina MJ, Rosen TA, Tominaga M, Brake AJ, Julius D (1999) A capsaicin-receptor homologue with a high threshold for noxious heat. Nature 398(6726):436–441
196. Nilius B, Biro T, Owsianik G (2014) TRPV3: time to decipher a poorly understood family member! J Physiol 592(Pt 2):295–304
197. Sokabe T, Fukumi-Tominaga T, Yonemura S, Mizuno A, Tominaga M (2010) The TRPV4 channel contributes to intercellular junction formation in keratinocytes. J Biol Chem 285 (24):18749–18758
198. Barr MM, Sternberg PW (1999) A polycystic kidney-disease gene homologue required for male mating behaviour in C. elegans. Nature 401(6751):386–389

199. LopezJimenez ND, Cavenagh MM, Sainz E, Cruz-Ithier MA, Battey JF, Sullivan SL (2006) Two members of the TRPP family of ion channels, PKD1L3 and PKD2L1, are co-expressed in a subset of taste receptor cells. J Neurochem 98(1):68–77
200. The polycystic kidney disease 1 gene encodes a 14 kb transcript and lies within a duplicated region on chromosome 16. The European Polycystic Kidney Disease Consortium (1994). Cell 77(6):881–894
201. Hughes J, Ward CJ, Peral B, Aspinwall R, Clark K, San Millan JL, Gamble V, Harris PC (1995) The polycystic kidney disease 1 (PKD1) gene encodes a novel protein with multiple cell recognition domains. Nat Genet 10(2):151–160
202. Mochizuki T, Wu G, Hayashi T, Xenophontos SL, Veldhuisen B, Saris JJ, Reynolds DM, Cai Y, Gabow PA, Pierides A, Kimberling WJ, Breuning MH, Deltas CC, Peters DJ, Somlo S (1996) PKD2, a gene for polycystic kidney disease that encodes an integral membrane protein. Science 272(5266):1339–1342
203. Yamaguchi T, Hempson SJ, Reif GA, Hedge AM, Wallace DP (2006) Calcium restores a normal proliferation phenotype in human polycystic kidney disease epithelial cells. J Am Soc Nephrol 17(1):178–187
204. Guo L, Schreiber TH, Weremowicz S, Morton CC, Lee C, Zhou J (2000) Identification and characterization of a novel polycystin family member, polycystin-L2, in mouse and human: sequence, expression, alternative splicing, and chromosomal localization. Genomics 64 (3):241–251
205. Wu G, Hayashi T, Park JH, Dixit M, Reynolds DM, Li L, Maeda Y, Cai Y, Coca-Prados M, Somlo S (1998) Identification of PKD2L, a human PKD2-related gene: tissue-specific expression and mapping to chromosome 10q25. Genomics 54(3):564–568
206. Chen XZ, Vassilev PM, Basora N, Peng JB, Nomura H, Segal Y, Brown EM, Reeders ST, Hediger MA, Zhou J (1999) Polycystin-L is a calcium-regulated cation channel permeable to calcium ions. Nature 401(6751):383–386
207. Koulen P, Cai Y, Geng L, Maeda Y, Nishimura S, Witzgall R, Ehrlich BE, Somlo S (2002) Polycystin-2 is an intracellular calcium release channel. Nat Cell Biol 4(3):191–197
208. Hanaoka K, Qian F, Boletta A, Bhunia AK, Piontek K, Tsiokas L, Sukhatme VP, Guggino WB, Germino GG (2000) Co-assembly of polycystin-1 and −2 produces unique cation-permeable currents. Nature 408(6815):990–994
209. Huang AL, Chen X, Hoon MA, Chandrashekar J, Guo W, Trankner D, Ryba NJ, Zuker CS (2006) The cells and logic for mammalian sour taste detection. Nature 442(7105):934–938
210. Ishimaru Y, Inada H, Kubota M, Zhuang H, Tominaga M, Matsunami H (2006) Transient receptor potential family members PKD1L3 and PKD2L1 form a candidate sour taste receptor. Proc Natl Acad Sci U S A 103(33):12569–12574
211. Inada H, Kawabata F, Ishimaru Y, Fushiki T, Matsunami H, Tominaga M (2008) Off-response property of an acid-activated cation channel complex PKD1L3–PKD2L1. EMBO Rep 9(7):690–697
212. Kataoka S, Yang R, Ishimaru Y, Matsunami H, Kinnamon JC, Finger TE (2008) The candidate sour taste receptor, PKD2L1, is expressed by type III taste cells in the mouse. Chem Senses 33(3):243–254
213. Ishii S, Misaka T, Kishi M, Kaga T, Ishimaru Y, Abe K (2009) Acetic acid activates PKD1L3-PKD2L1 channel – a candidate sour taste receptor. Biochem Biophys Res Commun 385(3):346–350
214. Horio N, Yoshida R, Yasumatsu K, Yanagawa Y, Ishimaru Y, Matsunami H, Ninomiya Y (2011) Sour taste responses in mice lacking PKD channels. PLoS One 6(5):e20007
215. Nelson TM, Lopezjimenez ND, Tessarollo L, Inoue M, Bachmanov AA, Sullivan SL (2010) Taste function in mice with a targeted mutation of the *Pkd1l3* gene. Chem Senses 35 (7):565–577
216. Lee N, Chen J, Sun L, Wu S, Gray KR, Rich A, Huang M, Lin JH, Feder JN, Janovitz EB, Levesque PC, Blanar MA (2003) Expression and characterization of human transient receptor potential melastatin 3 (hTRPM3). J Biol Chem 278(23):20890–20897

217. Grimm C, Kraft R, Sauerbruch S, Schultz G, Harteneck C (2003) Molecular and functional characterization of the melastatin-related cation channel TRPM3. J Biol Chem 278 (24):21493–21501
218. Vriens J, Owsianik G, Hofmann T, Philipp SE, Stab J, Chen X, Benoit M, Xue F, Janssens A, Kerselaers S, Oberwinkler J, Vennekens R, Gudermann T, Nilius B, Voets T (2011) TRPM3 is a nociceptor channel involved in the detection of noxious heat. Neuron 70(3):482–494
219. Oberwinkler J, Lis A, Giehl KM, Flockerzi V, Philipp SE (2005) Alternative splicing switches the divalent cation selectivity of TRPM3 channels. J Biol Chem 280 (23):22540–22548
220. Wagner TF, Loch S, Lambert S, Straub I, Mannebach S, Mathar I, Dufer M, Lis A, Flockerzi V, Philipp SE, Oberwinkler J (2008) Transient receptor potential M3 channels are ionotropic steroid receptors in pancreatic beta cells. Nat Cell Biol 10(12):1421–1430
221. Straub I, Krügel U, Mohr F, Teichert J, Rizun O, Konrad M, Oberwinkler J, Schaefer M (2013) Flavanones that selectively inhibit TRPM3 attenuate thermal nociception in vivo. Mol Pharmacol 84(5):736–750
222. Straub I, Mohr F, Stab J, Konrad M, Philipp SE, Oberwinkler J, Schaefer M (2013) Citrus fruit and fabacea secondary metabolites potently and selectively block TRPM3. Br J Pharmacol 168(8):1835–1850
223. Erlund I (2004) Review of the flavonoids quercetin, hesperetin, and naringenin. Dietary sources, bioactivities, bioavailability, and epidemiology. Nutr Res 24(10):851–874

Top Med Chem (2017) 23: 115–150
DOI: 10.1007/7355_2015_100

Published online: 29 March 2014

# Olfactory Transduction Channels and Their Modulation by Varieties of Volatile Substances

**Hiroko Takeuchi and Takashi Kurahashi**

**Abstract** Olfaction starts at the sensory cilia of the olfactory receptor cell. One of the unusual features of this signal transduction is that the underlying ion channels are directly exposed to the external environment, so that the channel and olfactory senses can be modified by externally applied chemicals even by the airborne stimuli. In the human history, such properties have long been used as an olfactory masking that erases unpleasant smells present in the environment. It has been shown that a part of masking is responsible for direct suppression by odorants of olfactory signal transduction channels. It has also been shown that similar suppression by off-flavors included in foods and beverages brings negative effects on pleasant scents and flavors by suppressing original odor of products. In this chapter, we focus on the olfactory signal conversion system and its modulation by diverse types of volatile substances.

**Keywords** Channel inhibition, Cl(Ca) channel, CNG channel, Olfactory cilia, Olfactory masking

## Contents



H. Takeuchi (✉) and T. Kurahashi
Graduate School of Frontier Biosciences, Osaka University, 1-3 Machikaneyma-cho, Toyonaka, Osaka, Japan
e-mail: hiroko@bpe.es.osaka-u.ac.jp

# 1 Outline of the Olfactory Signal Transduction

## *1.1 Olfactory Transduction Machinery in the Cilia*

In the environment surrounding our lives, wide varieties of odorant molecules hang in the air all the time. The odorant molecules are mostly carbon hydrates and usually have the molecular weight of 300–400. Most of these chemicals have intrinsic fragrances. For instance, the smell of rose can be characterized by the principal ingredient of geraniol. Banana is specified by amyl acetate or butyl acetate. Sense of smell starts at olfactory receptor cells (ORCs) that are situated in the olfactory epithelium (OE). The receptor cell is also called olfactory receptor neurons (ORNs), presumably due to their spiking ability with an axon. The former abbreviation, ORC, is used especially when the homology with photoreceptor cells is considered; both olfactory receptor and photoreceptor cells use similar transduction machinery employing the cyclic nucleotide as second messengers. At the ORC, chemical information of the odorant molecule is converted into the biological electrical signals through activation of transduction channels. Depolarization caused by the transduction channels is graded, and when they reach once to the threshold potential, they in turn generate a train of action potentials (spikes) through activation of voltage-dependent channels. The spike train is transmitted to the olfactory bulb along the long axon of the ORC. The signals are integrated roughly at the olfactory bulb once and further transmitted to the deeper parts of the brain. Interestingly, in parallel to the recognition processes in the olfactory cortex, olfactory information projects near to the hippocampus and corpus amygdala. Therefore, there is an intimate relationship between olfaction and functions of these areas including memory, emotion, etc.

Figure 1 illustrates a schematic organ structure that mediates an initial step for olfactory perception. First, odorant molecules enter into the nasal cavity and they dissolve into the mucus that is gushed from the Bowman's gland [1]. This

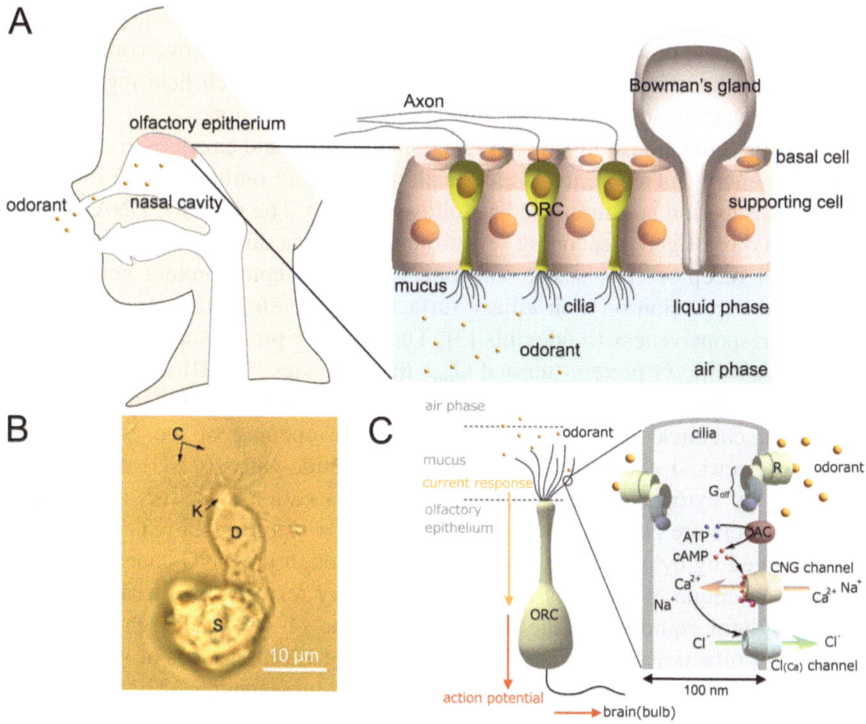


**Fig. 1** Olfactory sensation. (**a**) Olfactory organ. *Left*: sagittal plane of the head. *Right*: schematic diagram of the olfactory epithelium. *ORC* olfactory receptor cell. Note that the olfactory cilia are directly exposed to the external environment. Volatile chemicals (odorant molecules) are inhaled from the outside. Odorant molecules are dissolved in the mucus in accordance with its air–water partition coefficient and binding affinity to the OBP. (**b**) Photograph of isolated ORC. *C* cilia, *K* knob, *D* dendrite, *S* soma. (**c**) Scheme of molecular transduction cascade of ORC. *R* receptor protein, *AC* adenylyl cyclase, *CNG channel* cyclic nucleotide-gated channel, $Cl_{(Ca)}$ *channel* calcium-activated chloride channel, *ATP* adenosine triphosphate, *cAMP* cyclic nucleotide monophosphate

molecular dissolution is determined by the distribution coefficient between gaseous layer liquid phases (dissociation constant). Odorant molecules that are partitioned into the mucus are thought to bind to the olfactory binding protein (OBP) [2]. OBP thus increases the activity of odorant in the mucus and also OBP/odorant complexes are used for the clearance of odorants. Olfactory epithelium is composed of several types of cells: ORCs, basal cells (precursors for ORCs), and supporting cells.

The ORC shows a bipolar morphology; from one side of the cell body, a single dendrite extends toward the surface of the epithelium. The apical part of the dendrite is swollen and it is called the olfactory knob or terminal swelling. The site for the olfactory signal transduction is olfactory cilia that are extended from the knob of the ORCs. This functionary important region displays a fine cylindrical structure having submicron diameter structure (100–200 nm, length 10–100 μm). Because of such structure, understanding of the olfactory transduction system has been restricted for long period of time, especially in the quantitative aspect. During the past decades, however, some of such difficulties have been overcome, and the fundamental knowledge has been accumulated in the research field regarding the molecular mechanisms mediating olfactory energy conversion.

Olfactory cilia are equipped with functional proteins and enzymes to convert the chemical energy of odorants to electrical signals. The outline of this molecular signal transduction is summarized as follows (Fig. 1). The genomic DNA prepares multiple types of genes encoding seven-transmembrane domain odorant receptors. Individual receptor cells select only one type of receptor protein gene for the functional expression on their ciliary surface, and therefore, they express heterogeneous responsiveness to odorants [3]. The receptor proteins are coupled to the olfaction-specific G protein (termed $G_{olf}$) that activates type III adenylyl cyclase (AC). This enzyme converts cytoplasmic ATP to cyclic AMP (cAMP). These sequential chemical reactions finally lead to the opening of ion channels that underlie electrical excitation. First, cyclic nucleotide-gated (CNG) channels are activated by cytoplasmic cAMP. This channel is cation selective, so the main charge careers are $Na^{+}$ and $Ca^{2+}$. Subsequently, Ca-activated Cl ($Cl_{(Ca)}$) channels are activated by cytoplasmic $Ca^{2+}$ that flowed through the CNG channel. This Cl channel is unique in that they induce an excitatory response due to a strong $Cl^{-}$ uptake system equipped in the plasma membrane [4–15]. Although the natural ligands for olfaction have large diversity exceeding 100 thousands of varieties, the signal thus converges into only one type of second messenger, cAMP, through the enzymatic cascade equipped in the sensory cilia.

**Structures of CNG and $Cl_{(Ca)}$ Channels** The CNG channel has four subunits, each with six transmembrane domains and one cAMP-binding site. According to this, individual CNG channels have four binding sites for cAMP. The subunits of CNG channels are not homogeneous. The stoichiometry of subunits in the wild-type channels is thought to be CNGA2 × 2, CNGA4 × 1, and CNGB1b × 1. The molecular structure of the $Cl_{(Ca)}$ channel is still controversial. Bestrophin-2 (mBest2) and TMEM16B (ANO2) are candidates for $Cl_{(Ca)}$ channel [16–18].

## *1.2 Spatial Distribution of Sensitivity to the Odorant*

Spatial distribution of sensitivity to the odorant was examined with the puff application of odorant under the whole-cell patch clamp recording (Fig. 2) [19]. Small amounts of odorant were puff-applied locally to various regions of the isolated newt ORCs. Because of the movement of the cilia, mapping was made mainly on the apical region of the dendrite and on the cell body. The maximum response was evoked, when the stimulus was given to the apical part of the dendrite. Identical doses of odorant given to the cell body induced much less current. Furthermore, a longer response delay was seen, when the stimulus was applied to the cell body rather than to the apical dendrite. This difference may be attributed to the diffusion of the odorant to the receptive site. Indeed, the time delay (ca. 200 ms)

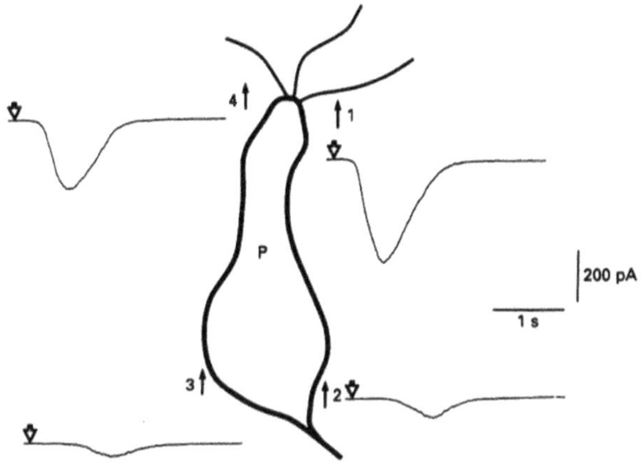


**Fig. 2** Spatial distribution of sensitivity to the odorant. *Black arrows* show sites of odorant application. *White arrows* show the start time of odorant stimulation (50 ms). Odorant was *n*-amyl acetate [19]

between these responses agrees well with the diffusion time course derived from the diffusion coefficient of odorants in water. The results from such experiment clearly show that the sensitivity to odorants of ORCs is strongly polarized to the apical region. It has now been established that cilia are the site for odorant reception.

In order to measure activities of transduction channels located in the cilia, ORCs are isolated from the OE and are subjected to electrical recordings. Usually, urodele (tailed amphibians; newts, tiger salamander) are used for the electrophysiological research, because they have ORCs relatively bigger than those in other vertebrates and have been used as one of the best model animals since the late twentieth century [19–23]. For instance, newt ORCs exhibit the cell body size of 15–20 μm, while in human and mice, the cell body size is 5–10 μm. The properties and underlying molecular mechanisms that include transduction, adaptation, signal amplification, and olfactory masking seem to be the same as those in mammals.

### *1.3 Identification of the Second Messenger cAMP and Third Messenger $Ca^{2+}$ in ORCs*

It has long been hypothesized that olfactory signal transduction is mediated by two parallel pathways [14, 24]. A generally accepted view was that the ORC selected either cAMP or $IP_3$ as a second messenger depending on the species of odorant molecules. In the cAMP pathway, odorant binding to the receptor protein triggers an activation of $G_{olf}$ and in turn activates AC, which produces cAMP intracellularly. Cytoplasmic cAMP activates CNG channels directly. In addition, the resultant $Ca^{2+}$ influx triggers the opening of $Cl_{(Ca)}$ channels. In contrast, the $IP_3$ pathway was thought to use completely independent molecules. The signal of receptor activation is transmitted to PLC via a G protein ($G_q$). As a result, the cytoplasm $IP_3$, concentration increases, which finally leads to the opening of plasma membrane cationic channels. Plasma membrane ion channels that are activated by cytoplasmic $IP_3$ are unusual, but their presence has been reported for ORCs in a

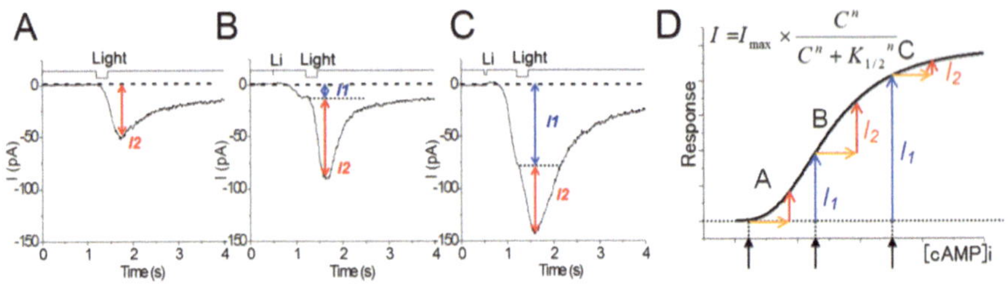


**Fig. 3** Summation of responses induced by $IP_3$ odorant and light. $I_1$ shows $IP_3$ odorant-induced current, $I_2$ shows the cAMP response induced by caged photolysis. If both pathways were independent, amplitudes of $I_2$ were constant through (**a**)–(**c**). (**d**) Explanation for the different $I_2$ depending on different basal level of cAMP [15]

wide variety of animals from invertebrates (lobster [25]) to vertebrates (catfish [26], *Xenopus* [27], bullfrog [28, 29], and rat [30]). The signaling pathway involving cAMP has been well established [5, 7, 8, 14, 22], but the involvement of $IP_3$ has long been controversial [8] (see also [30, 31]). The reason for this was that especially at the single-cell level, there was very limited information regarding responses to odorants that have been shown to produce $IP_3$ exclusively (tentatively termed "$IP_3$ odorants"). The technical limitation has been mainly the low probability of single-cell sensitivities to $IP_3$ odorants [23, 32]. Moreover, the fine structure of sensory cilia (0.2 μm diameter) has made experimental manipulations extremely difficult. In 2003, Takeuchi and Kurahashi showed that sensory responses induced by $IP_3$ odorants are actually generated by an increase in cytoplasmic cAMP (Fig. 3). The activity of transduction channels in olfactory cilia was recorded, while the $[cNMP]_i$ was freely manipulated through the photolysis of caged compounds. This allowed them to examine cross-interactions between $IP_3$ odorants and cytoplasmic cNMPs directly in real time. Responses induced by both stimulants were homologous in their characteristic properties and showed cross-adaptation with each other. Furthermore, both responses were additive in a manner as predicted precisely by the theory that signal transduction is mediated by cAMP (Fig. 3).

Although the olfactory signal transduction employs second (cAMP) and third ($Ca^{2+}$) messengers as soluble factors in the cytoplasm, these molecules seem not to travel for far distances from the site of generation [33]. Both CNG [34] and $Cl_{(Ca)}$ channels [33] are localized to the cilia, and within the cilia, distribution of channels are almost homogeneous when measured in the newt and mouse, while in frogs that have longer cilia, the density is reported to be maximum at around the 20 % distance from the proximal part of the cilia and is gradually reduced depending on the distance toward the tip [35]. Since the transduction channels are localized to the cilia that are exposed directly to the external environment, this sensory system is influenced by the external chemicals, either naturally or artificially.

# 2 Electrical Properties of the Transduction Channels

Electrical responses of the ORCs are generated by the sequential opening of CNG and $Cl_{(Ca)}$ channels. Since individual channels show cooperativities and since their relation is sequential, the resulted output exhibits high cooperativity. Such boosting system establishes a high nonlinear amplification of the transduction system that is not observed in homologous systems seen in biological system (e.g., photoreceptor cells).

## 2.1 CNG Channel

In 1987, Nakamura and Gold found with inside-out configuration of the patch clamp that the electrical conductance of olfactory cilia was increased, when cAMP was applied to the cytoplasmic side of the membrane patch [36]. The cAMP-induced conductance changes were shown to be similar to the cationic conductance activated by the odorant [19, 37]. The unitary event of conductance was analyzed and their distributions were shown to be localized to the cilia, where olfactory signal transduction takes place [37–39]. Because production by odorants of cAMP had been reported by Lancet [40] in advance, the discovery of conductance change by cytoplasmic cAMP filled a missing link between the odorant-induced chemical reaction and the electrical excitation of the receptor cell. In this transduction cascade, odorant information is transmitted as chemical signals all the way to the production of cAMP. The opening of the channel pore and ion fluxes then generates electrical signals. Thus the CNG channels play an important role for the energy conversion.

### 2.1.1 Ion Permeability and Open Probability of the CNG Channel

Ion permeability of CNG channels was obtained by patch clamp recording with ramp pulses (Fig. 4) [19, 37]) as follows:

Odor application; $P_{Li}:P_{Na}:P_K:P_{Rb}:P_{Cs} = 1.25:1:0.98:0.84:0.80$.
cAMP application; $P_{Li}:P_{Na}:P_K:P_{Rb}:P_{Cs} = 0.93:1:0.93:0.91:0.72$.

Furthermore, $P_{Ca}/P_{Na}$ was shown to be 6.5 [41]. These results suggest that the CNG channel permeates varieties of cations.

Unitary event of the CNG channels shows steplike current fluctuation caused by transition between open and close statutes. Single-channel conductance is approximately 30 pS [38, 39, 42], when external divalent cations are omitted, and its open probability becomes 0.7–0.8 when fully liganded [39, 43–45].

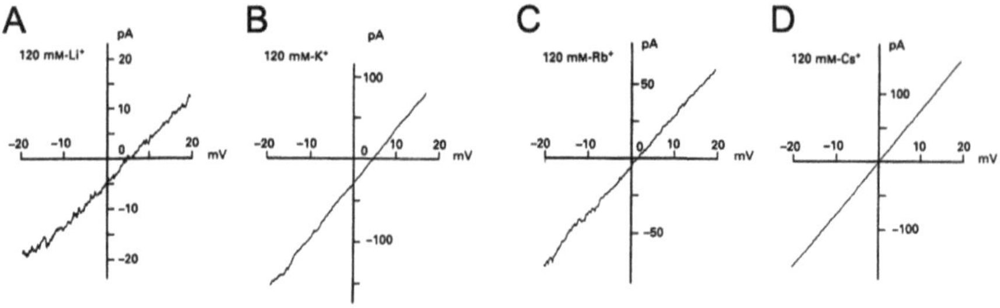


**Fig. 4** The I–V relations of the conductance activated by 10 mmol/L *n*-amyl acetate recorded in cells bathed in solutions containing various alkali metal ions. (**a**) 120 mmol/L $Li^+$. (**b**) 120 mmol/L $K^+$. (**c**) 120 mmol/L $Rb^+$. (**d**) 120 mmol/L $Cs^+$. Records were obtained from different cells. The position of the reversal potential provides an index of permeability ratio. The more the reversal potential shifts to positive direction, the more the external ions have permeability. The pipette contained 120 mmol/L $Cs^+$ solution [19]

### 2.1.2 High-Density Expression of CNG Channels on the Cilia

Odorant sensitivity of ORCs is localized to the cilia. Therefore, it is natural to consider that CNG channels are expressed densely on the ciliary membrane. The density of the channel was estimated by observing single-channel events and their dose dependence in inside-out membrane patches obtained from the cilia (Fig. 5). To obtain the membrane patch, the diameter of pipette tip was adjusted to less than 0.2 μm.

At low concentration (e.g., 1 μmol/L), the open probability of individual CNG channels is extremely low, and therefore, we can observe steplike unitary events caused by the transition between open and closed states of the channel molecule. As concentration of cAMP applied is increased, the open probability is also increased and the current becomes noisy [38, 39]. The channel density estimated with noise analyses of such data is 1,750 channel/μm$^2$ (toad), when the membrane patch was obtained from the cilia. On the other hand, the cell body showed the density of 6 channels/μm$^2$.

### 2.1.3 Broad Expression of CNG Channels Within the Cilia

In an early study employing electrophysiological technique using the ciliary bundle, Lowe and Gold (1993) showed that cyclic nucleotide (CN) sensitivities are distributed evenly along the entire cilia [46]. They made a whole-cell recording configuration on the tiger salamander ORCs and applied local UV stimulation along the cilia to photolyze the cytoplasmic caged cAMP that opens CNG channels. The amplitude of light-induced responses showed a linear relation with the area of cilia that were illuminated. With the electron microscopy (EM), however, immune staining against the subunit of CNG channel, CNGA2, was found to be dominant to the tip of the cilia [47]. In 2006, Flannery et al. used the detached ciliary

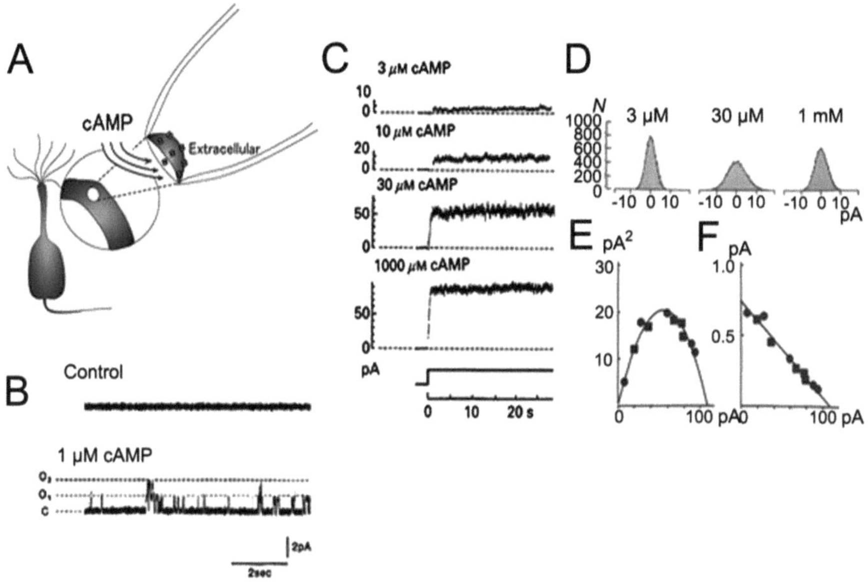


**Fig. 5** Inside-out recording from the ciliary membrane and estimation of the density of CNG channels. (**a**) Scheme of inside-out configuration. The tip of recording pipette is adjusted to less than the diameter of cilium. (**b**) Single-channel current from the patch membrane [38]. (**c**) cAMP-induced currents in various concentrations of cAMP. (**d**) Amplitude distribution of cAMP-induced current fluctuation around the mean current. Amplitude in the presence of 3 μmol/L, 30 μmol/L, and 1 mmol/L cAMP. (**e**) Relation between the variance and the mean current amplitude. The parabola shows the relation, $\sigma^2 = iI - I^2/N$. (**f**) Relation between $\sigma^2/I$ and $I$. The straight line shows the relation, $\sigma^2 = i - I/N$ [39]

preparation with their cytoplasmic side exposing to the bathing solution [35]. They applied cAMP from the opening of the proximal part of the cilia and the current time course was analyzed with a cAMP diffusion model. They suggested that the current development was well fitted, when CNG channels were estimated to be clustered at the middle part of the cilia. These controversies arose mainly from the technical limitations in a direct measurement of CN sensitivities in living cilia.

Since the diameter of the cilia is 0.1–0.2 μm, cytoplasmic dialysis within the cilia is not realistic. In some experiments, CNG channels in the cilia have been investigated with an inside-out mode of patch clamp method [48], but, the success rates for such kind of experiments are extremely low for systematic analysis of the channel. More realistic experiments are recordings with a whole-cell recording configuration. Simultaneously, caged cAMP is introduced into the cytoplasm, which defuses to the entire cell including fine cilia, and strong UV light is applied to photolyze caged cAMP into cAMP that opens the CNG channel that is localized to the ciliary membrane (Fig. 6). Large cells obtained from urodele can be persistent for such hard manipulations. Usually, voltage clamp condition is used to monitor the activity of the channels. However, one has to pay attention to the space clamp of

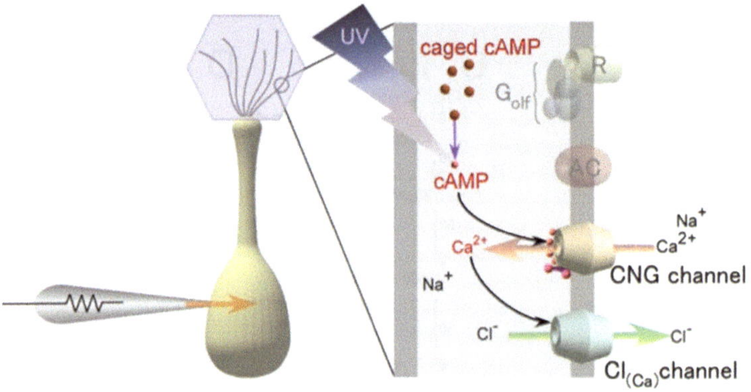


**Fig. 6** Scheme of the experimental protocol using caged photolysis under current recording

this fine tubing, which is considerably reduced depending on the opening probability of the channels.

In 2008, Takeuchi and Kurahashi addressed to this question by using a local uncaging of cytoplasmic caged compound with submicron-scale UV laser stimulation that was combined with simultaneous electrical recordings [49]. The single cilium was finely visualized with the laser scanning confocal microscope (LSM) system after loading with the fluorescent compound, lucifer yellow. When the CN sensitivities were mapped with local UV spots along the single cilium, cells responded to stimuli at any point of the cilia (Fig. 7).

Furthermore, data showed that local responses are independent from each other in small responses establishing a linear response summation. At the same time, strong stimulus onto only a very small region (1 μm) was found to induce a huge inward current exceeding 100 pA, which was equivalent to the opening of 700–2300 ion channels. The large local response indicates a presence of strong amplification operated by a high-density distribution of the transduction channels for the local excitation.

Although the transduction uses cytoplasmic soluble factor, it is suggested that signal conversion of odorant information takes place at the cilia locally, presumably due to the limited spread of cAMP and $Ca^{2+}$ [49].

## 2.2 $Cl_{(Ca)}$ Channel

In 1991, Kleene and Gesteland reported that application of $Ca^{2+}$ into the cytoplasm of the cilia increased the electrical conductance of the ciliary membrane [50]. The conductance change was mediated by the increase of Cl permeability of the membrane. Later, this conductance was shown to be significant even in the odorant-induced current [22]. The conductance is actually mediated by ion channels gated by cytoplasmic Ca ($Cl_{(Ca)}$ channel). $Ca^{2+}$ that opens the Cl channel enters

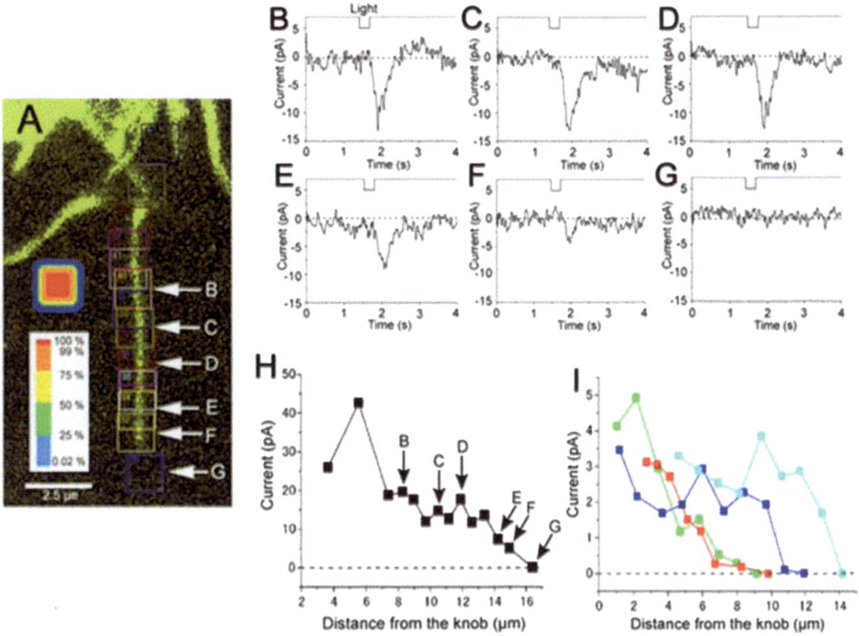


**Fig. 7** Spatial distribution of the CNG channel. Mapping of CN sensitivity along the single newt cilium. (**a**) Fluorescent image of a single cilium. To avoid showing complex figure, only ROIs are illustrated. Stimulus intensity is shown independently with a colored scale. (**b–g**) Waveforms of the current induced by the local laser irradiation. Cell was loaded with 1 mmol/L caged cGMP. (**h**) Relationship between the distance from the knob and local current responses in A. A large response observed in the second proximal region is likely to be induced the illumination onto two cilia. (**i**) Relationship between the distance from the knob and local current responses from four cells [49]

through the CNG channel. Since cAMP is a second messenger in the ORC, this Ca $^{2+}$ is sometimes called a "third messenger" in the olfactory signal transduction. In general in neurons, the Cl channel operates inhibitory signals, because $Cl^-$ has negative charges and the cytoplasmic concentration of $Cl^-$ is extremely low providing an equilibrium potential for $Cl^-$ to be comparable to that of $K^+$. However, Kurahashi and Yau (1993) showed that the $Cl^-$ current in olfactory cilia is actually excitatory [22]. Ache and Zhaninazarov (1995) also showed that the equilibrium potential for $Cl^-$ is near 0 mV [24]. As has been mentioned before, the first step of electrical signal generation is caused at the level of CNG channels. Therefore, it is interpreted that $Cl^-$ current component contributes to the signal amplification. Also, it is likely that alternative functions of cation and anion components can preserve response of the olfactory system even in the changed ionic strength surrounding cilia.

In the case that CNG and Cl channels are spatially separated to each other, the degree of nonlinear amplification could be variable depending on the strength of stimulus. In 2009, Takeuchi and Kurahashi measured distribution of the $Cl_{(Ca)}$

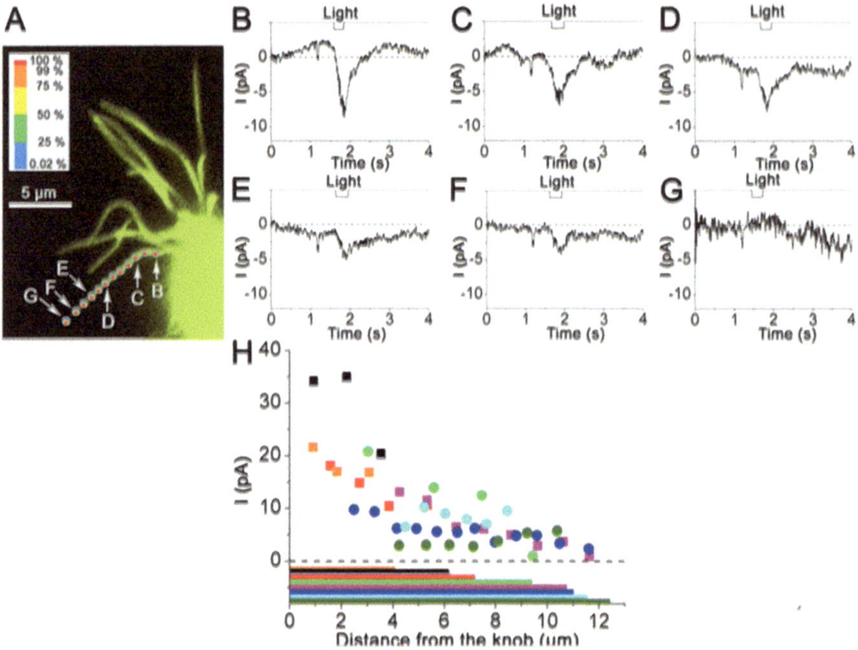


**Fig. 8** Spatial distribution of $[Ca^{2+}]_i$ sensitivity along the single cilium. (**a**) Fluorescent image of a single cilium and the locations of the UV laser stimuli. Stimulus intensity is depicted as a color scaling that is independently shown with a scale bar. (**b–g**) Waveforms of the current induced by local laser irradiation. (**h**) Relation between the distance from the knob and local current responses from eight cells. The lowest horizontal color bars indicate the length of the cilia of corresponding colored plots [49]

channel in living olfactory single cilium using a submicron local $[Ca^{2+}]_i$ elevation with the laser photolysis [33]. The data showed that $Cl_{(Ca)}$ channels are expressed broadly along the cilia (Fig. 8).

## 2.3 *Excitatory and Inhibitory Responses of the Olfactory Receptor Cell*

### 2.3.1 Excitatory Response: Nonlinear Signal Amplification

It has been widely accepted that the olfactory transduction system shows a high cooperativity establishing nonlinear signal amplification [21, 46]. In 2002, Takeuchi and Kurahashi reported that cAMP responses of the olfactory cilia showed a high cooperativity of Hill coefficient of about 5, similar to that of the odorant response [51]. Therefore, it is highly likely that nonlinear boosting of system is achieved solely at the channel level, not by the enzymatic cascade

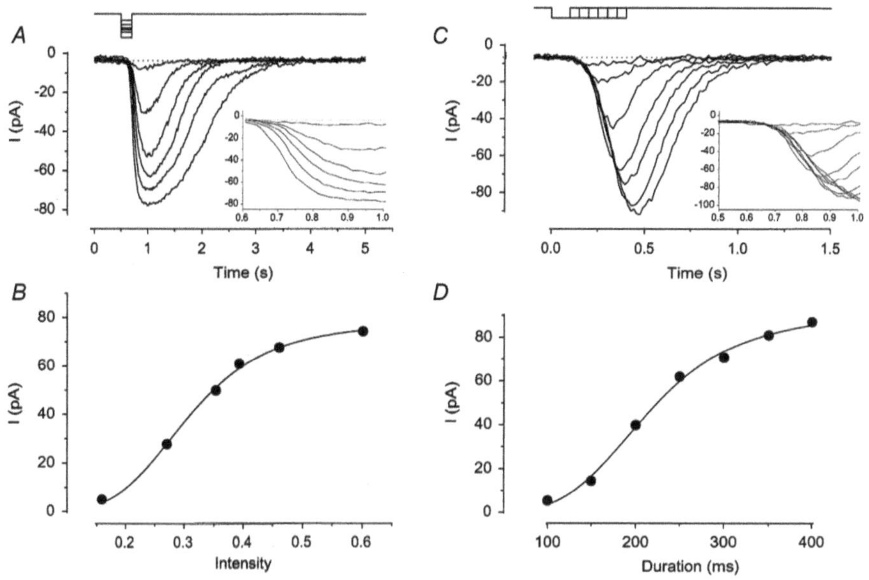


**Fig. 9** Dose–response relation with high cooperativity. (**a**, **b**) Light-intensity dependence. (**c**, **d**) Duration dependence [51]

locating upstream to adenylyl cyclase. The signaling process between the receptor and adenylyl cyclase would be almost linear [52, 53].

The response amplitude was dependent on both the intensity and duration of light illumination (Fig. 9). With a fixed duration, an increase in the light intensity increased the current amplitude monotonically. One noticeable feature was that the slope of rising phase became steeper as the light intensity was increased (Fig. 9a, inset). The relation between response amplitude and intensity could be fitted by the Hill equation with high cooperativity (Hill coefficient, $n_H = 5$). The amplitude of current responses was also dependent on duration of the light stimuli at fixed intensity (Fig. 9c). The relation between current amplitude and light duration was also fitted by the Hill equation (Fig. 9d) with a large Hill coefficient ($n_H = 4$). In contrast to the experiment in Fig. 9a, however, the slope of the rising phase was constant (Fig. 9c, inset). This is because the rising phase of cAMP-induced current is specified by the steepness of $[cAMP]_i$ increase that is dependent on the strength of the light. The time course, therefore, represents the dose–response relation of this system, which again shows a high cooperativity (Fig. 10).

Signal amplification of the transduction system can be compared in homologous systems (Fig. 11). It has been shown that the fundamental mechanism for signal transduction is similar between olfactory and photoreceptor cells. In rod photoreceptor cells, molecular amplification has been well documented by measuring activities and time courses of enzymes [54–56]. In contrast in ORC, the volume of the intra-ciliary space is extremely limited for the quantitative biochemical measurements. In addition, responsiveness of ORCs to odorants is heterogeneous,

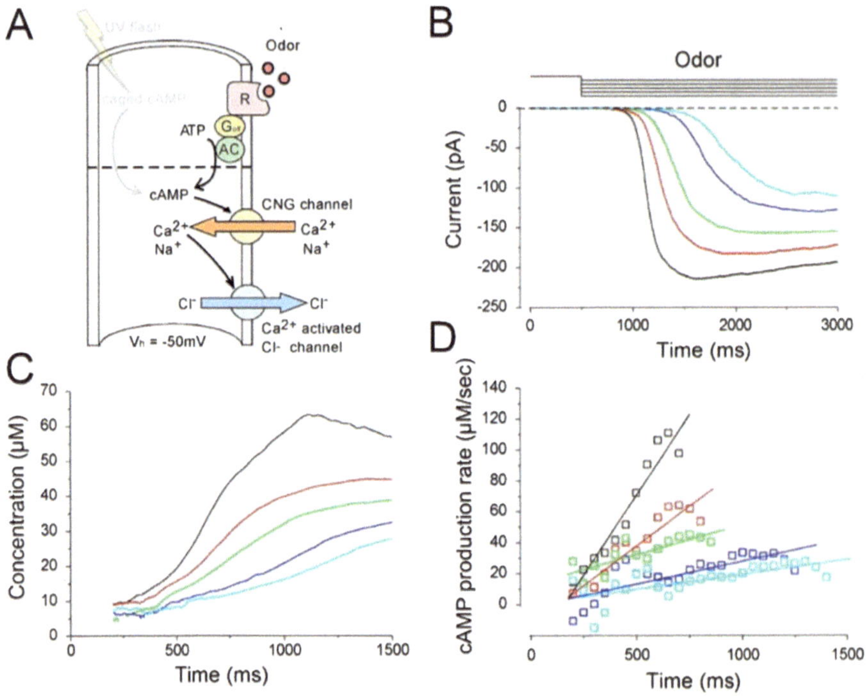


**Fig. 10** Odor-induced current, [cAMP]$_i$ change, and cAMP production rate during ligand application under the physiological condition. (**a**) Experimental logic. Experiments were conducted to drive native transduction pathway under the physiological condition. (**b**) Current responses induced by different intensity of long odor steps, while the duration was kept constant (3 s). (**c**) Cytoplasmic cAMP concentration. Curves were obtained by converting curves in B with the inverse function of the Hill equation. (**d**) cAMP production rate obtained as a derivative of curve C. Average slopes within 100 ms time windows were obtained from curves in C for every 50 ms. The rising phase was fitted by the straight line with the least square method [34]

which makes further difficult to obtain concentration of products in the cilia (Fig. 11). In 2005, Takeuchi and Kurahashi obtained intra-ciliary cAMP dynamics by comparing odorant-induced responses with the caged cAMP-induced current that was quantified in terms of the production for cytoplasmic cAMP [34]. It was shown that in olfaction cAMP production was extremely small (200,000 molecules/s/cell at the maximum, Fig. 11), in contrast to the cGMP hydrolysis in the rod (>100,000 molecules/photon). The observed numbers indicate that the ORC has lower amplification at the enzymatic cascade. Seemingly, such low amplification is a disadvantage for the signal transduction, but this unique mechanism would be essential to reduce the loss of ATP that is broadly used for the activities of cells (Fig. 11). Apparently, transduction by a smaller number of second-messenger formations would be achieved by the fine ciliary structure that has a high surface–volume ratio. In addition, it is speculated that this low amplification at their

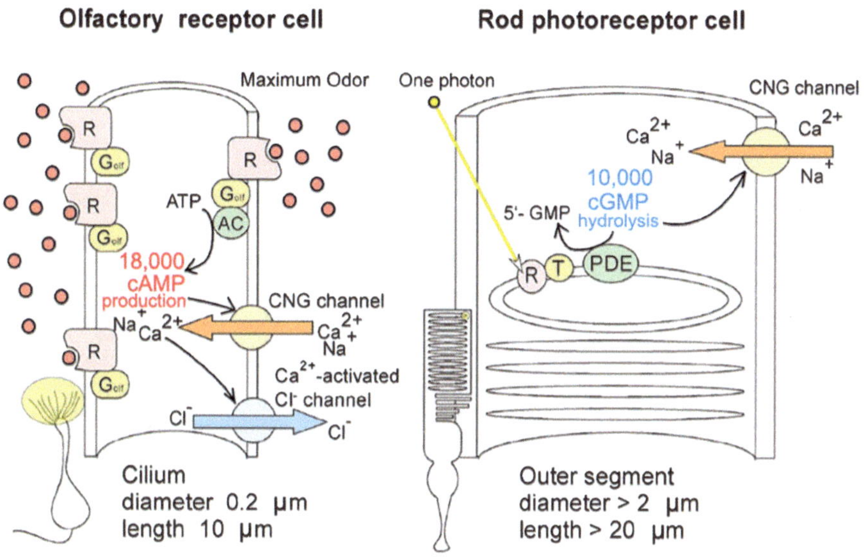


**Fig. 11** Scheme of signal amplification site. *Left*: ORC. *Right*: rod photoreceptor cell

enzymatic processes may be the reason why the ORC has acquired high amplification at the final stage of transduction channels using $Ca^{2+}$ as a third messenger.

### 2.3.2 Inhibitory Response from Inside of the Cell: Calcium-Dependent Adaptation

The adaptive properties of the ORCs have been pointed out even from very early studies of electrophysiological recording. The unit recording of spike discharges from ORCs shows a gradual decline during the long-lasting odor stimulation or application of repeated odorant stimuli [57]. Later, properties and mechanisms expressing such adaptation were investigated in further detail with whole-cell recording of ORCs. The size of odorant-induced current is gradually decreased with time during the long odorant stimulation (Fig. 12a,b [41]). Furthermore, current responses induced by repetitive stimulation were also decreased with time (Fig.12c,d [21]). In double-pulse experiments, the amplitude of secondary current becomes bigger, as the interstimulus interval becomes longer (Fig. 12d [21]).

There are two possibilities that explain the reduction of current during a long odor exposure and in double-pulse stimulation. One is a reduction in the absolute amplitude and the other is a shift of dose–response relation maintaining the maximum current constant. As adaptation of sensory system, the latter is thought to be appropriate, because such change in dose–response relation is directly related to an expansion of the dynamic range. In the odorant-induced current in ORC, it has been shown directly that dose–response relations are shifted to higher doses under

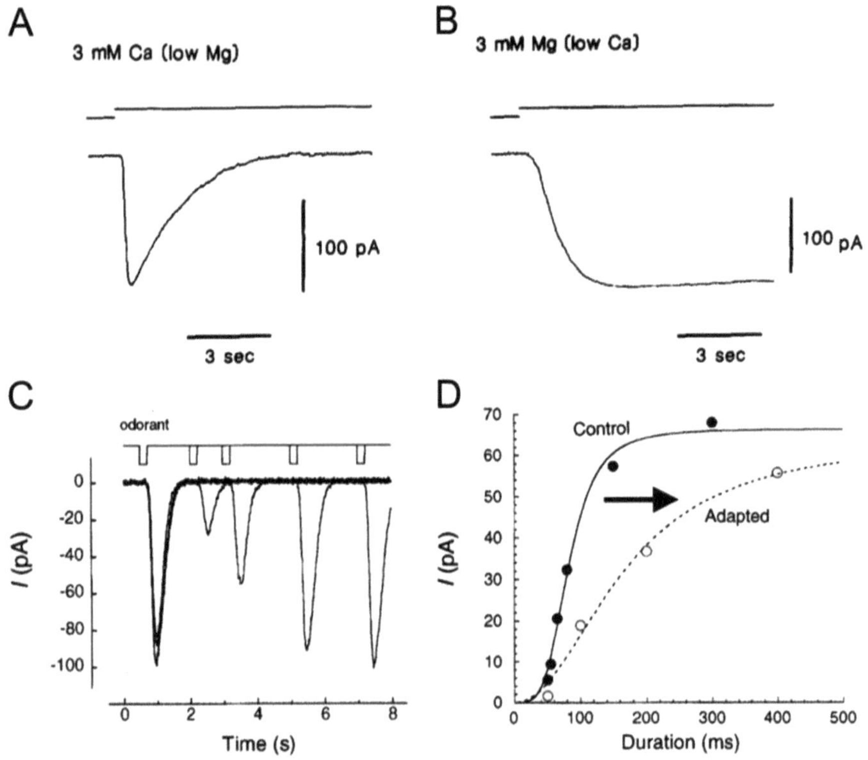


**Fig. 12** Odorant adaptation. Waveforms of odorant-induced currents recorded in 3 mmol/L $Ca^{2+}$ (**a**) and 3 mmol/L $Mg^{2+}$ (**b**). Adaptation is abolished, when the external Ca is replaced with Mg. Also, note that the time course for the rising phase becomes less steep in 0 Ca. This is explained from the fact that the rising phase is regulated by time-dependent increase in $[cAMP]_i$ in the cilia and the $[cAMP]_i$–response relation becomes nonlinearly amplified in the presence of Ca [41]. (**c**) Response reductions by a conditioning pulse and their recovery. (**d**) Relations between stimulus and response of ORCs under control and adapted conditions [21]

adapted conditions [21]. The data therefore suggest that adaptation is responsible for establishing a wider dynamic range.

It has been shown that olfactory adaptation is observed only, when $Ca^{2+}$ is present in external media. Falling phase of an inward current induced from odorant stimulation returns back to the baseline in normal $Ca^{2+}$ condition. On the other hand, under the absence of $Ca^{2+}$ in extracellular solution, the current stays at the constant level as long as the odorant is present in external media [19]. Furthermore, action of $Ca^{2+}$ has also been examined by several approaches [21]. Now, the most likely mechanism is that adaptation is regulated by a negative feedback to CNG channel by the influx of $Ca^{2+}$, presumably via a calmodulin [58, 59] and other proteins. Subunit CNGA4 may play an important role for adaptation [60]. Knockout of calmodulin does not cause significant changes in the adaptation [61]. This may indicate that there is alternative mechanisms mediating adaptation and may compensate for the lack of calmodulin.

#### 2.3.3 Inhibitory Response from Outside of the Cell: Olfactory Masking

Olfactory masking has been used to erase the unpleasant smells in long period of the human history. In our daily life, off-flavor substances in foods and beverages not only induce exogenous smells but also reduce the intrinsic flavors of foods and beverages even with a very low concentration. In many cases, a very low effective concentration has made it difficult to identify underlying off-flavor substances. In the next part, this phenomenon is expounded precisely with the latest knowledge.

## 3 Odorant Suppression of Transduction Channels

### *3.1 Inhibition by Odorants of the Transduction Channel*

In 1994, it was shown that certain types of volatile substances were capable of inhibiting the current that is generated by odorants [20]. When the odorant itself induces the response in a cell, such inhibition can be recognized as self-suppression. The suppression is observed consistently among cells examined. Ache's group [62] also reports that odorant responses are inhibited through the activities of phosphoinositide 3-kinase ($PI_3K$). In this case, suppression of response is heterogeneous among cells and therefore, such inhibition may be independent from odorant actions to be described here.

Since the kinetics of suppression is so rapid when examined with amyl acetate, the target of stimulants has been thought to be the CNG channel rather than suppression to upstream proteins that include the receptor protein, G protein, and adenylyl cyclase; inhibition of these molecules would be influenced as a delayed response [51]. To support such idea, Yamada and Nakatani (2001) showed that the response, induced by the phosphodiesterase inhibitor (IBMX) and membrane-permeable analog of cAMP (8-Br cAMP), was suppressed by externally applied volatile substances [63]. The idea was further confirmed later by experiments with recombinant CNG channels expressed in *Xenopus* oocytes. Furthermore, responses, induced by the photolysis of cytoplasmic caged cAMP in the ORC's cilia, were suppressed by odorants [64]. Such channel suppression is caused by a wide variety of odorants and the degree of suppression is variable depending on the species of chemicals (Fig. 13, Table 1).

In a report by Takeuchi et al (2009), various kinds of masking agents used in perfumery were shown to inhibit olfactory CNG channels, and the degree of suppression was proportional to their ability for olfactory masking in human [33]. These agents are commonly used to erase the original smell emitted from the underlying chemical substances. For instance, aroma of shampoo is designed to

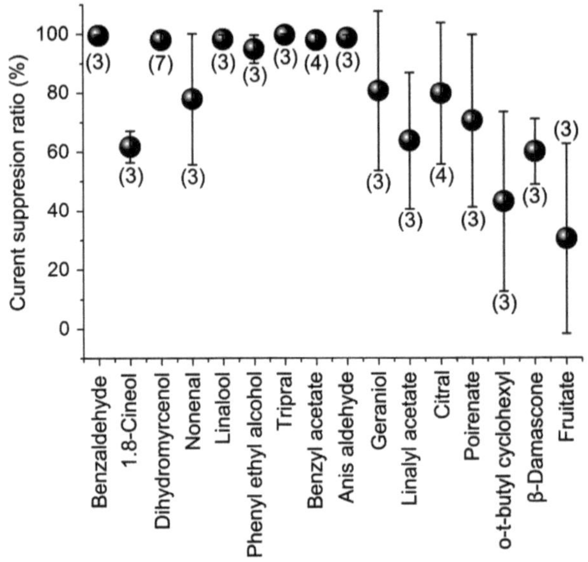


**Fig. 13** Current suppression ratio (SR) of 16 odorants. Odorant concentration in Ringer's solution was 0.1%. SR = 0 indicates no effect. *Error bars* indicate SD. *Numbers in parentheses* indicate the numbers of cells examined ([33], Fig. 1d)

erase smells emitted from human body and smells emitted even from detergents themselves and other compounds included in the product. Most of such masking agents produce also their own scents by activating receptor cells, presumably because suppression is induced with high concentrations (see Sect. 5).

## 3.2 Suppression of Signal Transduction by Off-Flavors in Foods and Beverages

In a recent work, more potent suppressors for the CNG channel were found in off-flavor substances included in wines [65]. One of the well-known off-flavors is 2,4,6-trichloroanisole (TCA) [73]. It has long been pointed out that TCA degrades the quality of wines even with very slight contamination (1 ppt = 5 pmol/L). A major cause of deterioration was initially responsible for the cork, and therefore, it was called "corked wine" or "cork taint." Chemically, TCA is produced from phenol. When the cork was sterilized with a Cl-breached compound, phenol included in wines is converted into 2,4,6,trichlorophenol (TCP). Then, this substance is further converted into TCA by the action of fungi living in the cork (Fig. 14). After these processes were revealed, some wineries avoided using Cl-containing sterilization and started to use the screw cap. Although the fraction of corked wine has been reduced by such carefulness, some fraction of wine has still been tainted. Now, generation of TCA is thought to be originating from a wide variety of sources (Fig. 14).

**Table 1** Odorants/compounds blocking the olfactory CNG channel

| Odorant/substance | Structure | Log *D* at pH 7.4[a] | Channel | $EC_{50}$ [μmol/L] or block (%) at single conc. | References |
|---|---|---|---|---|---|
| Amyl acetate | 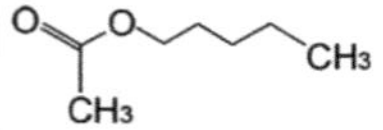 | 2.52 | Native CNG | 61.8% (0.1% solution) | [20] |
| | | | Homomeric CNGA2 | | [64] |
| Isoamyl acetate | 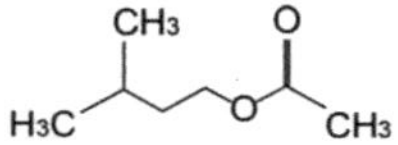 | 1.99 | Heterooligomeric CNGA2 + A4 + B1 | 27–57% (0.5 μl/ml) | [64] |
| Anisole | 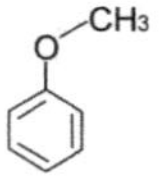 | 2.01 | Heterooligomeric CNGA2 + A4 + B1 | 48–81% (0.5 μl/ml) | [64] |
| Limonene | 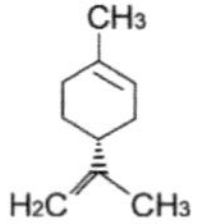 | 4.40 | NA | NA | [20] |
| Cineole | 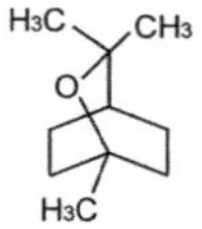 | 2.85 | Heterooligomeric CNGA2 + A4 + B1 | 8.9–61% (0.5 μl/ml) | [20, 33, 64] |
| Geraniol | 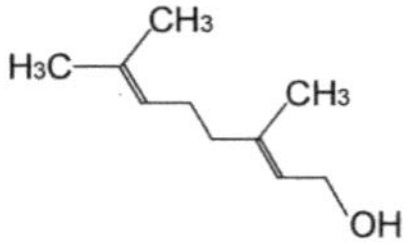 | 3.41 | Native CNG | 80.6% (0.1% solution) | [33] |
| | | | Native CNG | 29 μmol/L ($EC_{50}$) | [65] |

(continued)

**Table 1** (continued)

| Odorant/substance | Structure | Log *D* at pH 7.4[a] | Channel | $EC_{50}$ [μmol/L] or block (%) at single conc. | References |
|---|---|---|---|---|---|
| TCA | 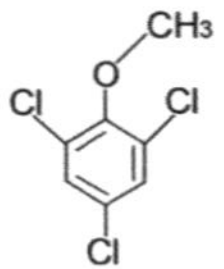 | 3.87 | Native CNG | 0.19 μmol/L ($EC_{50}$) | [65] |
| TBA | 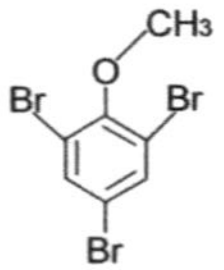 | 4.40 | Native CNG | 1.4 μmol/L ($EC_{50}$) | [65] |
| FM1-43 | 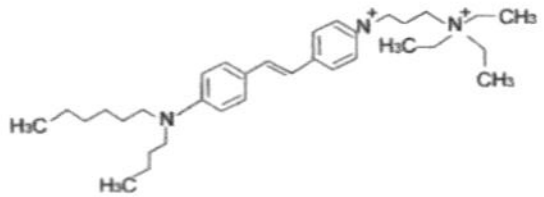 | NA | Native CNG | NA | [66] |
| L-cis diltiazem | 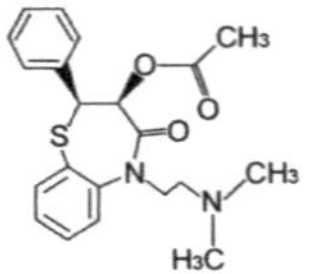 | 2.06 | Native CNG | 5.8 μmol/L ($EC_{50}$) | [65] |
| Dequalinium | 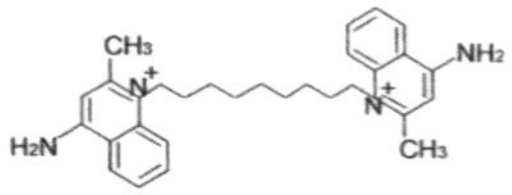 | NA | CNGA 2 (state-dependent block) | 1.3 μmol/L ($EC_{50}$) | [67, 68] |

| | | | | | |
|---|---|---|---|---|---|
| Tetracaine | 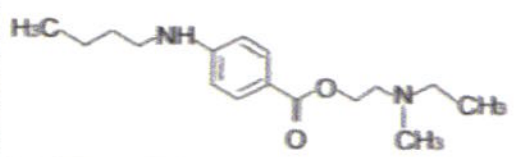 | 2.26 | Subunit 1 (state-dependent block) | 102.4 μmol/L ($K_D$) | [69] |
| Pseudechetoxin (PsTx) | 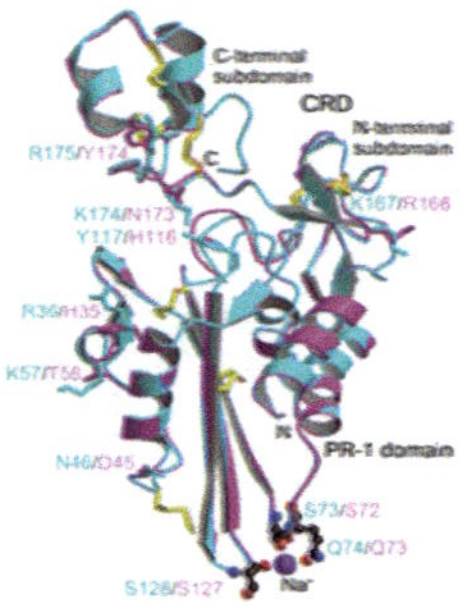 | NA | CNGA1, 2 (pore blocker) | | [70–72] |

[a]Log *D* was obtained from Chemspider. www.chemspider.com/

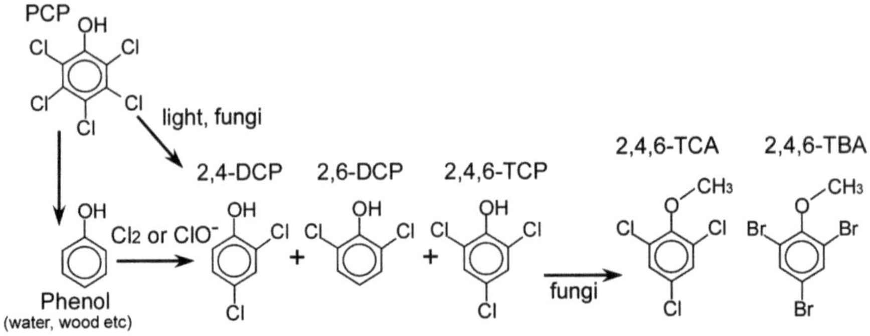


**Fig. 14** Generation of TCA in nature. Note that the fungi are involved in the synthesis. *PCP* pentachlorophenol; *2,4-DCP* 2,4-dichlorophenol; *2,6-DCP* 2,6-dichlorophenol; *2,4,6-TCP* 2,4,6-trichlorophenol; *2,4,6-TCA* 2,4,6-trichloroanisole; *2,4,6-TBA* 2,4,6-tribromoanisole

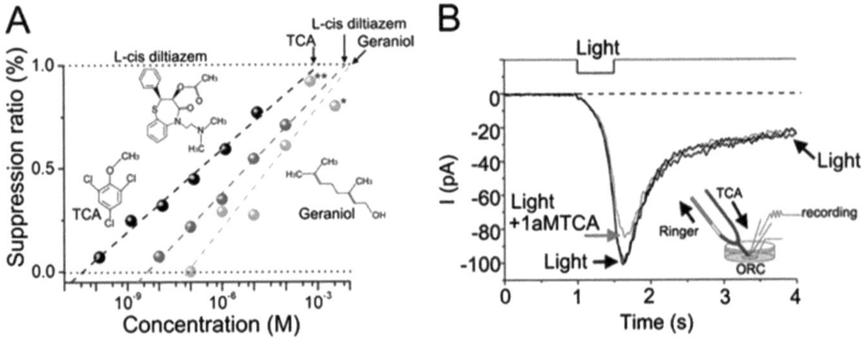


**Fig. 15** The least effective dose of TCA suppression. (**a**) Dose-suppression relation. TCA > > L-cis diltiazem > > geraniol. (**b**) Light-induced whole-cell current was suppressed by 1 amol/L TCA, applied via the U-tube system [65]. 1 mol/L caged cAMP was used

Thus, it has long been known that even a very small amount of TCA leads our sense of smell to an unpleasant direction. However, a question still remained: how does such a low concentration of chemical affect human olfaction? To answer to this question, Takeuchi et al. (2013) applied TCA to the ORC isolated from the newt. Although they failed to observe any response to a low concentration of TCA (e.g., sub-femtomolar), they recognized that the currents, induced by odorants (cineole) and by the photolysis of cytoplasmic caged cAMP, were suppressed by TCA [65]. The action of TCA in suppressing CNG channels was much more potent than that of the powerful masking agent (geraniol) and pharmacological agent (L-cis-diltiazem) (Fig. 15a). When TCA was delivered to the cell from a fast perfusion system (U-tube system), even an attomolar solution caused detectable reduction in the transduction current (Fig. 15b). It was shown that such high efficiency was, in part, achieved by a time integration of the effect. Although the transduction current is a mixture of CNG and $Cl_{(Ca)}$ components, the action seems to be strong at CNG channel; $Cl_{(Ca)}$ was relatively resistant. However, since $Cl_{(Ca)}$ is sequentially activated by the action of $Ca^{2+}$ that flows into the cytoplasm through CNG channels, abolishment of current through CNG channels completely abolishes the total transduction current. Also, because this sequential channel opening exhibits nonlinear signal transmission, the degree of suppression varies depending on the concentration of drugs [33].

## *3.3 Mechanisms of Channel Block*

Odorant suppression of the CNG channel shows several unique features. Based on such properties, it is speculated that the action of TCA and masking agents is mediated via a lipid bilayer of the membrane (Fig. 16). A similar mechanism has been thought to be involved in the action of local anesthesia [74], although the effective concentration is extremely lower for TCA.

### 3.3.1 Dose Dependence

When TCA was applied to the cell with a brief pulse (e.g., 1 s) of puff, the response suppression shows a dose dependence that spans a wide range of concentration (5–6 log unit) and is fitted to a straight line, when the dose is plotted in a logarithmic scale, rather than being fitted by Michaelis–Menten or Hill relation that describes

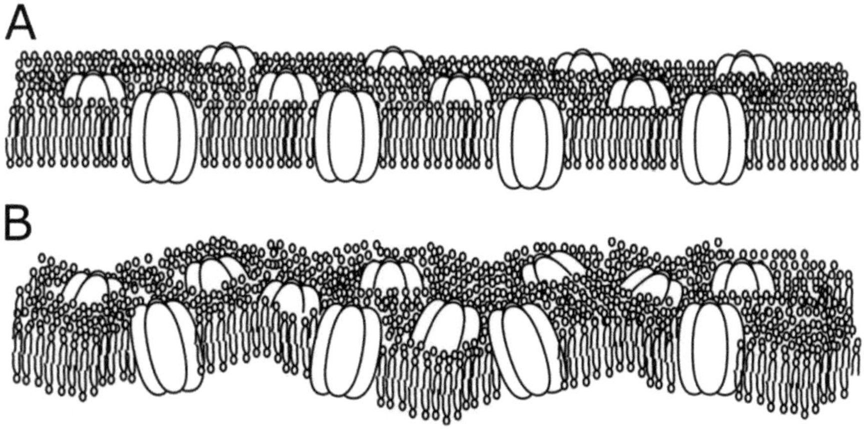


**Fig. 16** (**a**) Many CNG channels exist on the bilayer membrane in a normal condition. (**b**) Hypothesis; to suppress many CNG channels simultaneously, TCA may dissolve to the lipid bilayer of the ciliary membrane

ligand binding to the receptor site (Fig. 15a, TCA). This supports the notion that the underlying mechanisms are not a simple blockage.

### 3.3.2 Small Number of Molecules

As described before, TCA suppression can be observed even with 1 amol/L solution. 1 amol/L 1 mL solution applied includes approximately 600 TCA molecules. This solution was applied to the cell from the pipe having an opening diameter of about 200 μm. The stream directly hits the cilia, so, only a very small amount out of 600 molecules can impact the cilia. The transduction current of ORCs is carried by 100 thousands of CNG channels. This number was estimated from the channel density obtained from the inside-out version of the patch clamp applied to the cilium (see Fig. 5 and [38, 39]). This density was multiplied by the surface area of the cilia, because it has been shown that CNG channels distribute almost evenly along the entire cilia (see Fig. 7 and [49]). If suppression was caused by a direct binding of molecules to channels, a small number of channels blocked cannot represent the reduction of the current (Fig. 17). One percent reduction of the net current should be a function of 10,000 channels.

### 3.3.3 Time Integration

It is also shown that the suppression is dependent on exposure time in TCA solution. When the preexposure time before light stimulation (for photolysis of caged cAMP) is prolonged, the degree of suppression was also increased (Fig. 18). Furthermore, the recovery from suppression was slow displaying the half recovery of 10–20 s. It thus seems likely that the TCA effect has time integration that may contribute to the

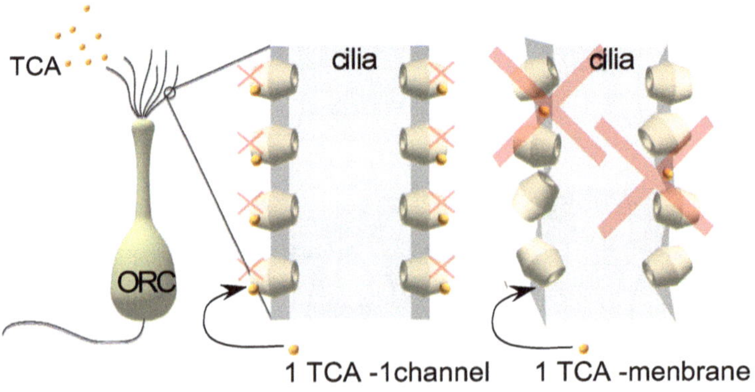


**Fig. 17** Two hypotheses; one TCA molecule may suppress many CNG channels simultaneously. It seems likely that TCA dissolve into membrane (lipid bilayer) first. The CNG channel block may be caused via structure change of lipid and channels

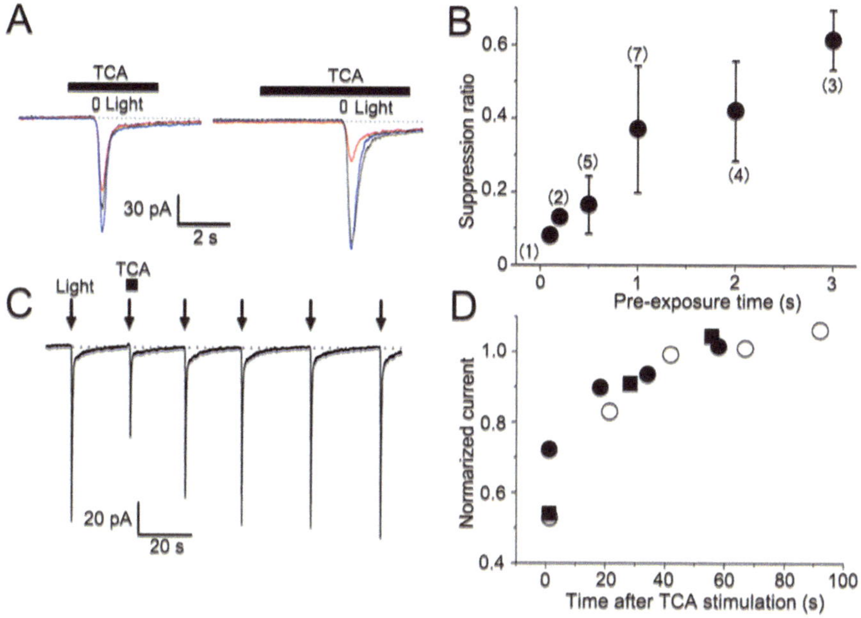


**Fig. 18** Time dependence of onset and offset of TCA suppression. (**a**, **b**) Onset time course of TCA suppression. (**c**, **d**) Offset time course of TCA suppression [65]

high efficiency. Such slow time course is consistent with an idea that TCA is gradually accumulated within the lipid bilayer before affecting ion channels.

### 3.3.4 Dependence on the Octanol/Water Partition Coefficient Log *D*

Although the efficiency varies largely, many types of chemicals can suppress CNG channels (Fig. 19a). The chemicals show large diversity in their structure. It is very unlikely that such diverse types of ligand suppress ion channels through the binding to the specific site.

Another property accompanying this suppression is its dependence on the hydrophobicity, the partition coefficient at octanol/water boundary of the chemicals (Fig. 19b). Especially in the biological systems, pH is strongly buffered to 7.4. So, partition coefficients at pH 7.4 (Log *D*, pH 7.4) are used in discussing hydrophobicity of chemicals that is present in biological organs. Strong dependence of the suppression on the partition coefficient indicates that the volatile substances act after integrating with the lipid bilayer.

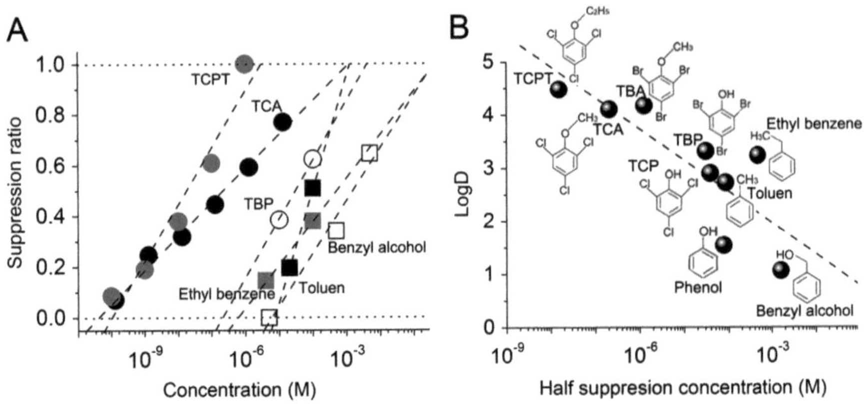


**Fig. 19** Current suppression relations. (**a**) Relation between SR and concentration by TCA analogs. (**b**) Relation between Log$D$ and half-suppression concentrations. $R = 0.84$. Modified from [65]

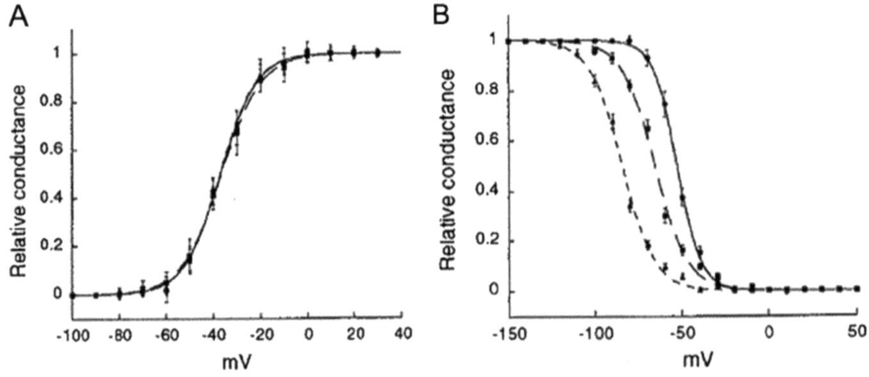


**Fig. 20** Effects of amyl acetate on activation and inactivation of voltage-dependent Na currents. (**a**) Activation curves (*filled circle* control, *filled square* 0.01 mmol/L amyl acetate, *filled triangle* 0.1 mmol/L amyl acetate). (**b**) Shift of inactivation curve of Na channel (*filled circle* control, *filled square* 0.01 mmol/L amyl acetate, *filled triangle* 0.1 mmol/L amyl acetate [75])

### 3.3.5 Diversity of Channels Suppressed and Similarity of Spectrum

Besides CNG channels, voltage-gated Na channel (Na), delayed K channel ($K_{delayed}$), Ca-activated K channel ($K_{Ca}$), T-type Ca channel ($Ca_T$), and L-type Ca channel ($Ca_L$) [75] are also sensitive to odorant suppression with different sensitivities, but the $Cl_{(Ca)}$ channel seems to be resistant [33, 65]. When compared with a fixed odorant concentration, the inhibition is most potent at CNG [76], followed by Na, Ca, $K_{Ca}$, and then $K_{delayed}$ channels. In case of Na channel, the reduction of amplitude is caused by the shift of the inactivation curve to the positive direction [75] (Fig. 20), while CNG channels do not show inactivation or desensitization [39].

As described, the degree of suppression on the CNG channel is variable depending on the type of chemicals. The rank order among substances in suppressing CNG channels matches that observed in Na channel [76]. This indicates that inhibition of these channels is regulated through a common mechanism.

# 4 Comparison of Channel Suppression with the Human Sensation

## *4.1 Masking Agents*

Takeuchi et al. also investigated the ability of CNG channel inhibitors to mask human olfactory sensation (Fig. 21) [33]. Several types of agents were selected. One of each was mixed with the vapor of isovaleric acid and panelists scored the degree of masking. In this kind of measurements, employment of expert panels for evaluating masking effects was important, because it was needed to extract the smell of isovaleric acid even in the mixture. People, who do not have an experience of evaluation, confuse the original smell of isovaleric acid in the mixture. The degree of masking varied among odorants and was correlated with the ability for suppressing the CNG channel (Fig. 21). It was important to pay attention to the solubility of chemicals. In human test, panels examine airborne stimuli that dissolve into the olfactory mucus layer, while in cell experiments chemicals are applied from the liquid phase with known concentrations. In experiments, chemicals that show similar solubility were selected.

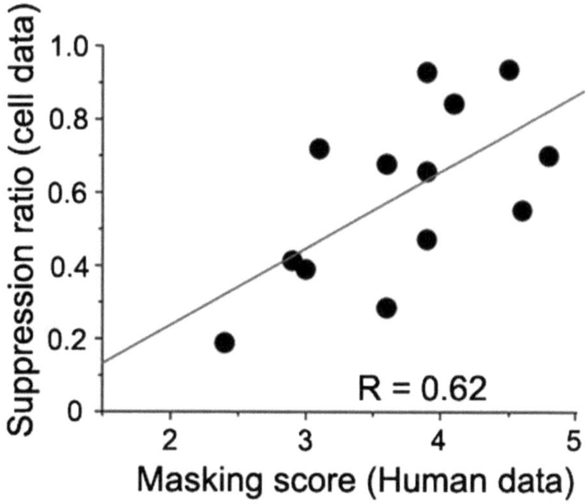


**Fig. 21** Relation between human masking score and channel suppression ratio. $R = 0.62$. Modified based upon data from [33]

## 4.2 Off-Flavors

Effect of TCA and related substances in suppressing the wine odor was also examined in panels. An essence in this kind of experiment is that the examiner has to avoid employing experts of wine evaluation, because they have their own criteria to identify TCA or bouchonne (cork taint in French). Panels selected are general experts for evaluating qualities of foods and beverages. They trained for and passed the panel selection test that has been conducted yearly. Among possible panelists, who do not hesitate to smell wines, 20 people were randomly selected. The concentration of recognition thresholds of TCA, TBA, and TCP for detecting the reduction in the wine odor were measured with the triangle test (Table 2). The rank order was TCA ≥ TCB >> TCP, which matched the rank order of these substances on suppressing CNG channel (Fig. 22).

**Table 2** Human sensory test by TCA relatives. Note that the effects are TCA ≥ TBA >> TCP [65]

| | | | |
|---|---|---|---|
| TCA | 3 ppt (14 pM) | 10 ppt (47 pM) | 14 ppt (71 pM) |
| Red wine | 7 | 12[a] | – |
| White wine | 5 | 3 | 15[a] |
| TBA | 10 ppt (29 pM) | 30 ppt (87 pM) | 10 ppt (290 pM) |
| Red wine | 4 | 7 | 13[a] |
| White wine | 6 | 5 | 12[a] |
| TCP | 30 ppb (152 nM) | 10 ppb (506 nM) | 300 ppb (1518 nM) |
| Red wine | 6 | 6 | 11[a] |
| White wine | 4 | 12[a] | – |

[a]According to binominal distribution, 11 or more correct judgments are significantly above chance at $p < 0.05$. $N = 20$

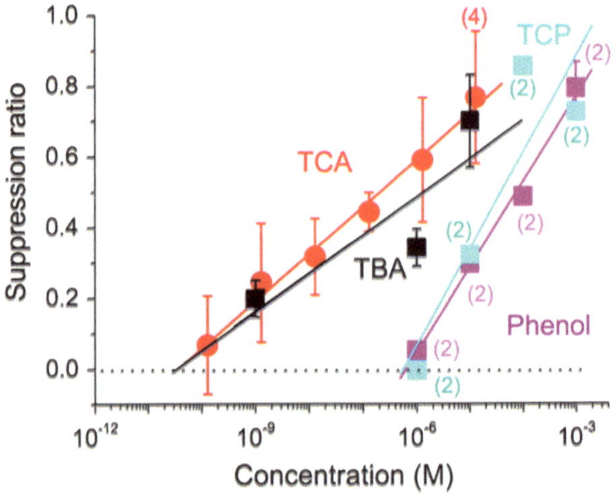


**Fig. 22** Current suppression by TCA relatives. Dose-suppression relation measured with cAMP-induced current. Note that the effects are TCA ≥ TBA >> TCP [65]

# 5 Masking and Deterioration of Foods and Beverages

Since the flavors of foods and beverages are strongly dependent on the olfactory sensation, suppression of olfactory transduction channels strongly affects the quality of those products. In addition to the extrinsic unpleasant smell emitted by off-flavors, such inhibition reduces the intrinsic pleasant odor. When examined with humans, such reduction is observed with a contamination for 10 ppt TCA or less amount. TCA of 10 ppt is equivalent to 50 pmol/L, and this value is further reduced, when they are evaporated into the air before getting into the nasal cavity. Furthermore, the concentration of TCA in the nasal mucus is determined by the partition at the air/water boundary. Experimentally, the final concentration through such processes has been shown to be as low as 5 fmol/L [65].

## *5.1 Induction of Smells by Masking Agents and Off-Flavors*

One may be puzzled why masking agents can generate sensation for their original aroma, while they suppress CNG channels. For instance, geraniol is a powerful inhibitor of the CNG channel [77, 78], but it induces a roselike aroma simultaneously. We assume that such sensation is generated by differences in the dose dependence and in temporal kinetics between excitation and inhibition. For most of odorants, suppression occurs at higher concentration than their emission of original aroma. In addition, the time course of suppression is rather simple; it occurs when the surrounding environment contains the inhibitor. Also, the effect disappears quickly, when the substance is removed (note that this is slightly different for TCA). But, excitation is mediated through activation of several enzymes and accumulation of cytoplasmic factors. These factors remain activated, until they are broken down and/or extruded. Such excitatory signals can be amplified further in the brain. It thus seems likely that the sense of smell is induced from masking agents as a result of differences in the effective dose and time dependence between the excitation and inhibition.

In case of TCA, it has been known that they induce musty unpleasant smells when included in wines and other beverages. It is natural to think that high concentration of TCA causes their intrinsic smell that is mediated through activation of receptor proteins. Such smell could be a musty smell that can be quickly recognized, as well as the reduction of intrinsic flavor induced by a suppressing effect of TCA.

The biggest question is why sub-femtomolar concentration of TCA causes different impression from normal wines. Low level of TCA (down to amol/L level) has been shown to inhibit CNG channels, which is equivalent to the reduction of olfactory sensation. Such effective concentration of TCA seems to be much lower than the value that causes excitation of the cell (e.g., 1 nmol/L–1 μmol/L).

The finding of suppression at very low concentration of TCA may serve several possibilities that have yet been considered.

The simplest possibility is that the human olfactory system may have receptors to detect sub-femtomolar concentrations of TCA. Alternatively, and beyond a simple partition between air and water, the olfactory mucus may contain some factors that increase the activity of TCA at the plasma membrane level. Odorant-binding protein found in the mucus may serve such effects [79]. Another possibility is that TCA may trigger some chemical reactions that increase some products that evoke the musty smell in the mucus. These hypotheses may be further investigated in experiments using ORCs and olfactory epithelium from human.

Also, it may be possible to assume that suppression itself causes sensation that may be accepted as an unpleasant smell. At a high-concentration contamination, original smell by TCA that is caused by receptor activation would easily be detected upon evaluation of products including wine. Usually, wine that includes the threshold level of TCA is evaluated with the triangle test that extracts even a very small difference in comparison with the quality of normal wine. So, the detected signal may not be the same as the original smell (musty smell) of TCA. Because the effect of TCA shows time integration, the low concentration effect may become obvious depending on the time exposed. This may indicate that the quality of products became worse during evaluation and tasting.

Theoretically, activation of receptor protein by a femtomolar ligand is unusual. Instead, it may be possible to assume that the inhibitory pathway may induce olfactory sensation. This idea is derived from the following observations: *a.* low effective concentration of TCA to induce sensation; *b.* agreements in concentrations between suppression and induction, both could be induced via the same mechanism; and *c.* similar sensory qualities induced by TCA, TBA, and TCP. Although the effective concentrations are different, the rank order matches that of channel suppression. Receptor-mediated olfactory sensation is variable in quality depending on the very small change in the chemical structure.

It is not a surprising fact that one feels an olfactory sensation upon the shutdown of the system. In visual system, similar phenomena that induce sensation by reduction of input have been well known. An example is an opponent color. After one looks at the red spot on a white field, a sudden removal of red spot causes a sensation for the green spot (complementary color) that is not actually presented.

TCA derivatives (TBA, TCP, and TCPT, a synthesized molecule) show similar effects for both inhibitions and induction of smell. Actions of TCA derivatives are unusual in terms of their high efficiency and induction of unpleasant smells at low concentrations. Of course, TCA derivatives induce different qualities of odor, when the concentration is increased, presumably by activating certain types of receptor proteins by their original binding to appropriate binding sites.

Another possibility for the reduction of olfactory sense may be derived from the macroscopic view of signal processing. Since the effect of TCA and other maskers are universal to the cells via suppression of CNG channel, the effect spans wide varieties of cell populations. When one considers an inhibition to the inhibitory circuit in the brain, such wide effect may cause a broad noise in the olfactory signal processing. It has been reported that application of wide spectrum of noise causes reduction in detecting certain type of olfactory sensation [80]. Effects of CNG channel inhibitors may also serve inhibition of sensation through a similar manner.

## 5.2 TCA Is Included in a Wide Variety of Foods and Beverages

In the work by Takeuchi et al. (2013), TCA was found not only in wines and other related products that emit musty smell [81–85] but also in a wide variety of foods and beverages, especially when the qualities of products are judged to be low grade from expert evaluators [65].

People did not pay attention to the presence of TCA in those products, because most of such foods and beverages did not show musty smells that have been thought to be a typical feature of TCA contamination. Because action of TCA was found to be a reduction of flavor in the work by Takeuchi et al. (2013), they tried to find TCA in products that are rated to be a low quality [65]. Since the concentration of TCA included was estimated to be lower than the detection threshold by GC/MS, TCA was measured after concentrating the sample solution.

Furthermore, TCA was found not only in foods and beverages but also in packing film, paper, and storage housings. These findings may suggest that TCA is not only generated within the products but also transferred from surround environments to products. It is highly likely that TCA degrades the quality of products. The finding serves possible targets to be paid attention to, in order to avoid deterioration of products.

## 5.3 Evolutionary Aspects of This Phenomenon

It is interesting to consider possible evolutionary reasons why TCA is detected at such low concentrations. Since the target of TCA seems to be a lipid, it appears unlikely that the perception itself is a product of biological evolution. Presumably, TCA molecules attack ORCs like a very effective drug. It is highly likely that animals may have evolved to detect such signals as a very initial sign for the deterioration of foods and beverages. At the same time, it may be interesting to assume that fungi may have used this chemical conversion for their biological survival strategy. If foods and beverages are judged to be non-delicious, the fungi living in them can survive more.

# 6 Summary and Perspectives

Sense of smell is initiated by ion channels that are directly facing to the external environment. Therefore, this sense is naturally affected from the external side of the body. When the modifier is generated naturally in foods and beverages, it degrades the quality of products considerably (off-flavors). In parallel, the ion channel can even be manipulated with artificially designed chemicals through the air. In the human history, such features have long been employed to erase unpleasant smells that include body odors, environmental malodors, and even bad odors emitted from meals (olfactory masking). It is interesting to see that these seemingly different phenomena share, at least in part, the same molecular mechanism that is presumably mediated via integration of odorant molecules with the lipid bilayer of the plasma membrane. The knowledge may serve to develop chemicals that are used for improving the quality of our life.

# References

1. Menco BP, Morrison EE (2003) Morphology of the mammalian olfactory epithelium: form, fine structure, function, and pathology. In: Doty RL (ed) Handbook of olfaction and gustation. Marcel Dekker, New York, pp 17–49
2. Pevsner J, Sklar PB, Snyder SH (1986) Odorant-binding protein: localization to nasal glands and secretions. Proc Natl Acad Sci U S A 83(13):4942–4946
3. Buck L, Axel R (1991) A novel multigene family may encode odorant receptors: a molecular basis for odor recognition. Cell 65(1):175–187
4. Firestein S (2001) How the olfactory system makes sense of scents. Nature 413(6852): 211–218
5. Firestein S, Shepherd GM (1995) Interaction of anionic and cationic currents leads to a voltage dependence in the odor response of olfactory receptor neurons. J Neurophysiol 73(2):562–567
6. Frings S, Lynch JW, Lindemann B (1992) Properties of cyclic nucleotide-gated channels mediating olfactory transduction. Activation, selectivity, and blockage. J Gen Physiol 100(1): 45–67
7. Frings S, Reuter D, Kleene SJ (2000) Neuronal $Ca^{2+}$-activated $Cl^-$ channels – homing in on an elusive channel species. Prog Neurobiol 60(3):247–289
8. Gold GH (1999) Controversial issues in vertebrate olfactory transduction. Annu Rev Physiol 61:857–871
9. Gold GH, Nakamura T, Lowe G (1990) Studies on the mechanism of olfactory transduction in vertebrates. Neurosci Res Suppl 12:S127–134
10. Kaupp UB, Seifert R (2002) Cyclic nucleotide-gated ion channels. Physiol Rev 82(3):769–824
11. Kleene SJ (1999) Both external and internal calcium reduce the sensitivity of the olfactory cyclic-nucleotide-gated channel to CAMP. J Neurophysiol 81(6):2675–2682
12. Pifferi S, Menini A, Kurahashi T (2010) Chapter 8. Signal transduction in vertebrate olfactory cilia. In: Menini A (ed) The neurobiology of olfaction. CRC Press/Taylor & Francis, Boca Raton, pp 203–224
13. Reisert J, Matthews HR (2001) Response properties of isolated mouse olfactory receptor cells. J Physiol 530(Pt 1):113–122
14. Schild D, Restrepo D (1998) Transduction mechanisms in vertebrate olfactory receptor cells. Physiol Rev 78(2):429–466

15. Takeuchi H, Kurahashi T (2003) Identification of second messenger mediating signal transduction in the olfactory receptor cell. J Gen Physiol 122(5):557–567
16. Pifferi S, Dibattista M, Sagheddu C, Boccaccio A, Al Qteishat A, Ghirardi F, Tirindelli R, Menini A (2009) Calcium-activated chloride currents in olfactory sensory neurons from mice lacking bestrophin-2. J Physiol 587(Pt 17):4265–4279
17. Pifferi S, Pascarella G, Boccaccio A, Mazzatenta A, Gustincich S, Menini A, Zucchelli S (2006) Bestrophin-2 is a candidate calcium-activated chloride channel involved in olfactory transduction. Proc Natl Acad Sci U S A 103(34):12929–12934
18. Stephan AB, Shum EY, Hirsh S, Cygnar KD, Reisert J, Zhao H (2009) ANO2 is the cilial calcium-activated chloride channel that may mediate olfactory amplification. Proc Natl Acad Sci U S A 106(28):11776–11781
19. Kurahashi T (1989) Activation by odorants of cation-selective conductance in the olfactory receptor cell isolated from the newt. J Physiol 419:177–192
20. Kurahashi T, Lowe G, Gold GH (1994) Suppression of odorant responses by odorants in olfactory receptor cells. Science 265(5168):118–120
21. Kurahashi T, Menini A (1997) Mechanism of odorant adaptation in the olfactory receptor cell. Nature 385(6618):725–729
22. Kurahashi T, Yau KW (1993) Co-existence of cationic and chloride components in odorant-induced current of vertebrate olfactory receptor cells. Nature 363(6424):71–74
23. Lowe G, Gold GH (1991) The spatial distributions of odorant sensitivity and odorant-induced currents in salamander olfactory receptor cells. J Physiol 442:147–168
24. Ache BW, Zhainazarov A (1995) Dual second-messenger pathways in olfactory transduction. Curr Opin Neurobiol 5(4):461–466
25. Fadool DA, Ache BW (1992) Plasma membrane inositol 1,4,5-trisphosphate-activated channels mediate signal transduction in lobster olfactory receptor neurons. Neuron 9(5):907–918
26. Restrepo D, Miyamoto T, Bryant BP, Teeter JH (1990) Odor stimuli trigger influx of calcium into olfactory neurons of the channel catfish. Science 249(4973):1166–1168
27. Schild D, Lischka FW, Restrepo D (1995) InsP3 causes an increase in apical [Ca2+]i by activating two distinct current components in vertebrate olfactory receptor cells. J Neurophysiol 73(2):862–866
28. Suzuki N (1994) IP3-activated ion channel activities in olfactory receptor neurons from different vertebrate species. In: Olfaction and taste XI. Springer, Berlin, pp 173–177
29. Kashiwayanagi M, Shimano K, Kurihara K (1996) Existence of multiple receptors in single neurons: responses of single bullfrog olfactory neurons to many cAMP-dependent and independent odorants. Brain Res 738(2):222–228
30. Kaur R, Zhu XO, Moorhouse AJ, Barry PH (2001) IP3-gated channels and their occurrence relative to CNG channels in the soma and dendritic knob of rat olfactory receptor neurons. J Membr Biol 181(2):91–105
31. Vogl A, Noe J, Breer H, Boekhoff I (2000) Cross-talk between olfactory second messenger pathways. Eur J Biochem 267(14):4529–4535
32. Chen S, Lane AP, Bock R, Leinders-Zufall T, Zufall F (2000) Blocking adenylyl cyclase inhibits olfactory generator currents induced by "IP(3)-odors". J Neurophysiol 84(1):575–580
33. Takeuchi H, Ishida H, Hikichi S, Kurahashi T (2009) Mechanism of olfactory masking in the sensory cilia. J Gen Physiol 133(6):583–601
34. Takeuchi H, Kurahashi T (2005) Mechanism of signal amplification in the olfactory sensory cilia. J Neurosci 25(48):11084–11091
35. Flannery RJ, French DA, Kleene SJ (2006) Clustering of cyclic-nucleotide-gated channels in olfactory cilia. Biophys J 91(1):179–188
36. Nakamura T, Gold GH (1987) A cyclic nucleotide-gated conductance in olfactory receptor cilia. Nature 325(6103):442–444
37. Kurahashi T (1990) The response induced by intracellular cyclic AMP in isolated olfactory receptor cells of the newt. J Physiol 430:355–371

38. Kurahashi T, Kaneko A (1991) High density cAMP-gated channels at the ciliary membrane in the olfactory receptor cell. Neuroreport 2(1):5–8
39. Kurahashi T, Kaneko A (1993) Gating properties of the cAMP-gated channel in toad olfactory receptor cells. J Physiol 466:287–302
40. Lancet D (1986) Vertebrate olfactory reception. Annu Rev Neurosci 9:329–355
41. Kurahashi T, Shibuya T (1990) Ca2(+)-dependent adaptive properties in the solitary olfactory receptor cell of the newt. Brain Res 515(1-2):261–268
42. Zufall F, Firestein S (1993) Divalent cations block the cyclic nucleotide-gated channel of olfactory receptor neurons. J Neurophysiol 69(5):1758–1768
43. Kleene SJ (1997) High-gain, low-noise amplification in olfactory transduction. Biophys J 73(2):1110–1117
44. Larsson HP, Kleene SJ, Lecar H (1997) Noise analysis of ion channels in non-space-clamped cables: estimates of channel parameters in olfactory cilia. Biophys J 72(3):1193–1203
45. Reisert J, Bauer PJ, Yau KW, Frings S (2003) The Ca-activated Cl channel and its control in rat olfactory receptor neurons. J Gen Physiol 122(3):349–363
46. Lowe G, Gold GH (1993) Nonlinear amplification by calcium-dependent chloride channels in olfactory receptor cells. Nature 366(6452):283–286
47. Matsuzaki O, Bakin RE, Cai X, Menco BP, Ronnett GV (1999) Localization of the olfactory cyclic nucleotide-gated channel subunit 1 in normal, embryonic and regenerating olfactory epithelium. Neuroscience 94(1):131–140
48. Hamill OP, Marty A, Neher E, Sakmann B, Sigworth FJ (1981) Improved patch-clamp techniques for high-resolution current recording from cells and cell-free membrane patches. Pflugers Arch 391(2):85–100
49. Takeuchi H, Kurahashi T (2008) Distribution, amplification, and summation of cyclic nucleotide sensitivities within single olfactory sensory cilia. J Neurosci 28(3):766–775
50. Kleene SJ, Gesteland RC (1991) Calcium-activated chloride conductance in frog olfactory cilia. J Neurosci 11(11):3624–3629
51. Takeuchi H, Kurahashi T (2002) Photolysis of caged cyclic AMP in the ciliary cytoplasm of the newt olfactory receptor cell. J Physiol 541(Pt 3):825–833
52. Ben-Chaim Y, Cheng MM, Yau KW (2011) Unitary response of mouse olfactory receptor neurons. Proc Natl Acad Sci U S A 108(2):822–827
53. Bhandawat V, Reisert J, Yau KW (2005) Elementary response of olfactory receptor neurons to odorants. Science 308(5730):1931–1934
54. Lamb TD, Pugh EN Jr (1992) A quantitative account of the activation steps involved in phototransduction in amphibian photoreceptors. J Physiol 449:719–758
55. Pugh EN Jr, Lamb TD (1990) Cyclic GMP and calcium: the internal messengers of excitation and adaptation in vertebrate photoreceptors. Vision Res 30(12):1923–1948
56. Pugh EN Jr, Nikonov S, Lamb TD (1999) Molecular mechanisms of vertebrate photoreceptor light adaptation. Curr Opin Neurobiol 9(4):410–418
57. Getchell TV, Shepherd GM (1978) Adaptive properties of olfactory receptors analysed with odour pulses of varying durations. J Physiol 282:541–560
58. Bradley J, Bonigk W, Yau KW, Frings S (2004) Calmodulin permanently associates with rat olfactory CNG channels under native conditions. Nat Neurosci 7(7):705–710
59. Chen TY, Yau KW (1994) Direct modulation by Ca(2+)-calmodulin of cyclic nucleotide-activated channel of rat olfactory receptor neurons. Nature 368(6471):545–548
60. Munger SD, Lane AP, Zhong H, Leinders-Zufall T, Yau KW, Zufall F, Reed RR (2001) Central role of the CNGA4 channel subunit in Ca2+-calmodulin-dependent odor adaptation. Science 294(5549):2172–2175
61. Song Y, Cygnar KD, Sagdullaev B, Valley M, Hirsh S, Stephan A, Reisert J, Zhao H (2008) Olfactory CNG channel desensitization by Ca2+/CaM via the B1b subunit affects response termination but not sensitivity to recurring stimulation. Neuron 58(3):374–386
62. Ukhanov K, Corey EA, Brunert D, Klasen K, Ache BW (2010) Inhibitory odorant signaling in Mammalian olfactory receptor neurons. J Neurophysiol 103(2):1114–1122

63. Yamada H, Nakatani K (2001) Odorant-induced hyperpolarization and suppression of cAMP-activated current in newt olfactory receptor neurons. Chem Senses 26(1):25–34
64. Chen TY, Takeuchi H, Kurahashi T (2006) Odorant inhibition of the olfactory cyclic nucleotide-gated channel with a native molecular assembly. J Gen Physiol 128(3):365–371
65. Takeuchi H, Kato H, Kurahashi T (2013) 2,4,6-trichloroanisole is a potent suppressor of olfactory signal transduction. Proc Natl Acad Sci U S A 110(40):16235–16240
66. Breunig E, Kludt E, Czesnik D, Schild D (2011) The styryl dye FM1-43 suppresses odorant responses in a subset of olfactory neurons by blocking cyclic nucleotide-gated (CNG) channels. J Biol Chem 286(32):28041–28048
67. Qu W, Moorhouse AJ, Lewis TM, Pierce KD, Barry PH (2005) Mutation of the pore glutamate affects both cytoplasmic and external dequalinium block in the rat olfactory CNGA2 channel. Eur Biophys J 34(5):442–453
68. Rosenbaum T, Gordon-Shaag A, Islas LD, Cooper J, Munari M, Gordon SE (2004) State-dependent block of CNG channels by dequalinium. J Gen Physiol 123(3):295–304
69. Fodor AA, Gordon SE, Zagotta WN (1997) Mechanism of tetracaine block of cyclic nucleotide-gated channels. J Gen Physiol 109(1):3–14
70. Brown RL, Haley TL, West KA, Crabb JW (1999) Pseudechetoxin: a peptide blocker of cyclic nucleotide-gated ion channels. Proc Natl Acad Sci U S A 96(2):754–759
71. Brown RL, Lynch LL, Haley TL, Arsanjani R (2003) Pseudechetoxin binds to the pore turret of cyclic nucleotide-gated ion channels. J Gen Physiol 122(6):749–760
72. Brown RL, Strassmaier T, Brady JD, Karpen JW (2006) The pharmacology of cyclic nucleotide-gated channels: emerging from the darkness. Curr Pharm Des 12(28):3597–3613
73. Buser HR, Zanier C, Tanner H (1982) Identification of 2,4,6-trichloroanisole as a potent compound causing cork taint in wine. J Agric Food Chem 30(2):359–362
74. Lundbaek JA (2008) Lipid bilayer-mediated regulation of ion channel function by amphiphilic drugs. J Gen Physiol 131(5):421–429
75. Kawai F, Kurahashi T, Kaneko A (1997) Nonselective suppression of voltage-gated currents by odorants in the newt olfactory receptor cells. J Gen Physiol 109(2):265–272
76. Kishino Y, Kato H, Kurahashi T, Takeuchi H (2011) Chemical structures of odorants that suppress ion channels in the olfactory receptor cell. J Physiol Sci 61(3):231–245
77. Sanhueza M, Bacigalupo J (1999) Odor suppression of voltage-gated currents contributes to the odor-induced response in olfactory neurons. Am J Physiol 277(6 Pt 1):C1086–C1099
78. Stotz SC, Vriens J, Martyn D, Clardy J, Clapham DE (2008) Citral sensing by Transient [corrected] receptor potential channels in dorsal root ganglion neurons. PLoS One 3(5), e2082
79. Pevsner J, Hou V, Snowman AM, Snyder SH (1990) Odorant-binding protein. Characterization of ligand binding. J Biol Chem 265(11):6118–6125
80. Weiss T, Snitz K, Yablonka A, Khan RM, Gafsou D, Schneidman E, Sobel N (2012) Perceptual convergence of multi-component mixtures in olfaction implies an olfactory white. Proc Natl Acad Sci U S A 109(49):19959–19964
81. Curtis RF, Land DG, Griffiths NM, Gee M, Robinson D, Peel JL, Dennis C, Gee JM (1972) 2,3,4,6-Tetrachloroanisole association with musty taint in chickens and microbiological formation. Nature 235(5335):223–224
82. Engel C, de Groot AP, Weurman C (1966) Tetrachloroanisole: a source of musty taste in eggs and broilers. Science 154(3746):270–271
83. Miki A, Isogai A, Utsunomiya H, Iwata H (2005) Identification of 2,4,6-trichloroanisole (TCA) causing a musty/muddy off-flavor in sake and its production in rice koji and moromi mash. J Biosci Bioeng 100(2):178–183
84. Qi F, Xu B, Chen Z, Ma J, Sun D, Zhang L, Wu F (2009) Ozonation catalyzed by the raw bauxite for the degradation of 2,4,6-trichloroanisole in drinking water. J Hazard Mater 168(1):246–252
85. Spadone JC, Takeoka G, Liardon R (1990) Analytical investigation of Rio off-flavor in green coffee. J Agric Food Chem 38(1):226–233

Top Med Chem (2017) 23: 151–174
DOI: 10.1007/7355_2016_101

Published online: 24 September 2014

# Chemosensory G Protein-Coupled Receptors (GPCR) in Blood Leukocytes

**Patrick Marcinek, Christiane Geithe, and Dietmar Krautwurst**

**Abstract** Chemosensory taste and smell perceptions are induced by adequate stimuli from our chemical environment interacting with their ca. 400 different odorant receptor types or 25 bitter taste receptor types, and sweet and umami receptor dimers in the sensory cells of the olfactory or gustatory epithelia, respectively. Beyond an expression in their canonical chemosensory epithelia, there is increasing evidence for an ectopic expression of at least some 90 olfactory and taste G protein-coupled receptors in a variety of non-chemosensory epithelia and cells, including our cellular immune system. Here we review the evidence for the expression of chemosensory receptors in different types of blood cells, and discuss their putative immunological functions and roles as targets for receptor- and immune cell-specific bioactives, such as foodborne flavor chemicals, or allelochemicals in general.

**Keywords** Blood leukocytes, Key food odorants, Olfactory, Receptors, Taste

## Contents



P. Marcinek, C. Geithe, and D. Krautwurst (✉)
Deutsche Forschungsanstalt fuer Lebensmittelchemie, Program Area III-Physiology/Chemoreception of Food Ingredients, Leibniz Institut, Lise-Meitner-Str. 34, 85354 Freising-Weihenstephan, Germany
e-mail: dietmar.krautwurst@lrz.tum.de

# 1 Introduction

Our chemical senses olfaction and taste sensitively respond to their adequate chemical stimuli and enable us to detect, recognize, discriminate, and hedonically evaluate them. The molecular basis for this is set by the expression of a vast variety of biological target molecules, for instance, by the expression of different families of chemosensory G protein-coupled receptors (GPCR) (Fig. 1) such as odorant receptors (OR) [1–3], trace amine-associated receptors (TAAR) [4–6], bitter taste receptors (TAS2R) [7–10], sweet and umami taste receptors (TAS1R) [11, 12], and free fatty acid receptors (FFAR) [13–15]. These chemosensory GPCR are typically expressed within the sensory cells of the olfactory epithelium in the nose or the taste buds of the tongue and have evolved to decode nature's signatures of smell and taste, enabling, for instance, the hedonic detection and energy monitoring and differentiation of food, or to prevent ingestion of poisonous substances [16, 17].

Most of what we eat has passed a rigorous chemosensory evaluation preprandially and, upon swallowing, has been allowed to enter our system. Not surprisingly, many chemosensory-relevant food ingredients, e.g., aroma or flavor compounds or their metabolites, can be detected postprandially in the blood [18–20] trafficking together with our immune cells. Thus, it is widely accepted that our cellular immune system responds to certain food ingredients or metabolites [21–26]. However, and until only recently, their molecular targets on our immune cells remained largely unknown.

Among a large and diverse set of surface receptors, leukocytes express a large variety of seven-transmembrane helix GPCR [27–29], for example, chemokine receptors [30, 31], free fatty acid receptors (FFAR) [25, 32–34], bradykinin receptors [35], prostanoid receptors [36], complement receptors [37], or formyl peptide receptors [38], to detect bacterial infections, to guide cells to injured tissue, or to orchestrate an immune response. For instance, just in T cells, B cells, and polymorphonuclear granulocytes (neutrophils, PMN), some 40 GPCR are highly expressed [27, 29]. Moreover, there is growing, yet mostly RNA-based evidence of an untypical, so-called "ectopic" expression of olfactory and taste chemosensory GPCR in a variety of peripheral, nonolfactory- and nontaste-related tissues [39–42] including leukocytes [34, 43–46].

Initially, only few studies reported on gene expression of some odorants or taste GPCR in leukocytes in general or in erythroid cells, for example, K562 erythroleukemia cells, by means of reverse transcription polymerase chain reaction (RT-PCR) [47], microarray RNA hybridization [48], or RNA sequencing [49]. These data were, however, inconsistent, which may have been due to method-intrinsic fluctuations [50], different RNA quality, or an insufficient number

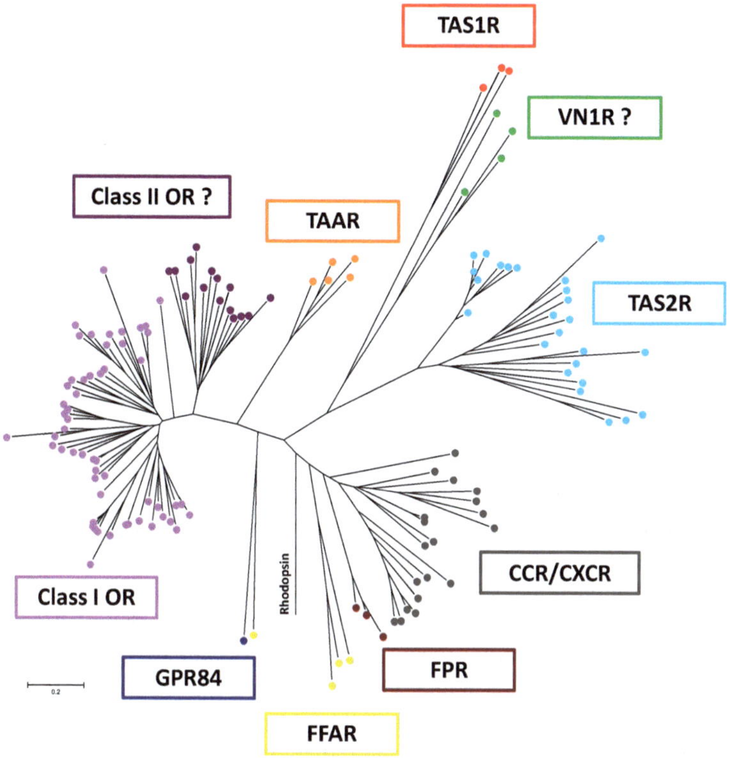


**Fig. 1** Evolutionary relationship of members of chemosensory GPCR families potentially expressed in human leukocytes. *Class I OR* class I odorant receptors, *Class II OR* class II odorant receptors, *TAAR* trace amine-associated receptors, *TAS1R* sweet and umami taste receptors type 1, *TAS2R* bitter taste receptors type 2, *VN1R* vomeronasal type 1 receptors, *FFAR* free fatty acid receptors, *GPR84* inflammation-related, medium-chain fatty acid-sensing receptor, *CCR/CXR* chemokine receptors, *FPR* formyl peptide receptors, *?* under debate. We established an evolutionary history of 134 deduced amino acid sequences, with human rhodopsin as an out-group, using the neighbor-joining method and computed evolutionary distances using the maximum composite likelihood method (MEGA 6.06 software). The tree is drawn to scale (scale bar, 0.2 amino acid substitutions per site)

of individual samples investigated. Recently, several groups, including ours, independently pushed forward and took the identification of OR, TAAR, TAS1R, and TAS2R in leukocytes beyond RNA to the level of protein and cellular function in a variety of blood leukocytes [43, 46, 51–55].

The investigation of the role of olfactory and taste receptors in immune cells has just begun. However, our immune system is rigorously trained to tell apart nonself from self, being perfectly suited for this task with a plethora of chemosensory receptors [25, 33, 45, 51, 56, 57] that have evolved to detect organic nutrients or allelochemicals of microbiological, plant, or animal origin, which carry biologically relevant information or are genuine bioactives [58–61]. Also, chemicals that typically carry chemosensory information, for instance, as food-specific aroma or flavor [17, 62–64] are active on our immune cells [44, 51, 65–67], which during chemosensory and hedonic evaluation of food, post-inhalation, or postprandially, constantly encounter and react to these chemicals, at least in the vicinity of the border epithelia of the nose, lung, mouth, or gastrointestinal tract [44, 56, 65, 66, 68–71]. The capability of self – nonself discrimination may be the common theme between the chemical senses olfaction and taste and the immune system [72].

Here, we review the evidence for an expression in leukocytes of chemosensory GPCR that are otherwise typical for our chemical senses and suggest them as a major group of new targets for old and new immunological-relevant bioactives.

## 2 Olfactory Receptors in Leukocytes

Olfactory receptors enable various animal species to analyze the chemical world that surrounds them. From tracking down a food source to conveying social clues via pheromones, they are certainly among the most versatile of sensing mechanisms for a multitude of chemical stimuli in the animal kingdom. Not surprisingly, recent years saw the emergence of olfactory receptors as sensors of chemical compounds within the cellular immune system, which has evolved to cope with stimulus complexities. The odorant receptors (OR) and the trace amine-associated receptors (TAAR) have proven to be of particular interest in the context of leukocytes, while vomeronasal type I receptors (VN1R) yet remain immunological orphans. Not long after the discovery of OR in 1991 [1], the finding of transcription factor early B-cell factor (EBF1), being identical to the olfactory transcription factor OLF1 (O/E-1), suggested a common gene expression within the cellular immune system and the olfactory neuroepithelium [73–77] and adds to the notion of a putative functional role of OR in blood immune cells.

### *2.1 Odorant Receptors (OR)*

Odorant receptors (OR) are the largest family of GPCR and allow the sensible exploration of an environment based on its volatile chemicals [1], utilizing ca. 400 different OR types just in humans [3, 78]. It appears that the majority of OR forego a "one-ligand-one-receptor" setup in favor of more variability, allowing for one best agonist and several agonists with lesser potency and/or efficacy per OR [17]. This combinatorial approach [79, 80] allows for sensing a great variety of chemicals.

An intriguing analogy between immunity and olfaction is that both are capable to differentiate between "self" and "nonself" for the purpose of immunological defense or, in the case of olfaction, for intraspecies communication, for example, maternal-newborn communication [81, 82], or mating purposes [72], with both biological functions being triggered by chemical cues. In light of an ever-increasing evidence of ectopic expression of OR genes [39, 41, 42], mostly detected in rats and mice [83, 84] but also observed in various human tissues such as sperm [85], kidney [86], and erythroid cells [47], particularly the discovery of OR genes expressed in erythroid cells represented a strong hint for their possible chemoreceptive role in leukocytes. The complex functions of OR in our cellular immune system are far from being clarified; however, single studies have shed some light on putative OR functions in single types of leukocytes. For instance, pulmonary macrophages, which depend on chemical stimuli to initiate migration and immunologic activity [87], were shown to upregulate the expression of four class I OR and four class II OR genes, upon co-stimulation with interferon-γ (IFNγ) and bacterial lipopolysaccharide (LPS) in a mouse model [88]. The same study also found a significant increase in expression of monocyte chemotactic protein 1 (MCP1) upon stimulation with IFNγ, LPS, and octanal, a model agonist for OR activation. The odorant octanal potentiated MCP1 production substantially, which then promoted macrophage migration. Most recently, Clark et al. [89] reported on RNA expression of two class I OR but also one class II OR, in mouse $CD4^+$ T cells suggesting an odorant-specific function.

State of the art molecular biology techniques also led to the discovery of OR expression in human leukocytes. One study identified 24 OR genes and pseudogenes that were expressed in white blood cells utilizing large RNA-Seq datasets [49], while another study identified OR11H1, OR4M1, and OR52N5 on human peripheral blood mononuclear cells as diagnostic markers for traumatic brain injury-associated symptoms [48]. In short, there is a small but growing number of studies reporting on an expression of OR within leukocytes, though mostly showing RNA expression, which has been rarely validated at the protein level (Table 1).

In human leukocytes, particularly in PMN, our group recently demonstrated the RNA expression of most class I OR as well as protein expression of three OR from the three class I OR families [46]. However, we could not confirm the RNA expression of any class II OR in general or of those class II OR that had been reported so far, in particular [46] (Table 1). Class I OR, which are of an evolutionary marine heritage [90–92], have been suggested to predominantly recognize water-soluble ligands such as carboxylic acids [17]. Indeed, butyric acid, diacetyl, or delta-decalactone, being constituents of a butter aroma recombinate, activated specific and overlapping sets of recombinate class I OR in HEK-293 cells and, moreover, induced a GPCR-dependent migration of PMN along a concentration gradient [51].

Taken together, the listed studies (Table 1) strongly suggest a functional role of OR in human immunity.

**Table 1** Expression of odorant receptors (OR) (RNA, protein) in human leukocytes

| Receptor | Leukocyte type | Technique/s utilized | Reference |
|---|---|---|---|
| *Class I OR* | | | |
| OR51A2 | NK | RT-(q)PCR | [46] |
| OR51A4 | Mono, NK, B, T, PM | RT-(q)PCR | [46] |
| OR51A7 | B, T, PMN | RT-(q)PCR | [46] |
| OR51B2 | NK, B, T, PMN | RT-(q)PCR | [46] |
| OR51B4 | Mono, NK, B, T, PMN | RT-(q)PCR | [46] |
| OR51B5 | B, T, PMN | RT-(q)PCR | [46] |
| OR51B6 | B, T, PMN | RT-(q)PCR, ICC | [46] |
| OR51D1 | NK, B, T, PMN | RT-(q)PCR | [46] |
| OR51E1 | B, T, PMN | RT-(q)PCR | [46] |
| OR51E2 | B, T, PMN | RT-(q)PCR | [46] |
| OR51F1 | Mono, B, T, PMN | RT-(q)PCR | [46] |
| OR51F2 | Mono, B, T, PMN | RT-(q)PCR | [46] |
| OR51G1 | Mono, B, T, PMN | RT-(q)PCR | [46] |
| OR51G2 | Mono, B, T, PMN | RT-(q)PCR | [46] |
| OR51I1 | Mono, NK, B, T, PMN | RT-(q)PCR | [46] |
| OR51I2 | Mono, B, T, PMN | RT-(q)PCR | [46] |
| OR51J1 | Mono, NK, B, T, PMN | RT-(q)PCR | [46] |
| OR51L1 | B, T, PMN<br>PBMC | RT-(q)PCR<br>RNA-CHIP, RT-(q)PCR | [46]<br>[48] |
| OR51M1 | Mono, NK, B, T, PMN | RT-(q)PCR | [46] |
| OR51Q1 | Mono, NK, B, T, PMN | RT-(q)PCR | [46] |
| OR51S1 | Mono, B, T, PMN | RT-(q)PCR | [46] |
| OR51T1 | B, T, PMN | RT-(q)PCR | [46] |
| OR51V1 | B, T, PMN | RT-(q)PCR | [46] |
| OR52A1 | Ery<br>B, T, PMN | RT-PCR<br>RT-(q)PCR | [47]<br>[46] |
| OR52A4 | B, T, PMN | RT-(q)PCR, ICC | [46] |
| OR52A5 | B, T, PMN | RT-(q)PCR | [46] |
| OR52B2 | Mono, NK, B, T, PMN | RT-(q)PCR | [46] |
| OR52B4 | Mono, NK, B, T, PMN | RT-(q)PCR | [46] |
| OR52B6 | Mono, NK, B, T, PMN | RT-(q)PCR | [46] |
| OR52D1 | Mono, NK, B, T, PMN | RT-(q)PCR | [46] |
| OR52E2 | Mono, B, T, PMN | RT-(q)PCR | [46] |
| OR52E4 | B, T, PMN | RT-(q)PCR | [46] |
| OR52E5 | B, T, PMN | RT-(q)PCR | [46] |
| OR52E6 | Mono, B, T, PMN | RT-(q)PCR | [46] |
| OR52E8 | B, T, PMN | RT-(q)PCR | [46] |
| OR52H1 | Mono, B, T, PMN | RT-(q)PCR | [46] |
| OR52I1 | Mono, NK, B, T, PMN | RT-(q)PCR | [46] |
| OR52I2 | Mono, NK, B, T, PMN | RT-(q)PCR | [46] |
| OR52J3 | Mono, B, T, PMN | RT-(q)PCR | [46] |
| OR52K1 | Mono, NK, B, T, PMN | RT-(q)PCR | [46] |

(continued)

**Table 1** (continued)

| Receptor | Leukocyte type | Technique/s utilized | Reference |
|---|---|---|---|
| OR52K2 | Mono, NK, B, T, PMN | RT-(q)PCR | [46] |
| OR52L1 | B, T, PMN | RT-(q)PCR | [46] |
| OR52M1 | Mono, B, T, PMN | RT-(q)PCR | [46] |
| OR52N1 | Mono, B, T, PMN | RT-(q)PCR | [46] |
| OR52N2 | Mono, B, T, PMN | RT-(q)PCR | [46] |
| OR52N4 | **n.d.:** PMN (specPrim), Mono, NK, B, T (degPrim)<br>L | RT-(q)PCR<br>NGS | [46]<br>[49] |
| OR52N5 | **n.d.:** PMN (specPrim), Mono, NK, B, T (degPrim)<br>PBMC | RT-(q)PCR<br>RNA-CHIP, RT-(q)PCR | [46]<br>[48] |
| OR52R1 | NK, B, T, PMN | RT-(q)PCR | [46] |
| OR52W1 | Mono, B, T, PMN | RT-(q)PCR | [46] |
| OR56A1 | Mono, NK, B, T, PMN | RT-(q)PCR | [46] |
| OR56A3 | Mono, B, T, PMN | RT-(q)PCR | [46] |
| OR56A4 | B, T, PMN | RT-(q)PCR | [46] |
| OR56A5 | B, T, PMN | RT-(q)PCR | [46] |
| OR56B1 | T | RT-(q)PCR | [46] |
| OR56B4 | B, T, PMN | RT-(q)PCR, ICC | [46] |
| *Class II OR* | | | |
| OR2A4 | **n.d.**: PMN (specPrim), Mono, NK, B, T (degPrim)<br>L | RT-(q)PCR<br>NGS | [46]<br>[49] |
| OR2B6 | **n.d.**: PMN (specPrim), Mono, NK, B, T (degPrim)<br>L | RT-(q)PCR<br>NGS | [46]<br>[49] |
| OR2C1 | **n.d.** PMN (specPrim), Mono, NK, B, T (degPrim)<br>L | RT-(q)PCR<br>NGS | [46]<br>[49] |
| OR2C3 | **n.d.**: PMN (specPrim), Mono, NK, B, T (degPrim)<br>L | RT-(q)PCR<br>NGS | [46]<br>[49] |
| OR2J3 | **n.d.**: PMN (specPrim), Mono, NK, B, T (degPrim)<br>PBMC | RT-(q)PCR<br>RNA-CHIP, RT-(q)PCR | [46]<br>[48] |
| OR2L13 | **n.d.**: PMN (specPrim), Mono, NK, B, T (degPrim)<br>L | RT-(q)PCR<br>NGS | [46]<br>[49] |
| OR2W3 | **n.d.**: PMN (specPrim), Mono, NK, B, T (degPrim)<br>L | RT-(q)PCR<br>NGS | [46]<br>[49] |
| OR3A3 | **n.d.**: PMN (specPrim), Mono, NK, B, T (degPrim)<br>L | RT-(q)PCR<br>NGS | [46]<br>[49] |
| OR4D10 | **n.d.**: PMN (specPrim), Mono, NK, B, T (degPrim)<br>PBMC | RT-(q)PCR<br>RNA-CHIP, RT-(q)PCR | [46]<br>[48] |

(continued)

**Table 1** (continued)

| Receptor | Leukocyte type | Technique/s utilized | Reference |
|---|---|---|---|
| OR4M1 | **n.d.**: PMN (specPrim), Mono, NK, B, T (degPrim)<br>PBMC | RT-(q)PCR<br>RNA-CHIP, RT-(q)PCR | [46]<br>[48] |
| OR4Q3 | **n.d.**: PMN (specPrim), Mono, NK, B, T (degPrim)<br>PBMC | RT-(q)PCR<br>RNA-CHIP, RT-(q)PCR | [46]<br>[48] |
| OR6V1 | **n.d.**: PMN (specPrim), Mono, NK, B, T (degPrim)<br>Ery | RT-(q)PCR<br>NB | [46]<br>[47] |
| OR7D2 | **n.d.**: PMN (specPrim), Mono, NK, B, T (degPrim)<br>L | RT-(q)PCR<br>NGS | [46]<br>[49] |
| OR9A2 | **n.d.**: PMN (specPrim), Mono, NK, B, T (degPrim)<br>Ery | RT-(q)PCR<br>NB | [46]<br>[47] |
| OR11H1 | **n.d.**: PMN (specPrim), Mono, NK, B, T (degPrim)<br>PBMC | RT-(q)PCR<br>RNA-CHIP, RT-(q)PCR | [46]<br>[48] |

*Abbreviations*: *L* whole leukocytes, *PBMC* peripheral blood mononuclear cells, *NK* natural killer cells, *Mono* monocytes, *PMN* polymorphonuclear granulocytes, neutrophils, *T* T cells, *B* B cells, *Ery* erythroid cells, ***n.d.*** not detectable, *specPrim* gene-specific oligonucleotide primer, *degPrim* degenerate oligonucleotide primer, *RT* reverse transcriptase, *q* quantitative, *PCR* polymerase chain reaction, *ICC* immunocytochemistry, *NGS* next-generation sequencing, *RNA-CHIP* RNA hybridization, *NB* Northern Blot

## 2.2 Vomeronasal Type I Receptors (VN1R)

Vomeronasal type I receptors (VN1R) belong to a family of GPCR that is predominantly associated with the vomeronasal organ of various species [93, 94]. The vomeronasal organ's predominant function appears to be the detection of pheromones [95], though it is markedly reduced or nonfunctional in higher apes including *Homo sapiens sapiens* [96–98]. Adding to this notion is the fact that the human genome harbors approximately 200 VN1R pseudogenes but only five remnant, putative functional genes of this family [99–101]. Functional VN1R supposedly recognize small volatiles such as key food odorants, compounds from axillary sweat, or sulfated steroids [102–104]. Despite their many similarities to OR, the expression of VN1R in leukocytes has not been demonstrated yet.

## 2.3 Trace Amine-Associated Receptors (TAAR)

Trace amine-associated receptors (TAAR) recognize a wide range of endogenously produced monoamines that only appear in trace amounts, i.e., trace amines [4, 105, 106]. Through this, they participate in the modulation of human neurotransmission

[107]. Particularly TAAR1 has emerged as a modulator of dopaminergic activity becoming a promising target for the treatment of psychiatric disorders and substance abuse [108, 109].

However, recent discoveries also suggest an odorant receptor (OR)-like activity of TAAR in zebrafish and mice [110, 111]. Further parallels can be drawn between OR and TAAR by observing the far greater abundance of TAAR genes in fishes (13–109) than in tetrapods (9–22) [112] mimicking the OR's evolutionary marine heritage. In humans, we find seven functional TAAR genes (TAAR1, 2, 5, 6, 7, 8, and 9) [113] and two pseudogenes (TAAR3 and 4) [114], of which TAAR1 is the most investigated member of this receptor family to date. Furthermore, TAAR were found to be capable of detecting volatile amines in the urine of mice, thus enabling the detection of social chemical cues in a fashion very similar to VN1R and their detection of pheromones [5, 6]. Moreover, members of the human gut microbiome are capable of producing trace amines de novo in a clear display of allelopathy [115–117]. They also modify endogenous trace amines to establish cross species communication with their human host [118].

There is an increasing evidence for the genetic expression of TAAR in leukocytes of primates [54], mice [119, 120], and humans [43, 121, 122] (Table 2). Little is known, however, with respect to studies of functionality of TAAR – the scarcity of publications displays how recently the focus of the scientific community has shifted onto the topic of TAAR in immunity. Whatever little there is published still paints an interesting picture of how psychostimulants and biogenic amines may influence our cellular immune system.

**Table 2** Expression of trace amine-associated receptors (TAAR) (RNA, protein) in human leukocytes

| Receptor | Leukocyte type | Technique/s utilized | Reference |
|---|---|---|---|
| TAAR1 | L | RT-PCR | [121] |
| | PBMC | RT-PCR | [119] |
| | Mono, B,T, NK, PMN | RT-(q)PCR | [43] |
| | PMN | IB, ICC | [43, 46] |
| | B | IB | [54, 55] |
| | T | IB | [122] |
| TAAR2 | PBMC | RT-PCR | [119] |
| | Mono, B,T, NK, PMN | RT-(q)PCR | [43] |
| | PMN | IB, ICC | [43] |
| TAAR5 | Mono, B,T, NK, PMN | RT-PCR | [43] |
| TAAR6 | L | RT-PCR | [121] |
| | Mono, B,T, NK, PMN | RT-PCR | [43] |
| TAAR8 | L | RT-PCR | [121] |
| | **n.d.**: Mono, B,T, NK, PMN | RT-PCR | [43] |
| TAAR9 | L | RT-PCR | [121] |
| | Mono, B,T, NK, PMN | RT-PCR | [43] |

*Abbreviations*: *L* whole leukocytes, *PBMC* peripheral blood mononuclear cells, *NK* natural killer cells, *Mono* monocytes, *PMN* polymorphonuclear granulocytes, neutrophils, *T* T cells, *B* B cells, ***n.d.*** not detectable, *RT* reverse transcriptase, *q* quantitative, *PCR* polymerase chain reaction, *IB* immunoblot, *ICC* immunocytochemistry

Methamphetamine (meth) is an artificially produced strong stimulant of the central nervous system and is frequently used as a recreational drug. It can impact immunity on various levels, altering chemical and physical natural barriers, and may influence both innate and adaptive immunity [123]. Recent discoveries could directly correlate meth abuse with an increase of TAAR1 expression [54] and linked it with increased susceptibility toward HIV, both in astrocytes in vitro [124] and lymphocytes in patients [122].

A significant baseline production of TAAR in leukocytes was reported in conjunction with biogenic amines that are commonly found in food (octopamine, tyramine, and synephrine) [121]. This coincides with results reported by Babusyte et al. [43], which for the first time demonstrated the functional expression of both TAAR1 and TAAR2 in human leukocytes, presumably functioning as heterodimers. The same study demonstrated functional effects of tyramine, 2-phenylethylamine, and 3-iodothyronamine at plasma-typical "trace" concentrations in the pmol/L to nmol/L range, triggering the migration of PMN, IL-4 production in T cells, and IgE secretion in B cells [43, 125].

An interesting link between foodborne molecules and TAAR-mediated influence on the immune system was established by Nelson et al. [119], when they found human peripheral blood lymphocytes to upregulate the expression of TAAR1 and TAAR2 in response to the treatment of phytohemagglutinin, a lectin commonly found in plants, especially in certain legumes.

In conclusion, there is profound and increasing evidence for TAAR expression as well as for a variety of TAAR functions in our cellular immune system.

## 3 Taste Receptors in Leukocytes

While less versatile than olfactory receptors, taste receptors are nonetheless essential chemosensory molecules, as they allow us to analyze food and to avoid the ingestion of potentially harmful material. Moreover, both the TAS1 (sweet/umami) and TAS2 (bitter) receptor families have been associated with innate immunity [45], though not as extensively as the free fatty acid receptors (FFAR) [34]. There is accumulating evidence for a variety of taste receptor functions in leukocytes reacting to food ingredients or inhaled compounds that reach the border epithelia of the gastrointestinal tract or respiratory epithelia such as the lung or subsequently the blood.

### *3.1 Bitter Taste Receptors (TAS2R)*

Type 2 taste receptors (TAS2R) were discovered in 2000 [7, 8]. They detect bitter taste stimuli [62]. Currently, 43 human TAS2R genes, among them 25 protein-coding TAS2R genes, are known [126], and it has been speculated that their great diversity stems from selective pressure on the ability to recognize a multitude of often bitter-tasting poisons in the environment [10, 62]. Much like the olfactory receptors described above, TAS2R expression has also been found in ectopic tissues such as the liver, the heart, or the brain [49, 127, 128]. The lung, however, appears to be the predominant tissue, where TAS2R RNA expression and protein function have been linked to nontaste-related cellular functions. Particularly in the case of asthma, TAS2R protein functionality has been correlated with relief from symptoms, first in a mouse model [70], then, more recently, in both human cells and patients [69, 129]. Here, TAS2R not only show involvement in this inherited disease of the lung but have also proven functionality in upper respiratory innate immunity. They accurately identified bacteria-derived molecules [130] and induced defensive mechanisms such as the production of nitrous oxide and the excretion of antibacterial peptides [52, 131].

TAS2R proteins thus clearly appear to be involved in the intricate cellular network that makes up human immunity (Table 3). Mast cells, best known for their role in allergies, express a variety of TAS2R genes (Table 3). Upon stimulation, for example, with the antimalarial drug chloroquine, TAS2R gene products effectively mediated the inhibition of an IgE-induced release of histamine from mast cells underlining the TAS2R protein-mediated anti-inflammatory activity observed in asthma [44]. Malki et al. [46] showed TAS2R43 and TAS2R31 RNA and protein expression in neutrophils and demonstrated their involvement in mediating chemotaxis, triggered by plasma-typical saccharin concentrations in a siRNA-controlled study, suggesting a role of TAS2R in postprandial chemosensing of food ingredients by certain blood leukocytes.

In conclusion, several independent research groups unambiguously demonstrated gene expression for the majority of the 25 human bitter taste receptors in various types of human leukocytes (Table 3). For few TAS2R, however, protein expression and/or cellular function in leukocytes have been demonstrated so far (Table 3).

## 3.2 Sweet Taste Receptor (TAS1R2/TAS1R3)

The sweet taste receptor heterodimer TAS1R2/TAS1R3 of the taste buds responds to natural sugars and artificial sweeteners triggering the impression of sweetness in human gustation [12]. Beyond, its ectopic expression in nontaste-related tissues and cells in general [40, 42], and leukocytes in particular [45], has been reported (Table 4). As suggested for the bitter taste receptors, the TAS1R2/TAS1R3 heterodimer appears to be also involved in upper airway immunity [45, 132]. Here, Lee et al. [52] demonstrated the sweet taste receptor to be an antagonist to a TAS2R function blocking bitter taste receptor-mediated anti-microbial activity by attenuating its calcium-dependent signaling. Lee et al. also hypothesized that the sweet taste heterodimer measures glucose concentration in the mucosa. As microorganisms use the sugar in their environment for energy

**Table 3** Expression of TAS2R bitter taste receptors (RNA, protein) in human leukocytes

| Receptor | Leukocyte type | Technique/s utilized | Reference |
|---|---|---|---|
| TAS2R1 | Mono, NK, B, T, PMN | RT-(q)PCR | [46] |
| TAS2R3 | Mono, NK, B, T, PMN<br>MC | RT-(q)PCR<br>RT-(q)PCR | [46]<br>[44] |
| TAS2R4 | Mono, NK, B, T, PMN<br>L, Mono, Lym, PMN<br>MC | RT-(q)PCR<br>RT-(q)PCR<br>RT-(q)PCR | [46]<br>[69]<br>[44] |
| TAS2R5 | Mono, NK, B, T, PMN<br>L, Mono, Lym, PMN<br>MC | RT-(q)PCR<br>RT-(q)PCR<br>RT-(q)PCR | [46]<br>[69]<br>[44] |
| TAS2R7 | Mono, NK, B, T, PMN | RT-(q)PCR | [46] |
| TAS2R8 | Mono, NK, B, T, PMN | RT-(q)PCR | [46] |
| TAS2R9 | Mono, NK, B, T, PMN | RT-(q)PCR | [46] |
| TAS2R10 | Mono, NK, B, T, PMN<br>L, Mono, Lym, PMN<br>MC | RT-(q)PCR<br>RT-(q)PCR<br>RT-(q)PCR | [46]<br>[69]<br>[44] |
| TAS2R13 | Mono, NK, B, T, PMN<br>L, Mono, Lym, PMN<br>MC | RT-(q)PCR<br>RT-(q)PCR<br>RT-(q)PCR | [46]<br>[69]<br>[44] |
| TAS2R14 | NK, B, T, PMN<br>L, Mono, Lym, PMN<br>MC<br>L | RT-(q)PCR<br>RT-(q)PCR<br>RT-(q)PCR<br>NGS | [46]<br>[69]<br>[44]<br>[49] |
| TAS2R16 | Mono, NK, B, T, PMN | RT-(q)PCR | [46] |
| TAS2R19 | Mono, NK, B, T, PMN<br>L, Mono, Lym, PMN<br>MC | RT-(q)PCR<br>RT-(q)PCR<br>RT-(q)PCR | [46]<br>[69]<br>[44] |
| TAS2R20 | Mono, NK, B, T, PMN<br>L, Mono, Lym, PMN<br>MC<br>L | RT-(q)PCR<br>RT-(q)PCR<br>RT-(q)PCR<br>NGS | [46]<br>[69]<br>[44]<br>[49] |
| TAS2R30 | Mono, NK, B, T, PMN | RT-(q)PCR | [46] |
| TAS2R31 | Mono, NK, B, T, PMN<br>L, Mono, Lym, PMN | RT-(q)PCR<br>RT-(q)PCR | [46]<br>[69] |
| TAS2R38 | Mono, NK, B, T, PMN<br>PMN | RT-(q)PCR, ICC<br>IB, ICC | [46]<br>[53] |
| TAS2R39 | Mono, NK, B, T, PMN | RT-(q)PCR | [46] |
| TAS2R40 | Mono, NK, B, T, PMN<br>L | RT-(q)PCR<br>NGS | [46]<br>[49] |
| TAS2R41 | Mono, NK, B, T, PMN | RT-(q)PCR | [46] |
| TAS2R42 | Mono, NK, B, T, PMN | RT-(q)PCR | [46] |
| TAS2R43 | Mono, NK, B, T, PMN | RT-(q)PCR, ICC | [46] |
| TAS2R45 | Mono, NK, B, T, PMN<br>L, Mono, Lym, PMN | RT-(q)PCR<br>RT-(q)PCR | [46]<br>[69] |
| TAS2R46 | Mono, NK, B, T, PMN<br>L, Mono, Lym, PMN<br>MC | RT-(q)PCR<br>RT-(q)PCR<br>RT-(q)PCR | [46]<br>[69]<br>[44] |

(continued)

**Table 3** (continued)

| Receptor | Leukocyte type | Technique/s utilized | Reference |
|---|---|---|---|
| TAS2R50 | Mono, NK, B, T, PMN | RT-(q)PCR | [46] |
| | L, Mono, Lym, PMN | RT-(q)PCR | [69] |
| TAS2R60 | Mono, NK, B, T, PMN | RT-(q)PCR | [46] |
| | L | NGS | [49] |

*Abbreviations*: *L* whole leukocytes, *Lym* lymphocytes, *MC* mast cells, *NK* natural killer cells, *Mono* monocytes, *PMN* polymorphonuclear granulocytes, neutrophils, *T* T cells, *B* B cells, *RT* reverse transcriptase, *q* quantitative, *PCR* polymerase chain reaction, *NGS* next-generation sequencing, *IB* immunoblot, *ICC* immunocytochemistry

**Table 4** Expression of TAS1R sweet and umami taste receptors (RNA, protein) in human leukocytes

| TAS1R | | | |
|---|---|---|---|
| Receptor | Leukocyte type | Technique/s utilized | Reference |
| TAS1R1 | Mono, NK, B, T, PMN | RT-(q)PCR, ICC | [46] |
| TAS1R2 | Mono, NK, B, T, PMN | RT-(q)PCR, ICC | [46] |
| TAS1R3 | B | RT-(q)PCR, ISH | [133] |
| | Mono, NK, B, T, PMN | RT-(q)PCR, ICC | [46] |
| | L | NGS | [49] |
| | PBC | RT-(q)PCR | [134] |

*Abbreviations*: *L* whole leukocytes, *PBC* peripheral blood cells, *NK* natural killer cells, *Mono* monocytes, *PMN* polymorphonuclear granulocytes, neutrophils, *T* T cells, *B* B cells, *RT* reverse transcriptase, *q* quantitative, *PCR* polymerase chain reaction, *ISH* in situ RNA hybridization, *NGS* next-generation sequencing, *ICC* immunocytochemistry

consumption, a drop in glucose concentration would indicate infection, which would in turn remove the sweet taste roadblock to activate bitter taste-induced immunity.

Particularly the subunit TAS1R3, which dimerizes with both TAS1R1 and TAS1R2, has been shown to be expressed in peripheral B cells in pigs [133]. This is in line with Malki et al. [46] reporting on TAS1R1, TAS1R2, and TAS1R3 to be expressed in subsets of five different peripheral human leukocyte types including B cells (Table 4). Furthermore, in a siRNA-controlled study, the TAS1R2/TAS1R3 heterodimer proved necessary for neutrophil chemotaxis along a defined saccharin gradient [46]. In this study, notably, the sweet taste blocker lactisole induced neutrophil migration in a TAS1R3-dependent but TAS1R2-independent manner suggesting a TAS1R3 homomer function in these cells [46].

### *3.3 Umami Taste Receptor (TAS1R1/TAS1R3)*

The umami taste receptor is an amino acid receptor in taste buds [11]. As a heterodimer, TAS1R1/TAS1R3 shares its TAS1R3 subunit with the sweet taste receptor [11, 135]. However, unlike its sweet taste receptor equivalent, very little is known about TAS1R1/TAS1R3 with respect to nontaste-related functionality, though an ectopic expression has been reported [49]. In a variety of nontaste-related cells, TAS1R1/TAS1R3 has been suggested to be involved in nutrient and amino acid sensing interpreted as a mechanism for communicating amino acid availability to the cell [131, 136]. Malki et al. [46] demonstrated RNA expression of TAS1R1/TAS1R3 in five different types of leukocytes as well as protein expression in neutrophils, T cells, and B cells suggesting an immune-related function of TAS1R1/TAS1R3 in leukocytes (Table 4). A gut-associated lymphoid tissue that warrants immune sensing of nutrients, such as amino acids, are the Peyer's patches, which harbor a variety of leukocytes [71].

## 3.4 Free Fatty Acid Receptors (FFAR)

The human genome includes the genes for four free fatty acid receptors (FFAR): FFAR1 (also known as GPR40), FFAR2 (GPR43), FFAR3 (GPR41), FFAR4 (GPR120), and GPR84. While FFAR1 and 4 both recognize medium to long chain fatty acids, FFAR 2 and 3 register short chain fatty acids (SCFA) only [137]. GPR84 exclusively accepts medium chain fatty acids [138]. Currently, FFAR1 is known predominantly for its presence on the Islets of Langerhans [139] and in the brain [140] overtaking metabolic and neurologic duties, respectively. An immune-related function has not been demonstrated, yet. FFAR4 is as potent as FFAR1 with respect to its influence on the metabolism, recognizing omega-3 fatty acids, and implementing a powerful anti-inflammatory program utilizing monocytes/macrophages and adipocytes [141, 142]. FFAR4 has been discussed to exert antidiabetic and anti-obesity effects and FFAR in general are regarded as targets for novel drugs to treat metabolic disorders [143]. SCFA and their receptors FFAR2 and FFAR3 have long been known to influence gut immunity through their ability to measure the metabolic activity of the gut microbiome [25, 144, 145]. The gut also represents a possible site of interaction between FFAR2 in particular and the abovementioned TAAR, as the microbiome induces the production of serotonin, a TAAR agonist, through enterochromaffin cells by supplying SCFA [146]. FFAR2 is the free fatty acid receptor that is most associated to immunity in general. FFAR2-dependent neutrophil migration has been demonstrated in both mouse and human [32, 147]. FFAR2 has also been shown to form a heterodimer with M-ficolin on monocytes, providing signal transduction for the ficolin's pathogen recognition signal, such as lipopolysaccharides (LPS) and *N*-acetylglucosamin (GlcNAc), and leading to the production of the neutrophil chemoattractant factor IL-8 [148].

Finally, GPR84 has been related to immunity in various experimental setups. On mouse macrophages, it was shown to be upregulated in response to bacterial LPS [149], while human adipocytes upregulated its expression as a response to macrophage-produced tumor necrosis factor-α (TNFα) [150]. Suzuki et al. also demonstrated that, much like FFAR2, GPR84 also seems to induce IL-8 production upon LPS stimulation – whether this also occurs via the formation of a M-ficolin-FFAR2-like heterodimer remains to be seen [151]. The same study also could show a GPR84-induced release of TNF$\alpha$ from macrophages upon LPS stimulation [151].

Altogether, there is increasing evidence suggesting a role of free fatty acid receptors in the cellular immune system (for a review, see [34]).

## 4 Olfactory and Taste Signaling Pathway Components in Leukocytes

There is very recent evidence that canonical signal transduction molecules of olfactory and taste receptors express in blood leukocytes. Flegel et al. [49] found RNA expression of olfactory G protein G$\alpha_{olf}$ (GNAL), its guanine nucleotide exchange factor RIC8B, and adenylyl cyclase type III (ADCY3) in leukocytes by means of NGS. Malki et al. [46] detected RNA expression of G proteins GNAL, G$\gamma_{13}$ (GNG13) and G$\alpha_{gustducin}$ (GNAT3), and transient receptor potential channel TRPM5 in isolated and purified peripheral blood neutrophils by means of RT-qPCR. Moreover, they demonstrated GNAT3 to be necessary for a migration of isolated human neutrophils toward an artificial sweetener stimulus (saccharin) by means of siRNA knockdown of GNAT3 or pretreatment of neutrophils with pertussis toxin (PTX) [46]. Notably, 2-phenylethylamine, a TAAR agonist, activated chemotactic migration of neutrophils that was also largely abolished by PTX or siRNA knockdown of GNAT3 suggesting TAAR signaling via G$\alpha_i$-type proteins, at least in neutrophils [46]. Most recently, Clark et al. [89] detected RNA expression of GNAL and ADCY3 in CD4$^+$ T cells by means of RT-qPCR suggesting OR-specific odorants for a cAMP-dependent modulation of T-cell activity.

These reports, so far, add to the notion of a "chemosensory equipment" of at least subpopulations of blood leukocytes with olfactory and taste receptors and their signaling molecules [46].

## 5 Concluding Remarks

With some 40 GPCR being highly expressed in, for instance, PMN, T, or B cells, the recent identification of ca. 90 chemosensory GPCR in leukocytes has increased the number of known GPCR in these cells about threefold. As it stands, there is significant evidence suggesting the existence of cell type-specific "chemosensory"-type subpopulations of leukocytes. These subpopulations display interindividual variations in their equipment with certain olfactory and taste receptors as well as regulation of expression of these GPCR depending on, for instance, the health status of an individual. This suggests that these chemosensory GPCR as a major group of immune markers, as targets for receptor- and immune cell-specific bioactives, such as foodborne flavor chemicals, or allelochemicals in general and beyond, opens the perspective of a specific pharmacological modulation of our cellular immune system. Thus, the research on the role of olfactory and taste receptors in our immune system is likely to be looming large in the nearest future.

## References

1. Buck L, Axel R (1991) A novel multigene family may encode odorant receptors: a molecular basis for odor recognition. Cell 65(1):175–187
2. Mainland JD, Keller A, Li YR, Zhou T, Trimmer C, Snyder LL, Moberly AH, Adipietro KA, Liu WL, Zhuang H, Zhan S, Lee SS, Lin A, Matsunami H (2014) The missense of smell: functional variability in the human odorant receptor repertoire. Nat Neurosci 17(1):114–120
3. Olender T, Waszak SM, Viavant M, Khen M, Ben-Asher E, Reyes A, Nativ N, Wysocki CJ, Ge D, Lancet D (2012) Personal receptor repertoires: olfaction as a model. BMC Genomics 13:414
4. Borowsky B, Adham N, Jones KA, Raddatz R, Artymyshyn R, Ogozalek KL, Durkin MM, Lakhlani PP, Bonini JA, Pathirana S, Boyle N, Pu X, Kouranova E, Lichtblau H, Ochoa FY, Branchek TA, Gerald C (2001) Trace amines: identification of a family of mammalian G protein-coupled receptors. Proc Natl Acad Sci U S A 98(16):8966–8971
5. Liberles SD (2009) Trace amine-associated receptors are olfactory receptors in vertebrates. Ann N Y Acad Sci 1170:168–172
6. Liberles SD, Buck LB (2006) A second class of chemosensory receptors in the olfactory epithelium. Nature 442(7103):645–650
7. Matsunami H, Montmayeur JP, Buck LB (2000) A family of candidate taste receptors in human and mouse. Nature 404(6778):601–604
8. Chandrashekar J, Mueller KL, Hoon MA, Adler E, Feng L, Guo W, Zuker CS, Ryba NJ (2000) T2Rs function as bitter taste receptors. Cell 100(6):703–711
9. Adler E, Hoon MA, Mueller KL, Chandrashekar J, Ryba NJ, Zuker CS (2000) A novel family of mammalian taste receptors. Cell 100(6):693–702
10. Shi P, Zhang J, Yang H, Zhang YP (2003) Adaptive diversification of bitter taste receptor genes in Mammalian evolution. Mol Biol Evol 20(5):805–814
11. Nelson G, Chandrashekar J, Hoon MA, Feng L, Zhao G, Ryba NJ, Zuker CS (2002) An amino-acid taste receptor. Nature 416(6877):199–202
12. Nelson G, Hoon MA, Chandrashekar J, Zhang Y, Ryba NJ, Zuker CS (2001) Mammalian sweet taste receptors. Cell 106(3):381–390
13. Abdoul-Azize S, Selvakumar S, Sadou H, Besnard P, Khan NA (2014) Ca2+ signaling in taste bud cells and spontaneous preference for fat: unresolved roles of CD36 and GPR120. Biochimie 96:8–13
14. Galindo MM, Voigt N, Stein J, van Lengerich J, Raguse JD, Hofmann T, Meyerhof W, Behrens M (2012) G protein-coupled receptors in human fat taste perception. Chem Senses 37(2):123–139
15. Voigt N, Stein J, Galindo MM, Dunkel A, Raguse JD, Meyerhof W, Hofmann T, Behrens M (2014) The role of lipolysis in human orosensory fat perception. J Lipid Res 55(5):870–882
16. Breslin PA (2013) An evolutionary perspective on food and human taste. Curr Biol 23(9): R409–R418

17. Dunkel A, Steinhaus M, Kotthoff M, Nowak B, Krautwurst D, Schieberle P, Hofmann T (2014) Nature's chemical signatures in human olfaction: a foodborne perspective for future biotechnology. Angew Chem Int Ed Engl 53(28):7124–7143
18. Colburn WA, Bekersky I, Blumenthal HP (1981) Dietary saccharin kinetics. Clin Pharmacol Ther 30(4):558–563
19. Desage M, Schaal B, Soubeyrand J, Orgeur P, Brazier JL (1996) Gas chromatographic-mass spectrometric method to characterise the transfer of dietary odorous compounds into plasma and milk. J Chromatogr B Biomed Appl 678(2):205–210
20. Zeller A, Horst K, Rychlik M (2009) Study of the metabolism of estragole in humans consuming fennel tea. Chem Res Toxicol 22(12):1929–1937
21. Addai FK (2010) Natural cocoa as diet-mediated antimalarial prophylaxis. Med Hypotheses 74(5):825–830
22. Singh A, Holvoet S, Mercenier A (2011) Dietary polyphenols in the prevention and treatment of allergic diseases. Clin Exp Allergy 41(10):1346–1359
23. Sultan MT, Butt MS, Qayyum MM, Suleria HA (2014) Immunity: plants as effective mediators. Crit Rev Food Sci Nutr 54(10):1298–1308
24. Arnardottir HH, Freysdottir J, Hardardottir I (2013) Dietary fish oil increases the proportion of a specific neutrophil subpopulation in blood and total neutrophils in peritoneum of mice following endotoxin-induced inflammation. J Nutr Biochem 24(1):248–255
25. Maslowski KM, Vieira AT, Ng A, Kranich J, Sierro F, Yu D, Schilter HC, Rolph MS, Mackay F, Artis D, Xavier RJ, Teixeira MM, Mackay CR (2009) Regulation of inflammatory responses by gut microbiota and chemoattractant receptor GPR43. Nature 461 (7268):1282–1286
26. Tomasdottir V, Vikingsson A, Freysdottir J, Hardardottir I (2013) Dietary fish oil reduces the acute inflammatory response and enhances resolution of antigen-induced peritonitis. J Nutr Biochem 24(10):1758–1765
27. Kehrl JH (2004) G-protein-coupled receptor signaling, RGS proteins, and lymphocyte function. Crit Rev Immunol 24(6):409–423
28. Lattin J, Zidar DA, Schroder K, Kellie S, Hume DA, Sweet MJ (2007) G-protein-coupled receptor expression, function, and signaling in macrophages. J Leukoc Biol 82(1):16–32
29. Futosi K, Fodor S, Mocsai A (2013) Reprint of Neutrophil cell surface receptors and their intracellular signal transduction pathways. Int Immunopharmacol 17(4):1185–1197
30. Lira SA, Furtado GC (2012) The biology of chemokines and their receptors. Immunol Res 54 (1–3):111–120
31. Sharma M (2010) Chemokines and their receptors: orchestrating a fine balance between health and disease. Crit Rev Biotechnol 30(1):1–22
32. Vinolo MA, Ferguson GJ, Kulkarni S, Damoulakis G, Anderson K, Bohlooly YM, Stephens L, Hawkins PT, Curi R (2011) SCFAs induce mouse neutrophil chemotaxis through the GPR43 receptor. PLoS One 6(6), e21205
33. Yonezawa T, Kurata R, Yoshida K, Murayama MA, Cui X, Hasegawa A (2013) Free fatty acids-sensing G protein-coupled receptors in drug targeting and therapeutics. Curr Med Chem 20(31):3855–3871
34. Alvarez-Curto E, Milligan G (2016) Metabolism meets immunity: the role of free fatty acid receptors in the immune system. Biochem Pharmacol 114:3–13
35. Bockmann S, Paegelow I (2000) Kinins and kinin receptors: importance for the activation of leukocytes. J Leukoc Biol 68(5):587–592
36. Sandig H, Pease JE, Sabroe I (2007) Contrary prostaglandins: the opposing roles of PGD2 and its metabolites in leukocyte function. J Leukoc Biol 81(2):372–382
37. Ward PA (2008) Role of the complement in experimental sepsis. J Leukoc Biol 83 (3):467–470
38. Cui T, Tsolakis AV, Li SC, Cunningham JL, Lind T, Oberg K, Giandomenico V (2013) Olfactory receptor 51E1 protein as a potential novel tissue biomarker for small intestine neuroendocrine carcinomas. Eur J Endocrinol 168(2):253–261

39. Kang N, Koo J (2012) Olfactory receptors in non-chemosensory tissues. BMB Rep 45 (11):612–622
40. Behrens M, Meyerhof W (2011) Gustatory and extragustatory functions of mammalian taste receptors. Physiol Behav 105(1):4–13
41. Feldmesser E, Olender T, Khen M, Yanai I, Ophir R, Lancet D (2006) Widespread ectopic expression of olfactory receptor genes. BMC Genomics 7:121
42. Foster SR, Roura E, Thomas WG (2014) Extrasensory perception: odorant and taste receptors beyond the nose and mouth. Pharmacol Ther 142(1):41–61
43. Babusyte A, Kotthoff M, Fiedler J, Krautwurst D (2013) Biogenic amines activate blood leukocytes via trace amine-associated receptors TAAR1 and TAAR2. J Leukoc Biol 93 (3):387–394
44. Ekoff M, Choi JH, James A, Dahlen B, Nilsson G, Dahlen SE (2014) Bitter taste receptor (TAS2R) agonists inhibit IgE-dependent mast cell activation. J Allergy Clin Immunol 134 (2):475–478
45. Lee RJ, Cohen NA (2015) Taste receptors in innate immunity. Cell Mol Life Sci 72 (2):217–236
46. Malki A, Fiedler J, Fricke K, Ballweg I, Pfaffl MW, Krautwurst D (2015) Class I odorant receptors, TAS1R and TAS2R taste receptors, are markers for subpopulations of circulating leukocytes. J Leukoc Biol 97(3):533–545
47. Feingold EA, Penny LA, Nienhuis AW, Forget BG (1999) An olfactory receptor gene is located in the extended human beta-globin gene cluster and is expressed in erythroid cells. Genomics 61(1):15–23
48. Zhao W, Ho L, Varghese M, Yemul S, Dams-O'Connor K, Gordon W, Knable L, Freire D, Haroutunian V, Pasinetti GM (2013) Decreased level of olfactory receptors in blood cells following traumatic brain injury and potential association with tauopathy. J Alzheimers Dis 34(2):417–429
49. Flegel C, Manteniotis S, Osthold S, Hatt H, Gisselmann G (2013) Expression profile of ectopic olfactory receptors determined by deep sequencing. PLoS One 8(2), e55368
50. Schuchhardt J, Beule D, Malik A, Wolski E, Eickhoff H, Lehrach H, Herzel H (2000) Normalization strategies for cDNA microarrays. Nucleic Acids Res 28(10), E47
51. Geithe C, Andersen G, Malki A, Krautwurst D (2015) A butter aroma recombinate activates human class-i odorant receptors. J Agric Food Chem 63(43):9410–9420
52. Lee RJ, Kofonow JM, Rosen PL, Siebert AP, Chen B, Doghramji L, Xiong G, Adappa ND, Palmer JN, Kennedy DW, Kreindler JL, Margolskee RF, Cohen NA (2014) Bitter and sweet taste receptors regulate human upper respiratory innate immunity. J Clin Invest 124 (3):1393–1405
53. Maurer S, Wabnitz GH, Kahle NA, Stegmaier S, Prior B, Giese T, Gaida MM, Samstag Y, Hansch GM (2015) Tasting pseudomonas aeruginosa biofilms: human neutrophils express the bitter receptor T2R38 as sensor for the quorum sensing molecule N-(3-oxododecanoyl)-l-homoserine lactone. Front Immunol 6:369
54. Panas MW, Xie Z, Panas HN, Hoener MC, Vallender EJ, Miller GM (2012) Trace amine associated receptor 1 signaling in activated lymphocytes. J Neuroimmune Pharmacol 7 (4):866–876
55. Wasik AM, Millan MJ, Scanlan T, Barnes NM, Gordon J (2012) Evidence for functional trace amine associated receptor-1 in normal and malignant B cells. Leuk Res 36(2):245–249
56. Masui R, Sasaki M, Funaki Y, Ogasawara N, Mizuno M, Iida A, Izawa S, Kondo Y, Ito Y, Tamura Y, Yanamoto K, Noda H, Tanabe A, Okaniwa N, Yamaguchi Y, Iwamoto T, Kasugai K (2013) G protein-coupled receptor 43 moderates gut inflammation through cytokine regulation from mononuclear cells. Inflamm Bowel Dis 19(13):2848–2856
57. Wellendorph P, Johansen LD, Brauner-Osborne H (2009) Molecular pharmacology of promiscuous seven transmembrane receptors sensing organic nutrients. Mol Pharmacol 76 (3):453–465

58. Sbarbati A, Osculati F (2006) Allelochemical communication in vertebrates: kairomones, allomones and synomones. Cells Tissues Organs 183(4):206–219
59. Nielsen BL, Rampin O, Meunier N, Bombail V (2015) Behavioral responses to odors from other species: introducing a complementary model of allelochemics involving vertebrates. Front Neurosci 9:226
60. Mardani H, Sekine T, Azizi M, Mishyna M, Fujii Y (2015) Identification of safranal as the main allelochemical from saffron (Crocus sativus). Nat Prod Commun 10(5):775–777
61. Carvalho Ade O, Gomes VM (2011) Plant defensins and defensin-like peptides – biological activities and biotechnological applications. Curr Pharm Des 17(38):4270–4293
62. Meyerhof W, Batram C, Kuhn C, Brockhoff A, Chudoba E, Bufe B, Appendino G, Behrens M (2010) The molecular receptive ranges of human TAS2R bitter taste receptors. Chem Senses 35(2):157–170
63. Galindo-Cuspinera V, Winnig M, Bufe B, Meyerhof W, Breslin PA (2006) A TAS1R receptor-based explanation of sweet 'water-taste'. Nature 441(7091):354–357. Epub 2006 Apr 23
64. Roudnitzky N, Bufe B, Thalmann S, Kuhn C, Gunn HC, Xing C, Crider BP, Behrens M, Meyerhof W, Wooding SP (2011) Genomic, genetic and functional dissection of bitter taste responses to artificial sweeteners. Hum Mol Genet 20(17):3437–3449
65. Bonisch U, Bohme A, Kohajda T, Mogel I, Schutze N, von Bergen M, Simon JC, Lehmann I, Polte T (2012) Volatile organic compounds enhance allergic airway inflammation in an experimental mouse model. PLoS One 7(7), e39817
66. Serafino A, Sinibaldi Vallebona P, Andreola F, Zonfrillo M, Mercuri L, Federici M, Rasi G, Garaci E, Pierimarchi P (2008) Stimulatory effect of Eucalyptus essential oil on innate cell-mediated immune response. BMC Immunol 9:17
67. Schepetkin IA, Kushnarenko SV, Ozek G, Kirpotina LN, Utegenova GA, Kotukhov YA, Danilova AN, Ozek T, Baser KH, Quinn MT (2015) Inhibition of human neutrophil responses by the essential oil of Artemisia kotuchovii and its constituents. J Agric Food Chem 63 (20):4999–5007
68. Lechner J, von Baehr V (2015) Stimulation of proinflammatory cytokines by volatile sulfur compounds in endodontically treated teeth. Int J Gen Med 9:109–118
69. Orsmark-Pietras C, James A, Konradsen JR, Nordlund B, Soderhall C, Pulkkinen V, Pedroletti C, Daham K, Kupczyk M, Dahlen B, Kere J, Dahlen SE, Hedlin G, Melen E (2013) Transcriptome analysis reveals upregulation of bitter taste receptors in severe asthmatics. Eur Respir J 42(1):65–78
70. Deshpande DA, Wang WC, McIlmoyle EL, Robinett KS, Schillinger RM, An SS, Sham JS, Liggett SB (2010) Bitter taste receptors on airway smooth muscle bronchodilate by localized calcium signaling and reverse obstruction. Nat Med 16(11):1299–1304
71. Jung C, Hugot JP, Barreau F (2010) Peyer's patches: the immune sensors of the intestine. Int J Inflam 2010:823710
72. Doherty PC (2003) On the nose: shared themes for the sensory and immune self. Nat Immunol 4(11):1043–1045
73. Milatovich A, Qiu RG, Grosschedl R, Francke U (1994) Gene for a tissue-specific transcriptional activator (EBF or Olf-1), expressed in early B lymphocytes, adipocytes, and olfactory neurons, is located on human chromosome 5, band q34, and proximal mouse chromosome 11. Mamm Genome 5(4):211–215
74. Wang MM, Reed RR (1993) Molecular cloning of the olfactory neuronal transcription factor Olf-1 by genetic selection in yeast. Nature 364(6433):121–126
75. Wang SS, Betz AG, Reed RR (2002) Cloning of a novel Olf-1/EBF-like gene, O/E-4, by degenerate oligo-based direct selection. Mol Cell Neurosci 20(3):404–414
76. Wang SS, Lewcock JW, Feinstein P, Mombaerts P, Reed RR (2004) Genetic disruptions of O/E2 and O/E3 genes reveal involvement in olfactory receptor neuron projection. Development 131(6):1377–1388

77. Wang SS, Tsai RY, Reed RR (1997) The characterization of the Olf-1/EBF-like HLH transcription factor family: implications in olfactory gene regulation and neuronal development. J Neurosci 17(11):4149–4158
78. Malnic B, Godfrey PA, Buck LB (2004) The human olfactory receptor gene family. Proc Natl Acad Sci U S A 101(8):2584–2589
79. Malnic B, Hirono J, Sato T, Buck LB (1999) Combinatorial receptor codes for odors. Cell 96 (5):713–723
80. Nara K, Saraiva LR, Ye X, Buck LB (2011) A large-scale analysis of odor coding in the olfactory epithelium. J Neurosci 31(25):9179–9191
81. Doucet S, Soussignan R, Sagot P, Schaal B (2007) The "smellscape" of mother's breast: effects of odor masking and selective unmasking on neonatal arousal, oral, and visual responses. Dev Psychobiol 49(2):129–138
82. Schaal B (2010) Mammary odor cues and pheromones: mammalian infant-directed communication about maternal state, mammae, and milk. Vitam Horm 83:83–136
83. Zhang X, Rogers M, Tian H, Zhang X, Zou DJ, Liu J, Ma M, Shepherd GM, Firestein SJ (2004) High-throughput microarray detection of olfactory receptor gene expression in the mouse. Proc Natl Acad Sci U S A 101(39):14168–14173
84. Young JM, Trask BJ (2002) The sense of smell: genomics of vertebrate odorant receptors. Hum Mol Genet 11(10):1153–1160
85. Spehr M, Gisselmann G, Poplawski A, Riffell JA, Wetzel CH, Zimmer RK, Hatt H (2003) Identification of a testicular odorant receptor mediating human sperm chemotaxis. Science 299(5615):2054–2058
86. Wetzel CH, Oles M, Wellerdieck C, Kuczkowiak M, Gisselmann G, Hatt H (1999) Specificity and sensitivity of a human olfactory receptor functionally expressed in human embryonic kidney 293 cells and Xenopus Laevis oocytes. J Neurosci 19(17):7426–7433
87. Taylor PR, Martinez-Pomares L, Stacey M, Lin HH, Brown GD, Gordon S (2005) Macrophage receptors and immune recognition. Annu Rev Immunol 23:901–944
88. Li JJ, Tay HL, Plank M, Essilfie AT, Hansbro PM, Foster PS, Yang M (2013) Activation of olfactory receptors on mouse pulmonary macrophages promotes monocyte chemotactic protein-1 production. PLoS One 8(11), e80148
89. Clark AA, Nurmukhambetova S, Li X, Munger SD, Lees JR (2016) Odorants specifically modulate chemotaxis and tissue retention of CD4+ T cells via cyclic adenosine monophosphate induction. J Leukoc Biol
90. Mezler M, Fleischer J, Breer H (2001) Characteristic features and ligand specificity of the two olfactory receptor classes from Xenopus laevis. J Exp Biol 204(Pt 17):2987–2997
91. Hoover KC (2013) Evolution of olfactory receptors. Methods Mol Biol 1003:241–249
92. Niimura Y, Nei M (2005) Evolutionary dynamics of olfactory receptor genes in fishes and tetrapods. Proc Natl Acad Sci U S A 102(17):6039–6044
93. Rodriguez I, Greer CA, Mok MY, Mombaerts P (2000) A putative pheromone receptor gene expressed in human olfactory mucosa. Nat Genet 26(1):18–19
94. Young JM, Massa HF, Hsu L, Trask BJ (2010) Extreme variability among mammalian V1R gene families. Genome Res 20(1):10–18
95. Chamero P, Leinders-Zufall T, Zufall F (2012) From genes to social communication: molecular sensing by the vomeronasal organ. Trends Neurosci 35(10):597–606
96. Mast TG, Samuelsen CL (2009) Human pheromone detection by the vomeronasal organ: unnecessary for mate selection? Chem Senses 34(6):529–531. Epub 2009 May 28
97. Swaney WT, Keverne EB (2009) The evolution of pheromonal communication. Behav Brain Res 200(2):239–247. Epub 2008 Oct 11
98. Witt M, Hummel T (2006) Vomeronasal versus olfactory epithelium: is there a cellular basis for human vomeronasal perception? Int Rev Cytol 248:209–259
99. Rodriguez I, Mombaerts P (2002) Novel human vomeronasal receptor-like genes reveal species-specific families. Curr Biol 12(12):R409–R411

100. Giorgi D, Friedman C, Trask BJ, Rouquier S (2000) Characterization of nonfunctional V1R-like pheromone receptor sequences in human. Genome Res 10(12):1979–1985
101. Liman ER (2006) Use it or lose it: molecular evolution of sensory signaling in primates. Pflugers Arch 453(2):125–131
102. Boschat C, Pelofi C, Randin O, Roppolo D, Luscher C, Broillet MC, Rodriguez I (2002) Pheromone detection mediated by a V1r vomeronasal receptor. Nat Neurosci 5 (12):1261–1262
103. Shirokova E, Raguse JD, Meyerhof W, Krautwurst D (2008) The human vomeronasal type-1 receptor family--detection of volatiles and cAMP signaling in HeLa/Olf cells. FASEB J 22 (5):1416–1425
104. Wallrabenstein I, Gerber J, Rasche S, Croy I, Kurtenbach S, Hummel T, Hatt H (2015) The smelling of Hedione results in sex-differentiated human brain activity. Neuroimage 113:365–373
105. Bunzow JR, Sonders MS, Arttamangkul S, Harrison LM, Zhang G, Quigley DI, Darland T, Suchland KL, Pasumamula S, Kennedy JL, Olson SB, Magenis RE, Amara SG, Grandy DK (2001) Amphetamine, 3,4-methylenedioxymethamphetamine, lysergic acid diethylamide, and metabolites of the catecholamine neurotransmitters are agonists of a rat trace amine receptor. Mol Pharmacol 60(6):1181–1188
106. Lindemann L, Hoener MC (2005) A renaissance in trace amines inspired by a novel GPCR family. Trends Pharmacol Sci 26(5):274–281
107. Lam VM, Espinoza S, Gerasimov AS, Gainetdinov RR, Salahpour A (2015) In-vivo pharmacology of trace-amine associated receptor 1. Eur J Pharmacol 763(Pt B):136–142
108. Jing L, Li JX (2015) Trace amine-associated receptor 1: a promising target for the treatment of psychostimulant addiction. Eur J Pharmacol 761:345–352
109. Harmeier A, Obermueller S, Meyer CA, Revel FG, Buchy D, Chaboz S, Dernick G, Wettstein JG, Iglesias A, Rolink A, Bettler B, Hoener MC (2015) Trace amine-associated receptor 1 activation silences GSK3beta signaling of TAAR1 and D2R heteromers. Eur Neuropsychopharmacol 25(11):2049–2061
110. Hussain A, Saraiva LR, Ferrero DM, Ahuja G, Krishna VS, Liberles SD, Korsching SI (2013) High-affinity olfactory receptor for the death-associated odor cadaverine. Proc Natl Acad Sci U S A 110(48):19579–19584
111. Dewan A, Pacifico R, Zhan R, Rinberg D, Bozza T (2013) Non-redundant coding of aversive odours in the main olfactory pathway. Nature 497(7450):486–489
112. Hashiguchi Y, Nishida M (2007) Evolution of trace amine associated receptor (TAAR) gene family in vertebrates: lineage-specific expansions and degradations of a second class of vertebrate chemosensory receptors expressed in the olfactory epithelium. Mol Biol Evol 24 (9):2099–2107
113. Lindemann L, Ebeling M, Kratochwil NA, Bunzow JR, Grandy DK, Hoener MC (2005) Trace amine-associated receptors form structurally and functionally distinct subfamilies of novel G protein-coupled receptors. Genomics 85(3):372–385
114. Davenport AP, Alexander SP, Sharman JL, Pawson AJ, Benson HE, Monaghan AE, Liew WC, Mpamhanga CP, Bonner TI, Neubig RR, Pin JP, Spedding M, Harmar AJ (2013) International Union of Basic and Clinical Pharmacology. LXXXVIII G protein-coupled receptor list: recommendations for new pairings with cognate ligands. Pharmacol Rev 65 (3):967–986
115. Murooka Y, Azakami H, Yamashita M (1996) The monoamine regulon including syntheses of arylsulfatase and monoamine oxidase in bacteria. Biosci Biotechnol Biochem 60 (6):935–941
116. Chander H, Batish VK, Babu S, Singh RS (1989) Factors affecting amine production by a selected strain of *Lactobacillus bulgaricus*. J Food Sci 54(4):940–942
117. Williams BB, Van Benschoten AH, Cimermancic P, Donia MS, Zimmermann M, Taketani M, Ishihara A, Kashyap PC, Fraser JS, Fischbach MA (2014) Discovery and characterization of gut microbiota decarboxylases that can produce the neurotransmitter tryptamine. Cell Host Microbe 16(4):495–503

118. Yano JM, Yu K, Donaldson GP, Shastri GG, Ann P, Ma L, Nagler CR, Ismagilov RF, Mazmanian SK, Hsiao EY (2015) Indigenous bacteria from the gut microbiota regulate host serotonin biosynthesis. Cell 161(2):264–276
119. Nelson DA, Tolbert MD, Singh SJ, Bost KL (2007) Expression of neuronal trace amine-associated receptor (Taar) mRNAs in leukocytes. J Neuroimmunol 192(1–2):21–30
120. Muhlhaus J, Dinter J, Nurnberg D, Rehders M, Depke M, Golchert J, Homuth G, Yi CX, Morin S, Kohrle J, Brix K, Tschop M, Kleinau G, Biebermann H (2014) Analysis of human TAAR8 and murine Taar8b mediated signaling pathways and expression profile. Int J Mol Sci 15(11):20638–20655
121. D'Andrea G, Terrazzino S, Fortin D, Farruggio A, Rinaldi L, Leon A (2003) HPLC electrochemical detection of trace amines in human plasma and platelets and expression of mRNA transcripts of trace amine receptors in circulating leukocytes. Neurosci Lett 346(1–2):89–92
122. Sriram U, Cenna JM, Haldar B, Fernandes NC, Razmpour R, Fan S, Ramirez SH, Potula R (2016) Methamphetamine induces trace amine-associated receptor 1 (TAAR1) expression in human T lymphocytes: role in immunomodulation. J Leukoc Biol 99(1):213–223
123. Salamanca SA, Sorrentino EE, Nosanchuk JD, Martinez LR (2014) Impact of methamphetamine on infection and immunity. Front Neurosci 8:445
124. Cisneros IE, Ghorpade A (2014) Methamphetamine and HIV-1-induced neurotoxicity: role of trace amine associated receptor 1 cAMP signaling in astrocytes. Neuropharmacology 85:499–507
125. Babusyte A, Krautwurst D (2013) Tracing amines to their receptors: a synopsis in light of recent literature. Cell Biol Res Ther 2:1–3
126. Bachmanov AA, Beauchamp GK (2007) Taste receptor genes. Annu Rev Nutr 27:389–414
127. Foster SR, Blank K, See Hoe LE, Behrens M, Meyerhof W, Peart JN, Thomas WG (2014) Bitter taste receptor agonists elicit G-protein-dependent negative inotropy in the murine heart. FASEB J 28(10):4497–4508
128. Avau B, Depoortere I (2015) The bitter truth about bitter taste receptors: beyond sensing bitter in the oral cavity. Acta Physiol (Oxf)
129. Camoretti-Mercado B, Pauer SH, Yong HM, Smith DC, Deshpande DA, An SS, Liggett SB (2015) Pleiotropic effects of bitter taste receptors on [Ca2+]i mobilization, hyperpolarization, and relaxation of human airway smooth muscle cells. PLoS One 10(6), e0131582
130. Tizzano M, Gulbransen BD, Vandenbeuch A, Clapp TR, Herman JP, Sibhatu HM, Churchill ME, Silver WL, Kinnamon SC, Finger TE (2010) Nasal chemosensory cells use bitter taste signaling to detect irritants and bacterial signals. Proc Natl Acad Sci U S A 107 (7):3210–3215
131. Wauson EM, Zaganjor E, Lee AY, Guerra ML, Ghosh AB, Bookout AL, Chambers CP, Jivan A, McGlynn K, Hutchison MR, Deberardinis RJ, Cobb MH (2012) The G protein-coupled taste receptor T1R1/T1R3 regulates mTORC1 and autophagy. Mol Cell 47 (6):851–862
132. Workman AD, Palmer JN, Adappa ND, Cohen NA (2015) The role of bitter and sweet taste receptors in upper airway immunity. Curr Allergy Asthma Rep 15(12):72
133. Kiuchi S, Yamada T, Kiyokawa N, Saito T, Fujimoto J, Yasue H (2006) Genomic structure of swine taste receptor family 1 member 3, TAS1R3, and its expression in tissues. Cytogenet Genome Res 115(1):51–61
134. Priego T, Sanchez J, Pico C, Ahrens W, De Henauw S, Kourides Y, Lissner L, Molnar D, Moreno LA, Russo P, Siani A, Veidebaum T, Palou A (2015) TAS1R3 and UCN2 transcript levels in blood cells are associated with sugary and fatty food consumption in children. J Clin Endocrinol Metab 100(9):3556–3564
135. Chen QY, Alarcon S, Tharp A, Ahmed OM, Estrella NL, Greene TA, Rucker J, Breslin PA (2009) Perceptual variation in umami taste and polymorphisms in TAS1R taste receptor genes. Am J Clin Nutr 90(3):770S–779S
136. Wauson EM, Zaganjor E, Cobb MH (2013) Amino acid regulation of autophagy through the GPCR TAS1R1-TAS1R3. Autophagy 9(3):418–419

137. Hara T, Kimura I, Inoue D, Ichimura A, Hirasawa A (2013) Free fatty acid receptors and their role in regulation of energy metabolism. Rev Physiol Biochem Pharmacol 164:77–116
138. Wang J, Wu X, Simonavicius N, Tian H, Ling L (2006) Medium-chain fatty acids as ligands for orphan G protein-coupled receptor GPR84. J Biol Chem 281(45):34457–34464
139. Itoh Y, Kawamata Y, Harada M, Kobayashi M, Fujii R, Fukusumi S, Ogi K, Hosoya M, Tanaka Y, Uejima H, Tanaka H, Maruyama M, Satoh R, Okubo S, Kizawa H, Komatsu H, Matsumura F, Noguchi Y, Shinohara T, Hinuma S, Fujisawa Y, Fujino M (2003) Free fatty acids regulate insulin secretion from pancreatic beta cells through GPR40. Nature 422 (6928):173–176
140. Briscoe CP, Tadayyon M, Andrews JL, Benson WG, Chambers JK, Eilert MM, Ellis C, Elshourbagy NA, Goetz AS, Minnick DT, Murdock PR, Sauls HR Jr, Shabon U, Spinage LD, Strum JC, Szekeres PG, Tan KB, Way JM, Ignar DM, Wilson S, Muir AI (2003) The orphan G protein-coupled receptor GPR40 is activated by medium and long chain fatty acids. J Biol Chem 278(13):11303–11311
141. Li X, Ballantyne LL, Che X, Mewburn JD, Kang JX, Barkley RM, Murphy RC, Yu Y, Funk CD (2015) Endogenously generated omega-3 fatty acids attenuate vascular inflammation and neointimal hyperplasia by interaction with free fatty acid receptor 4 in mice. J Am Heart Assoc 4(4)
142. Oh DY, Talukdar S, Bae EJ, Imamura T, Morinaga H, Fan W, Li P, Lu WJ, Watkins SM, Olefsky JM (2010) GPR120 is an omega-3 fatty acid receptor mediating potent anti-inflammatory and insulin-sensitizing effects. Cell 142(5):687–698
143. Ichimura A, Hasegawa S, Kasubuchi M, Kimura I (2014) Free fatty acid receptors as therapeutic targets for the treatment of diabetes. Front Pharmacol 5:236
144. Kasubuchi M, Hasegawa S, Hiramatsu T, Ichimura A, Kimura I (2015) Dietary gut microbial metabolites, short-chain fatty acids, and host metabolic regulation. Nutrients 7(4):2839–2849
145. Tolhurst G, Heffron H, Lam YS, Parker HE, Habib AM, Diakogiannaki E, Cameron J, Grosse J, Reimann F, Gribble FM (2012) Short-chain fatty acids stimulate glucagon-like peptide-1 secretion via the G-protein-coupled receptor FFAR2. Diabetes 61(2):364–371
146. Reigstad CS, Salmonson CE, Rainey JF 3rd, Szurszewski JH, Linden DR, Sonnenburg JL, Farrugia G, Kashyap PC (2015) Gut microbes promote colonic serotonin production through an effect of short-chain fatty acids on enterochromaffin cells. FASEB J 29(4):1395–1403
147. Le Poul E, Loison C, Struyf S, Springael JY, Lannoy V, Decobecq ME, Brezillon S, Dupriez V, Vassart G, Van Damme J, Parmentier M, Detheux M (2003) Functional characterization of human receptors for short chain fatty acids and their role in polymorphonuclear cell activation. J Biol Chem 278(28):25481–25489
148. Zhang J, Yang L, Ang Z, Yoong SL, Tran TT, Anand GS, Tan NS, Ho B, Ding JL (2010) Secreted M-ficolin anchors onto monocyte transmembrane G protein-coupled receptor 43 and cross talks with plasma C-reactive protein to mediate immune signaling and regulate host defense. J Immunol 185(11):6899–6910
149. Lattin JE, Schroder K, Su AI, Walker JR, Zhang J, Wiltshire T, Saijo K, Glass CK, Hume DA, Kellie S, Sweet MJ (2008) Expression analysis of G protein-coupled receptors in mouse macrophages. Immunome Res 4:5
150. Nagasaki H, Kondo T, Fuchigami M, Hashimoto H, Sugimura Y, Ozaki N, Arima H, Ota A, Oiso Y, Hamada Y (2012) Inflammatory changes in adipose tissue enhance expression of GPR84, a medium-chain fatty acid receptor: TNFalpha enhances GPR84 expression in adipocytes. FEBS Lett 586(4):368–372
151. Suzuki M, Takaishi S, Nagasaki M, Onozawa Y, Iino I, Maeda H, Komai T, Oda T (2013) Medium-chain fatty acid-sensing receptor, GPR84, is a proinflammatory receptor. J Biol Chem 288(15):10684–10691

Top Med Chem (2017) 23: 175–194
DOI: 10.1007/7355_2014_78

Published online: 9 December 2014

# Neuronal Functions and Emerging Pharmacology of TAAR1

**Stefano Espinoza and Raul R. Gainetdinov**

**Abstract** Trace amine-associated receptor 1 (TAAR1) is a member of TAAR family of G protein-coupled receptors (GPCRs). The members of this class of receptors discovered in 2001 have been found in some tissues ranging from the central nervous system to the olfactory epithelium and in some peripheral organs. The best studied receptor, TAAR1, is activated by a class of compounds named trace amines (TAs) that include compounds such as β-phenylethylamine (PEA), *p*-tyramine, octopamine, and tryptamine normally present at low levels in the mammalian brain. Although TA levels have been associated with many neuropsychiatric disorders, only the discovery of TAAR1 validated their physiological role. TAAR1 can modulate monoamine neurotransmission and, in particular, dopamine systems. Several studies have demonstrated that TAAR1 knockout (TAAR1-KO) mice display a supersensitive dopaminergic system, while activation of TAAR1 can reduce dopaminergic hyperactivity obtained either with pharmacological tools or present in genetic mouse model. For these reasons, TAAR1 has been proposed as a novel therapeutic target for neuropsychiatric disorders such as schizophrenia, bipolar disorder, and addiction. Moreover, several peripheral functions of TAAR1 have been described recently indicating intriguing novel TAAR1 roles in system physiology. Here we will review brain and peripheral functions mediated by TAAR1 and other TAARs.

---

S. Espinoza
Department of Neuroscience and Brain Technologies, Istituto Italiano di Tecnologia, Genova, Italy

R.R. Gainetdinov (✉)
Department of Neuroscience and Brain Technologies, Istituto Italiano di Tecnologia, Genova, Italy

Skolkovo Institute of Science and Technology (Skoltech), Skolkovo, Moscow 143025, Russia

Faculty of Biology, St. Petersburg State University, St. Petersburg 199034, Russia
e-mail: raul.gainetdinov@iit.it

**Keywords** Dopamine, Neuropsychiatric disorders, TAAR1, Trace amines

**Contents**



# 1 Introduction

Trace amine-associated receptors (TAARs) are a family of G protein-coupled receptors (GPCRs) that have been discovered in 2001 [1, 2]. From their first description by two independent groups, TAARs attracted an enduring interest among physiologists and pharmacologists, since they have been shown to be involved in many different physiological processes ranging from regulation of brain functions to olfaction and, more recently, to immune system [3–6]. Among them, TAAR1 has been mostly characterized with its particular role in the regulation of brain functions, although new intriguing functions of this receptor in the periphery are emerging. The initial observation from TAAR1 knockout (TAAR1-KO) mice was further strengthened by several recent studies using selective TAAR1 full and partial agonists, suggesting an important role for TAAR1 in the regulation of dopaminergic system and indicating TAAR1 as new potential target for the treatment of neuropsychiatric disorders such as schizophrenia, bipolar disorder, ADHD, and addiction [7–12]. Since the amount of literature on TAARs is increasing monthly, this review will focus on TAAR1 and particularly on recent reports that highlight TAAR1 influence on monoamine systems and its possible role in psychiatric diseases. We will also give brief introduction of the history of TAAR discovery and provide description of potential functions of TAAR1 in the periphery.

# 2 Trace Amines

The term "trace amines" (TAs) refers to a class of endogenous compounds that have been found at low levels in mammalian brain [3, 13–16]. These non-catecholic amines are closely related to the classic monoamines (dopamine, noradrenaline, and serotonin) in terms of structure, synthesis, metabolism, and distribution, so that for a long time they have been considered only as side products with little physiological relevance [3, 14]. With the cloning of the family of TAARs in 2001, TAs' role has been reconsidered and a growing amount of studies are now focused on more precise physiological role for these compounds [3, 4, 14, 17]. Conventionally, TAs include molecules such as β-phenylethylamine (PEA), *p*-tyramine, octopamine, synephrine, and tryptamine, but if we consider as "trace amine" an endogenous amine that is able to activate a TAAR, then we should include also other compounds such as 3-methylamine (a TAAR5 agonist), 3-metoxythyramine (a TAAR1 agonist), and others [18, 19]. Generally, most of TAs are synthetized from the decarboxylation of the L-amino acid precursors by the action of the L-amino acid decarboxylase (AADC) [14], while octopamine is formed from the hydroxylation of *p*-tyramine by the enzyme dopamine β-hydroxylase. Regarding the metabolism, the major route of degradation is mediated by the action of the monoamine oxidases (MAO), with MAO-B being the prototypical enzyme for PEA degradation [20] and both MAO-A and MAO-B responsible for the metabolism of the other TAs [21, 22]. MAO-B metabolizes also dopamine, and MAO-B inhibitors are clinically used for the treatment of Parkinson's disease and depression. Interestingly, MAO-B-deficient mice show higher PEA extracellular levels with no change in dopamine levels [23], suggesting a potential role for PEA in the therapeutic efficacy of these drugs.

While in invertebrates TAs such as tyramine and octopamine serve as major neurotransmitters [14, 24, 25], in mammals the precise physiological functions of these compounds are still not clear [3]. Before the discovery of TAARs, TAs were described as false neurotransmitters and sympathomimetic substances [26–28]. These effects are believed to be due to the nonspecific action that they have on vesicular and plasma membrane monoamine transporters and occurs at high, nonphysiological concentrations. In fact, PEA, at high doses, is able to reverse the function of dopamine transporter (DAT) and to increase locomotor activity producing an amphetamine-like effect [14]. For this reason, PEA has been considered as the "endogenous amphetamine." Interestingly, when PEA, at the same stimulating doses, is administered to mice lacking the DAT (DAT-KO mice), it produces a decrease in locomotor activity [29] revealing the specific, non-DAT-related effect that PEA has likely via TAAR1 activation. Indeed, many reports showed that the TAAR1-selective activation produces a "calming" effect on a hyperactive dopamine system obtained by either a pharmacological treatment (cocaine) or present in genetic animal models (DAT-KO mice) (see next sections) [10, 11]. TAs, as demonstrated by many studies, are present at low levels in mammalian brains, with an extracellular concentration that normally is estimated in the low nanomolar range and is tightly regulated by AADC and MAO activity [14]. As described below, TAAR1 is activated by PEA with an $EC_{50}$ around 200–900 nmol/L, depending on the species and on the type of the assay, usually performed in heterologous cell system [1, 2, 30–36]. These discrepancies could be due to the nonphysiological cellular systems where these assays are performed and may reflect some deficiency of the transduction machinery in these artificial systems. For example, TAAR4 is also activated by PEA, with an $EC_{50}$ obtained in COS-7 cells in the low micromolar range [1]. Conversely, when PEA was used to stimulate TAAR4 in its natural environment, the olfactory sensory neurons, it emerged that PEA started to stimulate TAAR4 in the picomolar range [37], with a strikingly higher affinity than in transfected cells. Similarly, TAAR1 sensitivity to *p*-tyramine in inducing $K^+$ currents through Kir3 channels was much higher, when tested in dopaminergic neurons, compared to when they were expressed in heterologous systems [34]. Finally, PEA was able to induce different effects in leukocytes in a TAAR1-dependent manner at low nanomolar concentrations [38], further suggesting that these receptors would need their natural environment to fully express their native pharmacology.

# 3 Trace Amine-Associated Receptors

TAAR family consists of 9 genes in human (including 3 pseudogenes) and in chimpanzee (including 6 pseudogenes), while 19 and 16 genes (including 2 and 1 pseudogenes) are present in the rat and mouse genome, respectively [4, 39]. All these genes cluster in a narrow region of approximately 100–200 kilobases on the same chromosome, reminiscent of similar chromosomal organization of some members of the olfactory receptors [18]. All these genes are encoded by a single exon, except TAAR2, and are of a length of about 1 kilobase. Interestingly, the region where TAARs are located on chromosome 6 (the locus 6q23) has been associated with schizophrenia and bipolar disorder in linkage/association studies [40]. Regarding the in vivo expression and tissue distribution, different groups carried out several studies, with sometime conflicting results, most likely due to the different techniques used to detect TAARs expression. All TAARs, with the exception of TAAR1, have been found in the olfactory epithelium as described by Lieberles and Buck [18] and believed to serve as new class of olfactory receptors. In this study, no detection of TAAR transcripts in the brain and periphery was reported; however, several other studies found that at least some members of TAAR family are expressed in many organs and in the brain [1, 2, 34, 41]. The first two groups that cloned TAAR family found that TAAR1 is expressed in many brain regions and in peripheral organs such as the liver, kidney, spleen, pancreas, and heart [1, 2]. There is also evidence that other TAARs are expressed in the brain including TAAR5, TAAR6, and TAAR9 [3]. Intriguingly, TAAR6 polymorphisms have been associated with susceptibility to schizophrenia, and mutations found in this gene have been studied in correlation with antidepressant response and suicidal behavior [40, 42]. TAAR1 and TAAR2 are also expressed in human blood leukocytes, particularly in polymorphonuclear (PMN) T and B cells, and lower expression was found for TAAR5, TAAR6, and TAAR9 [38, 43, 44]. RNA extracted from rat heart demonstrated that TAAR1, TAAR2, TAAR3, TAAR4, and TAAR8a are expressed in this organ [3, 45].

# 4 TAAR1

At the moment, TAAR1 is the best studied member of the TAAR family. Since its expression was found in the brain regions, most of the studies focused on TAAR1 role in brain physiology and pathology. TAAR1, alongside with TAAR4, is the only TAAR subtype that is activated by TAs, and if we consider that TAAR4 in humans is a pseudogene, TAAR1 could be considered as the primary target for these amines in humans. Several studies, using different experimental approaches, demonstrated that TAAR1 is expressed in various brain regions. While this expression was not at such high levels as for some other GPCRs, it was found to be in key regions for monoamine systems [1–3, 34, 46]. RNA transcript for TAAR1 was found in mouse and rat brain in ventral tegmental area (VTA), dorsal raphe, substantia nigra, striatum, and frontal cortex. Expression in dopaminergic areas such as the substantia nigra and striatum has been found also in rhesus monkey brain [41]. In humans, fewer studies were performed, but high expression of TAAR1 was found at least in the amygdala [1]. Using a transgenic mouse model, where LacZ gene was inserted in TAAR1 gene to have a specific expression of the LacZ driven by TAAR1 promoter, Lindemann et al. confirmed that TAAR1 was present in dopaminergic and serotonergic areas such as VTA, amygdala, dorsal raphe, subiculum, and parahippocampal region [34]. In the periphery, TAAR1 was found in several organs including liver, spleen, kidney, gastrointestinal tract, pancreas, heart, and leukocytes [2, 3]. At subcellular levels, it is still not clear whether TAAR1 in its natural environment such as neurons is expressed at the membrane as most GPCRs or in the intracellular compartments. Difficulties in reliable evaluation of cellular distribution of TAAR1 are, in part, determined by technical limitations such as the lack of specific antibodies.

TAAR1 expression and functional studies in heterologous cell system such as HEK-293 cells have been challenging, since TAAR1 was found mainly in the intracellular compartments leading to a difficulty in its coupling to G protein to transduce the signal [2, 30, 31, 33, 36, 47]. TAAR1 is a Gαs-coupled receptor and its stimulation increases cAMP levels, but likely only with sufficient membrane expression it is possible to properly study its pharmacology. Many strategies have been used to improve TAAR1 membrane expression such as building a human/rat chimera [2] or adding a peptide from bovine rhodopsin at the N-terminus of the receptor [18]. In our lab, we fused the first nine amino acid of the β2-adrenergic receptor to the N-terminus, leading to a significant level of membrane expression sufficient to reliably study TAAR1 pharmacology [36]. TAAR1, as demonstrated by many laboratories under different experimental conditions, can be activated by PEA and *p*-tyramine, with PEA being more potent against mouse and human TAAR1 and *p*-tyramine being more potent against rat TAAR1 [1, 2, 4, 5, 30–33, 35, 36, 47]. TAAR1 has been reported also to have a weaker affinity for tryptamine and octopamine [1–3, 36]. When stimulated by an agonist, TAAR1 couples to Gαs and stimulates the production of cAMP. Using a BRET-based biosensor, it was possible to study the dynamics of cAMP fluctuations after TAAR1 stimulation, and by using this approach we observed that TAAR1 poorly desensitizes upon agonist binding [36]. While cAMP levels, produced by β2-adrenergic receptor, a prototypical GPCR, decrease after 5 min and then return to basal levels, TAAR1 agonist-stimulated cAMP decreases slightly only after 10–20 min. This evidence was confirmed by β-arrestin2 translocation studies after TAAR1 stimulation [36, 48]. PEA was able to recruit β-arrestin2-GFP but in a small proportion and at a lesser degree compared to other GPCRs such as β2-adrenergic receptor. This property of TAAR1 resembled D3 dopamine receptor that has been demonstrated to poorly recruit β-arrestin2 [49]. By using cAMP assay in rat TAAR1, Bunzow et al. performed a screening of several classes of compounds as regards to TAAR1 activity [2].

Apart from classic TAs, many other known compounds were able to activate TAAR1. Among them, amphetamines and their derivatives are of particular interest. In this assay, D-amphetamine, L-amphetamine, D-methamphetamine, and (±)-MDMA acted as TAAR1 agonists, and later reports confirmed these data on human, mouse, rat, and rhesus monkey TAAR1 expressed in different cellular systems [30, 31, 33, 35, 36, 41, 47, 50]. Interestingly, the concentrations of amphetamine found to be necessary to activate TAAR1 are in line with what was found in drug abusers [3, 51, 52]. Thus, it is likely that some of the effects produced by amphetamines could be mediated by TAAR1. Indeed, in a study in mice, MDMA effects were found to be mediated in part by TAAR1, in a sense that MDMA auto-inhibits its neurochemical and functional actions [46]. Based on this and other studies (see other section), it has been suggested that TAAR1 could play a role in reward mechanisms and that amphetamine activity on TAAR1 counteracts their known behavioral and neurochemical effects mediated via dopamine neurotransmission. On the other hand, whether TAAR1 mutations or functional deficits in humans are associated with drug addiction would be an interesting point to evaluate.

Other interesting endogenous substances that are TAAR1 agonist are thyronamines, compounds structurally related to thyroid hormones [53, 54]. 3-Iodothyronamine (T1AM) and its deiodinated relative thyronamine (T0AM) are potent full agonists of human, rat, and mouse TAAR1, and when administered to rats, they induce profound physiological effect such as hypothermia, alteration of metabolism, cardiac effects, and behavioral suppression [3, 54–56]. Similarly, metabolites of amiodarone that has a chemical structure similar to thyronamines and is used in clinic to treat arrhythmias are TAAR1 agonists [57]. It should be noted, however, that these compounds are highly nonselective and can interact, for example, with the functions of plasma membrane and vesicular monoamine transporters [58]. Bunzow et al. [2] have found also that the O-metabolites of catecholamines such as 3-methoxytyramine (3-MT), 4-methoxytyramine (4-MT), normetanephrine, and metanephrine can exert potent TAAR1 agonistic activity. Our group and others also confirmed that 3-MT and 4-MT are potent agonists against human TAAR1 [31, 36]. This observation is particularly intriguing, since 3-MT is a dopamine metabolite formed by the activity of catechol *o*-methyltransferase (COMT) and traditionally has been considered as a compound with no biological activity but only as reflection of extracellular dopamine levels [59]. Our lab studied the potential effect of 3-MT in mice and documented that

3-MT can induce a behavioral activation in a dopamine-independent manner [19]. Central infusion of 3-MT produced mild hyperactivity and a complex set of abnormal movements that were less pronounced in TAAR1-KO mice. Moreover, 3-MT could induce the phosphorylation of ERK and CREB in striatum in a TAAR1-dependent manner [19]. This study indicates that COMT is not simply a metabolizing enzyme, but it may also serve as the rate-limiting step for the production of this novel neuromodulator active at TAAR1. It would be interesting to evaluate the role of 3-MT in behavioral manifestations under conditions where its concentrations are particularly high such as in Parkinson's disease patients with dyskinesia after a long-term treatment with L-DOPA [60].

A recent intriguing study also suggested that food additive ractopamine used to feed livestock in the USA is a full agonist of TAAR1 [61]. It should be underlined that since TAs and most compounds with TAAR1 agonistic activity are not selective and bind also other receptors and transporters [2], it is difficult to study TAAR1 physiology in vivo by analyzing physiological effects of these compounds in normal mice [29]. Until recently, the use of TAAR1-KO mouse line was the only possibility to evaluate TAAR1 contribution in CNS physiology. However, in the last 4 years, several selective full and partial TAAR1 agonists were synthesized and characterized [9–11]. The studies performed with these compounds in various preclinical models supported the idea that TAAR1 could be a novel target for managing psychiatric disorders such as schizophrenia, bipolar disorder, and addiction (see below). Despite the large number of agonists available at the moment, only one antagonist was described. However, poor solubility and brain-blood barrier penetration of this compound gives the possibility to investigate only in vitro effects of TAAR antagonism [8]. To advance pharmacological innovation, two groups attempted to discover the molecular determinants responsible for ligand-receptor interaction for TAAR1 [62–64]. For this purpose, our group developed a theoretical model of the human TAAR1 developed by homology modeling and made docking studies with known TAAR1 agonists finding important amino acid residues for the activity of these ligands [63]. By comparing the derived hTAAR1 model with known models such as of the β2-adrenergic receptor and the 5-HT1a serotonergic receptor, we identified two residues (D103 and N286) as potential anchor points for the ligand recognition process. Also Grandy's laboratory, with mutagenesis studies, described the structural determinants responsible for TAAR1 ligand-binding pockets with respect to amphetamine and methamphetamine and identified a residue responsible for the species stereoselectivity toward D- and L-amphetamine [64]. These studies, along with the screening of new chemical entities, could help in the drug discovery process to find new TAAR1 ligands such as agonists and systemically available antagonists that would improve the comprehension of TAAR1 physiology.

# 5 TAAR1 as a Novel Drug Target for Psychiatric Disorders

The idea that TAAR1 could be involved in the pathophysiology of psychiatric disorders was initially suggested by the fact that TAs could activate TAAR1. Historically, dysregulated TAs levels have been linked to many human disorders such as schizophrenia, ADHD, Parkinson's disease, migraine, and depression [14, 16, 65, 66]. Clinical studies found elevated PEA plasma levels in schizophrenic patients and increased urinary excretion in paranoid schizophrenics [67, 68]. By looking to its expression pattern in the brain, TAAR1 is well positioned to modulate monoamine systems, and monoamines play a pivotal role in the pathophysiology of many psychiatric disorders [4, 34]. First evidence about the TAAR1 role in brain functions came from the study of the mouse line lacking this receptor. Up to now, three different TAAR1-KO mouse lines have been generated, and substantially similar phenotype has been reported as regards to a supersensitive dopaminergic system and other monoamine-related dysregulations [30, 34, 46]. TAAR1-KO mice do not demonstrate overt phenotype, breed normally, and do not show striking differences in most neurological and behavioral tests versus their wild-type (WT) littermates. However, these mice were found to be more sensitive to the amphetamine-induced behavioral and neurochemical effects. Amphetamine was able to produce an enhanced response in terms of locomotion as well as in terms of dopamine released in the striatum as measured in microdialysis experiments [30, 34, 46]. Another amphetamine compound, MDMA, has been shown to increase dopamine release in the frontal cortex, striatum, and nucleus accumbens at a greater extent in TAAR1-KO mice compared to WT mice [46]. Similar results were obtained for serotonin, with an enhanced release in the striatum and the nucleus accumbens, but not in the frontal cortex. Furthermore, TAAR1-KO mice showed a significant deficit in prepulse inhibition (PPI), indicating an impairment of sensorimotor gating that is known to be deficient in schizophrenic patients [30].

Another feature that links TAAR1 to schizophrenia is the D2 dopamine receptor supersensitivity. In particular, it has been found that TAAR1-KO mice have a greater proportion of D2 dopamine receptors in the high-affinity state compared to WT [30]. As for increased amphetamine responsiveness and PPI deficits, also an increase in D2 dopamine receptor-mediated striatal functions has been traditionally associated with schizophrenia [69]. A direct functional interaction between TAAR1 and D2 dopamine receptor is one of possible mechanisms that might be responsible for the modulation of dopaminergic system by TAAR1. In VTA dopaminergic neurons, stimulation of TAAR1 modulates D2 dopamine autoreceptor activity to decrease D2 receptor activity and promote D2 receptor desensitization [8, 9]. By studying the outward current mediated by D2 receptors, it has been demonstrated that in VTA slices from TAAR1-KO mice or in slices from WT mice treated with the selective TAAR1 antagonist EPPTB, quinpirole desensitization was prevented, and the quinpirole potency was increased by fourfold [8]. On the contrary, the application of *p*-tyramine decreased quinpirole potency. Interestingly, the same type of modulation was found between TAAR1 and 5-HT1A autoreceptors in the dorsal raphe, where TAAR1 is expressed and modulates 5-HT1A activity [9]. Both D2 dopamine and 5-HT1A serotonin autoreceptors are important for the regulation of mood, cognition, and motor behavior and for the response to antidepressant and antipsychotic drugs [70–72]. In the VTA and dorsal raphe, TAAR1 can also modulate the firing rate of dopaminergic and serotonergic neurons, with higher firing frequency in TAAR1-KO animals or in WT mice treated with the antagonist EPPTB being observed [8]. This modulation is independent on cAMP levels and D2 dopamine receptors but mediated by the $G_{\beta\gamma}$ subunits of the activated G protein and resulted in the modulation of the $K^+$ current mediated by the Kir3-type $K^+$ channel. D2 dopamine autoreceptor functions are also altered in the nucleus accumbens of TAAR1-KO mice [73]. In fact, in vivo microdialysis and fast-scan cyclic voltammetry (FSCV) studies have revealed that in the nucleus accumbens but not in the dorsal striatum, the basal dopamine release was increased and activation or blockade of TAAR1 could modulate this release. Moreover, with a FSCV paired-pulse approach, it was possible to directly demonstrate that D2 dopamine autoreceptor activity was reduced in TAAR1-KO animals, leading to higher level of second pulse-induced stimulated dopamine release [73]. There is also significant amount of evidence that TAAR1 could have an influence also on postsynaptic D2 dopamine receptors. In an in vitro study, by using a bioluminescence energy transfer (BRET)-based assay, it has been demonstrated that TAAR1 could form a heterodimer with the long isoform of the D2 dopamine receptor [74] that it is known to be mostly expressed at the postsynaptic sites. This heterodimer was sensitive to D2 receptor conformation, and haloperidol, a D2 antagonist, was able to decrease the complex formation. Similarly, haloperidol treatment increased the TAAR1 responses mediated by PEA. This functional interaction of the D2 postsynaptic receptors was also evident in vivo, since haloperidol-induced catalepsy and c-Fos expression in dorsal striatum were reduced in TAAR1-KO animals [74]. Another line of research also focused on a possible influence of TAAR1 on dopamine transporters [5, 33, 41]. Co-expression studies revealed the presence of TAAR1 and DAT in a subset of neurons in the substantia nigra, and experiments done in synaptosomal preparations and cells revealed that TAAR1 can modulate DAT functions [50, 75–77]. It is notably, however, that TAAR1 agonists can effectively block hyperactivity in mice lacking the DAT and no alterations in DA uptake kinetics were found after application of TAAR1 agonists or antagonist in FSCV studies in striatal and accumbal slices [73].

The recent development of TAAR1-selective ligands, particularly full and partial agonists, has been of extremely importance for a better understanding of TAAR1 physiology. Since all the other known TAAR1 ligands possess other important activities (e.g., on DAT), these new selective ligands provided the first opportunity to understand the consequence of TAAR1 activation in experimental animal models. As the absence of TAAR1 results in a supersensitive dopaminergic system, TAAR1 activation negatively regulates dopamine system, decreasing an excess of dopaminergic activation, either obtained with pharmacological treatment or present in genetic animal models [9–11]. Moreover, TAAR1 agonists influence also the serotonergic system, and for these reasons, they have been proposed as possible treatment for diseases such as schizophrenia, bipolar disorder, depression, and drug abuse [6, 10]. Two full agonists and two partial agonists have been tested in various studies, and although a substantially similar profile was found between these compounds, there are also some important differences between full and partial agonists. TAAR1 activation reduces hyperlocomotion induced by pharmacological treatment with the DAT inhibitor cocaine or with the NMDA receptor antagonists L-687,414 and PCP as well as spontaneous hyperlocomotion present in DAT-KO mice or in NR1 knockdown mice [9–11]. Both hyperdopaminergia and hypoglutamatergia are considered to model schizophrenia endophenotypes and represent the two main hypotheses for the etiology of schizophrenia. Thus, these data indicate that TAAR1 activation can reduce what are considered as the positive symptoms in different pharmacological and genetic model of schizophrenia [11]. Interestingly, both full and partial agonists can potentiate the effect of two atypical antipsychotics olanzapine and risperidone in these behavioral paradigms suggesting also that TAAR1 agonists are potential add-on treatment to current antipsychotics [11]. TAAR1 treatment seems not to produce extrapyramidal side effects, and partial TAAR1 agonist can in fact reduce haloperidol-induced catalepsy, suggesting that under certain conditions such as deficient dopamine transmission caused by D2 dopamine receptors blockade, TAAR1 partial agonist might behave as an antagonist [11]. In contrast to olanzapine and other atypical antipsychotics that lead to weight gain, these compounds do not possess this side effect, but rather they can reduce the weight gain induced by a chronic treatment with olanzapine.

Furthermore, pharmacological magnetic resonance imaging (phMRI) study has revealed that, while differences exists, both full and partial TAAR1 agonists share a similar pattern with olanzapine in terms of brain region activation pattern including the prefrontal area, suggesting a potential role of TAAR1 in prefrontal cortical-related functions such as cognition. In fact, TAAR1 agonists can improve cognitive performance in the object retrieval paradigm in monkeys increasing the percentage of correct responses [11]. Similarly, in rats, TAAR1 agonist can revert the deficit induced by PCP in the attentional set-shifting test. Since TAAR1 is known to influence serotonin system, potential antidepressant and anxiolytic properties of these compounds have been also explored [9]. While only the partial agonists were effective in the forced swim test in rats, both the partial and full agonists showed antidepressant effect in monkeys in the differential reinforcement of low-rate behavior paradigm [10, 11]. Moreover, TAAR1 activation induced anxiolytic-like behaviors in the stress-induced hyperthermia test further indicating that TAAR1 modulation of serotonergic system could be of importance in mood disorders [10]. These data strongly support the idea that TAAR1 could be considered as a new multifaceted target to treat neuropsychiatric diseases such as schizophrenia and mood disorders. Intriguingly, the partial TAAR1 agonist, while showing a similar profile to the full agonist, has at the same time some peculiar differences such as the ability to reduce the haloperidol-induced catalepsy. It also can increase the firing activity in VTA dopaminergic neurons like the selective TAAR1 antagonist EPPTB [8]. This suggests that under certain conditions a basal TAAR1 activation by natural ligands is present and that TAAR1 partial agonist could decrease or increase dopamine-related behavior depending on the rate of dopaminergic activity.

Moreover, partial TAAR1 agonist at high doses was able to promote wakefulness, like caffeine, as a stimulating compound further indicating this putative "stabilizing" property [10, 11]. Whether TAAR1 partial activation might be more useful for the treatment of mood and anxiety disorders and the full TAAR1 agonist in others such as schizophrenia remains to be tested, but it would be very interesting to further understand in detail the differences between these compounds.

A recent study explored the possibility that apomorphine, a prototypical D1 and D2 dopamine receptor nonselective agonist, might exert its behavioral actions in part via TAAR1 activation [78]. Following an initial observation by Bunzow et al., this study confirmed that apomorphine is a partial agonist at rat and mouse TAAR1 with little activity at human and cynomolgus monkey TAAR1. While the lack of TAAR1 did not influence the locomotor behavior induced by apomorphine at low doses, apomorphine-induced climbing behavior and stereotypies were reduced in TAAR1-KO mice. Interestingly, when WT mice were injected with a TAAR1 agonist in combination with a D1 and D2 dopamine receptor selective agonists, they could reproduce a level of climbing behavior similar to what was obtained with apomorphine. Since apomorphine-induced climbing has been used for decades as the screening test for new antipsychotics [79], this study suggests that not only dopamine receptors but also TAAR1 could be in part responsible for this apomorphine effect, and compounds with putative antipsychotic activity identified by using this test could have also TAAR1 activity.

# 6 Role of TAAR1 in Addiction

Since TAAR1 has a strong connection to the dopaminergic system, it has been suggested that TAAR1 could have a role in addiction. Moreover, the evidence that several amphetamines, known to be addictive substances, were able to activate TAAR1, led the speculation that at least some of their effects could be mediated via TAAR1. Addictive drugs modulate brain functions in several ways, but all of them seem to have a unifying property that is to enhance mesolimbic dopamine neurotransmission [80]. The major ways to modulate synaptic dopamine levels are either influence on neuronal firing, interference with the reuptake of dopamine through DAT, or alterations in the presynaptic regulation at the level of terminals [80]. As described above, there is evidence that TAAR1 can potentially influence all of these processes. Particularly, TAAR1 has been reported to influence the firing of VTA dopaminergic neurons [34] and alter the function of presynaptic D2 dopamine receptors in nucleus accumbens [73], the brain region particularly important for addiction. TAAR1-KO mice that generally have a supersensitive dopaminergic system seem more incline to addictive properties of substances of abuse. In a study by Achat-Mendes et al. [81], the psychomotor and rewarding properties of methamphetamine were evaluated in WT and TAAR1-KO mice. Both single and repeated treatment with methamphetamine was able to produce an enhanced locomotor response in TAAR1-KO mice. Moreover, in conditioned place preference

(CPP) experiments, TAAR1-KO mice acquired the methamphetamine-induced CPP earlier than WT and retained CPP longer as evaluated by extinction training [81]. Interestingly, no difference between WT and KO for CPP induced with morphine was observed.

Another study evaluated the potential involvement of TAAR1 in alcohol abuse [82]. Using a two-bottle choice paradigm, this study showed that TAAR1-KO mice have a greater preference and consume more ethanol than WT counterparts, without difference in consumption of sucrose solution. Similarly, the sedative-like effects after ethanol consumptions were enhanced and lasted longer. These data suggest a potential role for TAAR1 in alcohol abuse disorder indicating the necessity of further studies to evaluate effects of TAAR1-selective drugs in alcohol-induced behaviors.

More evidence exists regarding potential utility of TAAR1-based drugs in cocaine addiction. The first study that was performed few years ago focused on evaluation of effects of TAAR1 agonist on cocaine self-administration in rats [10]. Using this well-validated experimental model of drug addiction, Revel et al. [10] demonstrated that partial TAAR1 agonist, dose-dependently, reduced cocaine intake in rats with a history of cocaine self-administration. Importantly, TAAR1 partial agonist did not influence lever pressing behavior in control subjects. Recently, two articles have been published regarding cocaine abuse-related effects in rats. In the first study, both partial and full agonists were studied in connection with models of cocaine relapse [7]. Context-induced renewal of drug seeking is considered close to real-life situations, since addicts often go under relapse, because they re-experienced the same context associated to past drug intake [83]. Rats with a history of cocaine self-administration went into abstinence without extinction and then put back into the same context, where they had cocaine self-administration. While saline control animals showed robust relapse to drug seeking, the treatment with both partial and full agonists dose-dependently reduces drug seeking. Importantly, at the doses used, TAAR1 agonists had no influence on a lever pressing task maintained by food. In another model of cocaine-primed reinstatement, where after extinction rats were injected by single dose of cocaine to induce relapse, TAAR1 partial agonist was able to completely block the cocaine-primed reinstatement of cocaine seeking [7]. Regarding the mechanism of action, it is shown that TAAR1 activation reduces the dopamine release induced by cocaine, as measured by FCSV in the nucleus accumbens, without altering the DAT functions, suggesting an involvement of other mechanisms than direct interference with dopamine uptake such as the alteration of D2 receptors activity [7]. In another study, TAAR1 partial agonist has been used to reduce several cocaine-mediated behaviors [12, 84]. First, TAAR1 agonist administration reduced the expression of cocaine behavioral sensitization. Moreover, in a CPP paradigm, TAAR1 agonist was able to reduce the expression but not the development of the CPP. Thus, while when administered prior to cocaine conditioning, TAAR1 agonist did not modify the development of the CPP, it could reduce the expression of the already established CPP when administered prior to the test session. Also in a model of cocaine relapse, the cocaine-primed reinstatement of cocaine seeking, TAAR1 activation reduced the relapse of the cocaine-seeking behavior [12]. Altogether, these data indicate that TAAR1 activation reduces the sensitizing, rewarding, and reinforcing effects of cocaine and TAAR1 should be explored further as a potential target for the treatment of cocaine addiction.

## 7 Role of TAARs in Periphery

As described above, TAARs and in particular TAAR1 are expressed in many peripheral organs. TAs effects on cardiovascular system and their pressor action have been known for many years [14], but it is evident that these actions most likely have to be attributed to their "false transmitter" properties. However, the fact that TAAR1, as well as other TAAR members, is expressed in the heart raised many questions about its putative role in this organ. T1AM and T0AM are endogenous compounds found in brain extracts and also in periphery [45, 54]. They can activate potently TAAR1 in vitro and produce important physiological responses when injected to animals. As described in several studies, thyronamines induce a behavioral suppression with locomotor inhibition, ptosis, reduced metabolic rate, hypotension, and hypothermia [45, 54, 56, 85]. All these effects were dose-dependent and reversible in few hours after the administration of these compounds. Of particular importance, the cardiac effects produced by thyronamines T1AM and T0AM, when injected in mice, induced a drop in the heart rate and a similar response in isolated heart preparation [45, 54]. T1AM, in ex vivo experiments and in cardiomyocytes from rats, produced a dose-dependent negative chronotropic and inotropic effects further confirming thyronamine action on heart physiology [45]. Whether TAAR1 is solely responsible for these actions or other mechanisms are involved has still to be established, since thyronamines have also activity at the monoamine membrane transporter and vesicular monoamine transporter 2 [58]. Regarding temperature control, one study showed that the hypothermic response obtained by the administration of thyronamines and other TAAR1 ligands such as amphetamines was similar in WT and in TAAR1-KO mice, suggesting that the mechanism other than involving TAAR1 was responsible for this effect [86]. On the other hand, Millan group monitored the effect of MDMA at different time points in WT and TAAR1-KO animals [46]. In WT mice, MDMA induced a biphasic thermoregulatory response, with an initial hypothermia followed by a gradual hyperthermia. In contrast, TAAR1-KO mice experienced only a hyperthermic response, suggesting that TAAR1 could have a role in thermoregulation, although more studies are necessary to understand the precise mechanism of this effect.

The first evidence that certain TAARs are expressed in leukocytes comes from the study by Nelson et al. [87]. Further studies confirmed the presence of several TAAR members in human, mouse, and rhesus monkey leukocytes [38, 43, 44]. The fact that compounds that target monoamine receptors and transporters such as ecstasy (MDMA) could influence the functions of leukocytes and affect immune response [88, 89] led to the idea that TAARs might be involved in this process. By

Western blot technique, it has been shown that TAAR1 is expressed in normal and malignant B cells derived from patients with several diseases [43]. TAAR1 was more expressed in activated B cells compared to resting ones confirming already published data [87]. Moreover, several TAAR1 agonists induced cytotoxicity in these cells suggesting a potential use for these compounds in the treatment of blood diseases such as leukemias and lymphomas. Also in rhesus monkey lymphocytes, TAAR1 expression was increased after immune activation, and methamphetamine induced a TAAR1-dependent signaling through PKA and PKC phosphorylation [44]. A recent interesting study focused on human blood leukocytes and TAAR1-/TAAR2-mediated functions [38]. Krautwurst et al. found that TAAR1, TAAR2, TAAR5, TAAR6, and TAAR9 were expressed in different leukocyte types including PMN, T cells, B cells, NK cells, and monocytes [38]. Among them, TAAR1 and TAAR2 were the most abundant receptors, with a similar expression profile, and while all TAAR1-expressing cells co-expressed also TAAR2, there was some percentage (16%) of cells that expressed only TAAR2. Interestingly, PEA, tyramine, and T1AM were able to induce several activities at very low concentration, in the low nanomolar range, which reflects the endogenous levels of these TAs [38]. TAAR1/TAAR2 activation triggered chemotactic migration of PMN cells in a concentration-dependent manner. Importantly, when TAAR1 and TAAR2 were downregulated with siRNA, this response was largely abolished. PEA was also able to induce, in a TAAR1-/TAAR2-dependent way, the IL-4 secretion in T cells and to modulate the expression of several genes, with chemotactic chemokine CCL5 being the highest expressed gene, which plays a role in allergy. Finally, TAAR1/2 activation mediated the secretion of IgE from B cells. These data suggest that TAs could play a previously unappreciated important role in immune-mediated functions (through cells migration, cytokine, and IgE productions) at concentrations found normally in the blood that could be easily increased simply by the ingestions of some type of food. Thus, these observations suggest a role of TAARs in mechanisms involved in food-related allergy.

**Conclusions**

Since the discovery of TAARs and particularly TAAR1, many studies have been performed to understand their physiology. It is now evident that TAAR1 has a primary role in the modulation of monoaminergic systems, in particular dopamine. Both the studies on TAAR1-KO mice and the recent development of selective ligands demonstrate that TAAR1 generally behaves as a "brake" for dopamine neurotransmission decreasing a hyperactive dopaminergic system. It is interesting to note that partial TAAR1 agonists can act also as antagonists, depending on the context, and in these cases, as for haloperidol-induced catalepsy, they seem to counteract a hypofunctional dopamine signaling. Thus, it might be expected that TAAR1 partial agonists could behave as dopamine system stabilizer, although more studies are necessary to

(continued)

uncover the mechanisms of this phenomenon. This panel of actions on dopamine physiology indicates that TAAR1 could be a novel target to treat dopamine-related disorder such as schizophrenia and addiction. On the other hand, TAAR1 activity on serotonergic system suggests that TAAR1 is able to modulate also phenotypes related to mood disorders such as depression and bipolar disorder. While there is now strong evidence of TAAR1 role in several experimental models either in mouse, rat, or nonhuman primates, to fully validate TAAR1 role in brain physiology, it will be of great importance to have a proof of concept in humans. Another interesting point would be to extend this line of research to other aspects related to brain physiology, such as cognition, since it is likely that TAAR1 could be involved also in cognition-related brain functions. Finally, important TAAR functions in periphery are emerging, and detailed descriptions of these mechanisms will be necessary to uncover these intriguing new roles of TAARs.

## References

1. Borowsky B, Adham N, Jones KA, Raddatz R, Artymyshyn R, Ogozalek KL, Durkin MM, Lakhlani PP, Bonini JA, Pathirana S, Boyle N, Pu X, Kouranova E, Lichtblau H, Ochoa FY, Branchek TA, Gerald C (2001) Trace amines: identification of a family of mammalian G protein-coupled receptors. Proc Natl Acad Sci U S A 98(16):8966–8971
2. Bunzow JR, Sonders MS, Arttamangkul S, Harrison LM, Zhang G, Quigley DI, Darland T, Suchland KL, Pasumamula S, Kennedy JL, Olson SB, Magenis RE, Amara SG, Grandy DK (2001) Amphetamine, 3,4-methylenedioxymethamphetamine, lysergic acid diethylamide, and metabolites of the catecholamine neurotransmitters are agonists of a rat trace amine receptor. Mol Pharmacol 60(6):1181–1188
3. Grandy DK (2007) Trace amine-associated receptor 1-family archetype or iconoclast? Pharmacol Ther 116(3):355–390
4. Lindemann L, Hoener MC (2005) A renaissance in trace amines inspired by a novel GPCR family. Trends Pharmacol Sci 26(5):274–281
5. Miller GM (2012) Avenues for the development of therapeutics that target trace amine associated receptor 1 (TAAR1). J Med Chem 55(5):1809–1814
6. Sotnikova TD, Caron MG, Gainetdinov RR (2009) Trace amine-associated receptors as emerging therapeutic targets. Mol Pharmacol 76(2):229–235
7. Pei Y, Lee JA, Leo D, Gainetdinov RR, Hoener MC, Canales JJ (2014) Activation of the trace amine-associated receptor 1 prevents relapse to cocaine seeking. Neuropsychopharmacology 39:2299–2308
8. Bradaia A, Trube G, Stalder H, Norcross RD, Ozmen L, Wettstein JG, Pinard A, Buchy D, Gassmann M, Hoener MC, Bettler B (2009) The selective antagonist EPPTB reveals TAAR1-mediated regulatory mechanisms in dopaminergic neurons of the mesolimbic system. Proc Natl Acad Sci U S A 106:20081–20086
9. Revel FG, Moreau JL, Gainetdinov RR, Bradaia A, Sotnikova TD, Mory R, Durkin S, Zbinden KG, Norcross R, Meyer CA, Metzler V, Chaboz S, Ozmen L, Trube G, Pouzet B, Bettler B, Caron MG, Wettstein JG, Hoener MC (2011) TAAR1 activation modulates monoaminergic neurotransmission, preventing hyperdopaminergic and hypoglutamatergic activity. Proc Natl Acad Sci U S A 108(20):8485–8490

10. Revel FG, Moreau JL, Gainetdinov RR, Ferragud A, Velazquez-Sanchez C, Sotnikova TD, Morairty SR, Harmeier A, Groebke Zbinden K, Norcross RD, Bradaia A, Kilduff TS, Biemans B, Pouzet B, Caron MG, Canales JJ, Wallace TL, Wettstein JG, Hoener MC (2012) Trace amine-associated receptor 1 partial agonism reveals novel paradigm for neuropsychiatric therapeutics. Biol Psychiatry 72(11):934–942
11. Revel FG, Moreau JL, Pouzet B, Mory R, Bradaia A, Buchy D, Metzler V, Chaboz S, Groebke Zbinden K, Galley G, Norcross RD, Tuerck D, Bruns A, Morairty SR, Kilduff TS, Wallace TL, Risterucci C, Wettstein JG, Hoener MC (2013) A new perspective for schizophrenia: TAAR1 agonists reveal antipsychotic- and antidepressant-like activity, improve cognition and control body weight. Mol Psychiatry 18(5):543–556
12. Thorn DA, Jing L, Qiu Y, Gancarz-Kausch AM, Galuska CM, Dietz DM, Zhang Y, Li JX (2014) Effects of the trace amine associated receptor 1 agonist RO5263397 on abuse-related effects of cocaine in rats. Neuropsychopharmacology 39:2309–2316
13. Sandler M, Ruthven CR, Goodwin BL, Reynolds GP, Rao VA, Coppen A (1980) Trace amine deficit in depressive illness: the phenylalanine connexion. Acta Psychiatr Scand Suppl 280:29–39
14. Berry MD (2004) Mammalian central nervous system trace amines. Pharmacologic amphetamines, physiologic neuromodulators. J Neurochem 90(2):257–271
15. Boulton AA (1980) Trace amines and mental disorders. Can J Neurol Sci 7(3):261–263
16. Branchek TA, Blackburn TP (2003) Trace amine receptors as targets for novel therapeutics: legend, myth and fact. Curr Opin Pharmacol 3(1):90–97
17. Berry MD (2007) The potential of trace amines and their receptors for treating neurological and psychiatric diseases. Rev Recent Clin Trials 2(1):3–19
18. Liberles SD, Buck LB (2006) A second class of chemosensory receptors in the olfactory epithelium. Nature 442(7103):645–650
19. Sotnikova TD, Beaulieu JM, Espinoza S, Masri B, Zhang X, Salahpour A, Barak LS, Caron MG, Gainetdinov RR (2010) The dopamine metabolite 3-methoxytyramine is a neuromodulator. PLoS One 5(10):e13452
20. Yang HY, Neff NH (1973) Beta-phenylethylamine: a specific substrate for type B monoamine oxidase of brain. J Pharmacol Exp Ther 187(2):365–371
21. Durden DA, Philips SR (1980) Kinetic measurements of the turnover rates of phenylethylamine and tryptamine in vivo in the rat brain. J Neurochem 34(6):1725–1732
22. Philips SR, Boulton AA (1979) The effect of monoamine oxidase inhibitors on some arylalkylamines in rate striatum. J Neurochem 33(1):159–167
23. Grimsby J, Toth M, Chen K, Kumazawa T, Klaidman L, Adams JD, Karoum F, Gal J, Shih JC (1997) Increased stress response and beta-phenylethylamine in MAOB-deficient mice. Nat Genet 17(2):206–210
24. Juorio AV, Philips SR (1976) Arylalkylamines in Octopus tissues. Neurochem Res 1 (5):501–509
25. Robertson HA, Juorio AV (1976) Octopamine and some related noncatecholic amines in invertebrate nervous systems. Int Rev Neurobiol 19:173–224
26. Parker EM, Cubeddu LX (1988) Comparative effects of amphetamine, phenylethylamine and related drugs on dopamine efflux, dopamine uptake and mazindol binding. J Pharmacol Exp Ther 245(1):199–210
27. Fuxe K, Grobecker H, Jonsson J (1967) The effect of beta-phenylethylamine on central and peripheral monoamine-containing neurons. Eur J Pharmacol 2(3):202–207
28. Raiteri M, Del Carmine R, Bertollini A, Levi G (1977) Effect of sympathomimetic amines on the synaptosomal transport of noradrenaline, dopamine and 5-hydroxytryptamine. Eur J Pharmacol 41(2):133–143
29. Sotnikova TD, Budygin EA, Jones SR, Dykstra LA, Caron MG, Gainetdinov RR (2004) Dopamine transporter-dependent and -independent actions of trace amine beta-phenylethylamine. J Neurochem 91(2):362–373

30. Wolinsky TD, Swanson CJ, Smith KE, Zhong H, Borowsky B, Seeman P, Branchek T, Gerald CP (2007) The trace amine 1 receptor knockout mouse: an animal model with relevance to schizophrenia. Genes Brain Behav 6(7):628–639
31. Wainscott DB, Little SP, Yin T, Tu Y, Rocco VP, He JX, Nelson DL (2007) Pharmacologic characterization of the cloned human trace amine-associated receptor 1 (TAAR1) and evidence for species differences with the rat TAAR1. J Pharmacol Exp Ther 320(1):475–485
32. Navarro HA, Gilmour BP, Lewin AH (2006) A rapid functional assay for the human trace amine-associated receptor 1 based on the mobilization of internal calcium. J Biomol Screen 11 (6):688–693
33. Miller GM, Verrico CD, Jassen A, Konar M, Yang H, Panas H, Bahn M, Johnson R, Madras BK (2005) Primate trace amine receptor 1 modulation by the dopamine transporter. J Pharmacol Exp Ther 313(3):983–994
34. Lindemann L, Meyer CA, Jeanneau K, Bradaia A, Ozmen L, Bluethmann H, Bettler B, Wettstein JG, Borroni E, Moreau JL, Hoener MC (2008) Trace amine-associated receptor 1 modulates dopaminergic activity. J Pharmacol Exp Ther 324(3):948–956
35. Lewin AH, Navarro HA, Mascarella SW (2008) Structure-activity correlations for beta-phenethylamines at human trace amine receptor 1. Bioorg Med Chem 16(15):7415–7423
36. Barak LS, Salahpour A, Zhang X, Masri B, Sotnikova TD, Ramsey AJ, Violin JD, Lefkowitz RJ, Caron MG, Gainetdinov RR (2008) Pharmacological characterization of membrane-expressed human trace amine-associated receptor 1 (TAAR1) by a bioluminescence resonance energy transfer cAMP biosensor. Mol Pharmacol 74(3):585–594
37. Zhang J, Pacifico R, Cawley D, Feinstein P, Bozza T (2013) Ultrasensitive detection of amines by a trace amine-associated receptor. J Neurosci 33(7):3228–3239
38. Babusyte A, Kotthoff M, Fiedler J, Krautwurst D (2013) Biogenic amines activate blood leukocytes via trace amine-associated receptors TAAR1 and TAAR2. J Leukoc Biol 93 (3):387–394
39. Lindemann L, Ebeling M, Kratochwil NA, Bunzow JR, Grandy DK, Hoener MC (2005) Trace amine-associated receptors form structurally and functionally distinct subfamilies of novel G protein-coupled receptors. Genomics 85(3):372–385
40. Duan J, Martinez M, Sanders AR, Hou C, Saitou N, Kitano T, Mowry BJ, Crowe RR, Silverman JM, Levinson DF, Gejman PV (2004) Polymorphisms in the trace amine receptor 4 (TRAR4) gene on chromosome 6q23.2 are associated with susceptibility to schizophrenia. Am J Hum Genet 75(4):624–638
41. Xie Z, Miller GM (2007) Trace amine-associated receptor 1 is a modulator of the dopamine transporter. J Pharmacol Exp Ther 321(1):128–136
42. Pae CU, Drago A, Kim JJ, Patkar AA, Jun TY, De Ronchi D, Serretti A (2010) TAAR6 variations possibly associated with antidepressant response and suicidal behavior. Psychiatry Res 180(1):20–24
43. Wasik AM, Millan MJ, Scanlan T, Barnes NM, Gordon J (2012) Evidence for functional trace amine associated receptor-1 in normal and malignant B cells. Leuk Res 36(2):245–249
44. Panas MW, Xie Z, Panas HN, Hoener MC, Vallender EJ, Miller GM (2012) Trace amine associated receptor 1 signaling in activated lymphocytes. J Neuroimmune Pharmacol 7 (4):866–876
45. Chiellini G, Frascarelli S, Ghelardoni S, Carnicelli V, Tobias SC, DeBarber A, Brogioni S, Ronca-Testoni S, Cerbai E, Grandy DK, Scanlan TS, Zucchi R (2007) Cardiac effects of 3-iodothyronamine: a new aminergic system modulating cardiac function. FASEB J 21 (7):1597–1608
46. Di Cara B, Maggio R, Aloisi G, Rivet JM, Lundius EG, Yoshitake T, Svenningsson P, Brocco M, Gobert A, De Groote L, Cistarelli L, Veiga S, De Montrion C, Rodriguez M, Galizzi JP, Lockhart BP, Coge F, Boutin JA, Vayer P, Verdouw PM, Groenink L, Millan MJ (2011) Genetic deletion of trace amine 1 receptors reveals their role in auto-inhibiting the actions of ecstasy (MDMA). J Neurosci 31(47):16928–16940

47. Reese EA, Bunzow JR, Arttamangkul S, Sonders MS, Grandy DK (2007) Trace amine-associated receptor 1 displays species-dependent stereoselectivity for isomers of methamphetamine, amphetamine, and para-hydroxyamphetamine. J Pharmacol Exp Ther 321(1):178–186
48. Barak LS, Ferguson SS, Zhang J, Caron MG (1997) A beta-arrestin/green fluorescent protein biosensor for detecting G protein-coupled receptor activation. J Biol Chem 272 (44):27497–27500
49. Kim KM, Gainetdinov RR, Laporte SA, Caron MG, Barak LS (2005) G protein-coupled receptor kinase regulates dopamine D3 receptor signaling by modulating the stability of a receptor-filamin-beta-arrestin complex. A case of autoreceptor regulation. J Biol Chem 280 (13):12774–12780
50. Xie Z, Westmoreland SV, Bahn ME, Chen GL, Yang H, Vallender EJ, Yao WD, Madras BK, Miller GM (2007) Rhesus monkey trace amine-associated receptor 1 signaling: enhancement by monoamine transporters and attenuation by the D2 autoreceptor in vitro. J Pharmacol Exp Ther 321(1):116–127
51. Asghar SJ, Tanay VA, Baker GB, Greenshaw A, Silverstone PH (2003) Relationship of plasma amphetamine levels to physiological, subjective, cognitive and biochemical measures in healthy volunteers. Hum Psychopharmacol 18(4):291–299
52. Bowyer JF, Young JF, Slikker W, Itzak Y, Mayorga AJ, Newport GD, Ali SF, Frederick DL, Paule MG (2003) Plasma levels of parent compound and metabolites after doses of either D-fenfluramine or D-3,4-methylenedioxymethamphetamine (MDMA) that produce long-term serotonergic alterations. Neurotoxicology 24(3):379–390
53. Hart ME, Suchland KL, Miyakawa M, Bunzow JR, Grandy DK, Scanlan TS (2006) Trace amine-associated receptor agonists: synthesis and evaluation of thyronamines and related analogues. J Med Chem 49(3):1101–1112
54. Scanlan TS, Suchland KL, Hart ME, Chiellini G, Huang Y, Kruzich PJ, Frascarelli S, Crossley DA, Bunzow JR, Ronca-Testoni S, Lin ET, Hatton D, Zucchi R, Grandy DK (2004) 3-Iodothyronamine is an endogenous and rapid-acting derivative of thyroid hormone. Nat Med 10(6):638–642
55. Tan ES, Groban ES, Jacobson MP, Scanlan TS (2008) Toward deciphering the code to aminergic G protein-coupled receptor drug design. Chem Biol 15(4):343–353
56. Frascarelli S, Ghelardoni S, Chiellini G, Vargiu R, Ronca-Testoni S, Scanlan TS, Grandy DK, Zucchi R (2008) Cardiac effects of trace amines: pharmacological characterization of trace amine-associated receptors. Eur J Pharmacol 587(1–3):231–236
57. Snead AN, Miyakawa M, Tan ES, Scanlan TS (2008) Trace amine-associated receptor 1 (TAAR1) is activated by amiodarone metabolites. Bioorg Med Chem Lett 18(22):5920–5922
58. Snead AN, Santos MS, Seal RP, Miyakawa M, Edwards RH, Scanlan TS (2007) Thyronamines inhibit plasma membrane and vesicular monoamine transport. ACS Chem Biol 2(6):390–398
59. Kehr W (1976) 3-Methoxytyramine as an indicator of impulse-induced dopamine release in rat brain in vivo. Naunyn Schmiedebergs Arch Pharmacol 293(3):209–215
60. Rajput AH, Fenton ME, Di Paolo T, Sitte H, Pifl C, Hornykiewicz O (2004) Human brain dopamine metabolism in levodopa-induced dyskinesia and wearing-off. Parkinsonism Relat Disord 10(4):221–226
61. Xuehong L, Grandy DK, Janowsky AJ (2014) Ractopamine, a livestock feed additive, is a full agonist at trace amine-associated receptor 1. J Pharmacol Exp Ther 350:124–129
62. Tallman KR, Grandy DK (2012) A decade of pharma discovery delivers new tools targeting trace amine-associated receptor 1. Neuropsychopharmacology 37(12):2553–2554
63. Cichero E, Espinoza S, Gainetdinov RR, Brasili L, Fossa P (2013) Insights into the structure and pharmacology of the human trace amine-associated receptor 1 (hTAAR1): homology modelling and docking studies. Chem Biol Drug Des 81(4):509–516
64. Reese EA, Norimatsu Y, Grandy MS, Suchland KL, Bunzow JR, Grandy DK (2014) Exploring the determinants of trace amine-associated receptor 1's functional selectivity for the stereoisomers of amphetamine and methamphetamine. J Med Chem 57(2):378–390

65. Sandler M, Reynolds GP (1976) Does phenylethylamine cause schizophrenia? Lancet 1 (7950):70–71
66. D'Andrea G, Terrazzino S, Fortin D, Cocco P, Balbi T, Leon A (2003) Elusive amines and primary headaches: historical background and prospectives. Neurol Sci 24(Suppl 2):S65–S67
67. Shirkande S, O'Reilly R, Davis B, Durden D, Malcom D (1995) Plasma phenylethylamine levels of schizophrenic patients. Can J Psychiatry 40(4):221
68. Potkin SG, Karoum F, Chuang LW, Cannon-Spoor HE, Phillips I, Wyatt RJ (1979) Phenylethylamine in paranoid chronic schizophrenia. Science 206(4417):470–471
69. Seeman P, Weinshenker D, Quirion R, Srivastava LK, Bhardwaj SK, Grandy DK, Premont RT, Sotnikova TD, Boksa P, El-Ghundi M, O'Dowd BF, George SR, Perreault ML, Mannisto PT, Robinson S, Palmiter RD, Tallerico T (2005) Dopamine supersensitivity correlates with D2High states, implying many paths to psychosis. Proc Natl Acad Sci U S A 102 (9):3513–3518
70. Millan MJ (2000) Improving the treatment of schizophrenia: focus on serotonin (5-HT)(1A) receptors. J Pharmacol Exp Ther 295(3):853–861
71. Barnes NM, Sharp T (1999) A review of central 5-HT receptors and their function. Neuropharmacology 38(8):1083–1152
72. Gardier AM, Malagie I, Trillat AC, Jacquot C, Artigas F (1996) Role of 5-HT1A autoreceptors in the mechanism of action of serotoninergic antidepressant drugs: recent findings from in vivo microdialysis studies. Fundam Clin Pharmacol 10(1):16–27
73. Leo D, Mus L, Espinoza S, Hoener MC, Sotnikova TD, Gainetdinov RR (2014) Taar1-mediated modulation of presynaptic dopaminergic neurotransmission: role of D2 dopamine autoreceptors. Neuropharmacology 81:283–291
74. Espinoza S, Salahpour A, Masri B, Sotnikova TD, Messa M, Barak LS, Caron MG, Gainetdinov RR (2011) Functional interaction between trace amine-associated receptor 1 and dopamine D2 receptor. Mol Pharmacol 80(3):416–425
75. Xie Z, Westmoreland SV, Miller GM (2008) Modulation of monoamine transporters by common biogenic amines via trace amine-associated receptor 1 and monoamine autoreceptors in human embryonic kidney 293 cells and brain synaptosomes. J Pharmacol Exp Ther 325 (2):629–640
76. Xie Z, Miller GM (2009) Trace amine-associated receptor 1 as a monoaminergic modulator in brain. Biochem Pharmacol 78(9):1095–1104
77. Xie Z, Miller GM (2008) Beta-phenylethylamine alters monoamine transporter function via trace amine-associated receptor 1: implication for modulatory roles of trace amines in brain. J Pharmacol Exp Ther 325(2):617–628
78. Sukhanov I, Espinoza S, Yakovlev DS, Hoener MC, Sotnikova TD, Gainetdinov RR (2014) TAAR1-dependent effects of apomorphine in mice. Int J Neuropsychopharmacol 17:1683–1693
79. Costall B, Naylor RJ, Nohria V (1978) Climbing behaviour induced by apomorphine in mice: a potential model for the detection of neuroleptic activity. Eur J Pharmacol 50(1):39–50
80. Sulzer D (2011) How addictive drugs disrupt presynaptic dopamine neurotransmission. Neuron 69(4):628–649
81. Achat-Mendes C, Lynch LJ, Sullivan KA, Vallender EJ, Miller GM (2012) Augmentation of methamphetamine-induced behaviors in transgenic mice lacking the trace amine-associated receptor 1. Pharmacol Biochem Behav 101(2):201–207
82. Lynch LJ, Sullivan KA, Vallender EJ, Rowlett JK, Platt DM, Miller GM (2013) Trace amine associated receptor 1 modulates behavioral effects of ethanol. Subst Abuse 7:117–126
83. Fuchs RA, Branham RK, See RE (2006) Different neural substrates mediate cocaine seeking after abstinence versus extinction training: a critical role for the dorsolateral caudate-putamen. J Neurosci 26(13):3584–3588
84. Thorn DA, Zhang C, Zhang Y, Li JX (2014) The trace amine associated receptor 1 agonist RO5263397 attenuates the induction of cocaine behavioral sensitization in rats. Neurosci Lett 566:67–71

85. Zucchi R, Chiellini G, Scanlan TS, Grandy DK (2006) Trace amine-associated receptors and their ligands. Br J Pharmacol 149(8):967–978
86. Panas HN, Lynch LJ, Vallender EJ, Xie Z, Chen GL, Lynn SK, Scanlan TS, Miller GM (2010) Normal thermoregulatory responses to 3-iodothyronamine, trace amines and amphetamine-like psychostimulants in trace amine associated receptor 1 knockout mice. J Neurosci Res 88:1962–1969
87. Nelson DA, Tolbert MD, Singh SJ, Bost KL (2007) Expression of neuronal trace amine-associated receptor (TAAR) mRNAs in leukocytes. J Neuroimmunol 192(1–2):21–30
88. Meredith EJ, Holder MJ, Chamba A, Challa A, Drake-Lee A, Bunce CM, Drayson MT, Pilkington G, Blakely RD, Dyer MJ, Barnes NM, Gordon J (2005) The serotonin transporter (SLC6A4) is present in B-cell clones of diverse malignant origin: probing a potential anti-tumor target for psychotropics. FASEB J 19(9):1187–1189
89. Chamba A, Holder MJ, Jarrett RF, Shield L, Toellner KM, Drayson MT, Barnes NM, Gordon J (2010) SLC6A4 expression and anti-proliferative responses to serotonin transporter ligands chlomipramine and fluoxetine in primary B-cell malignancies. Leuk Res 34(8):1103–1106

# Index

MIX
Papier aus verantwortungsvollen Quellen
Paper from responsible sources
FSC® C105338

If you have any concerns about our products,
you can contact us on
**ProductSafety@springernature.com**

In case Publisher is established outside the EU,
the EU authorized representative is:
**Springer Nature Customer Service Center GmbH**
**Europaplatz 3, 69115 Heidelberg, Germany**

Printed by Libri Plureos GmbH
in Hamburg, Germany